T0190473

Communications
in Computer and Information Science　　2062

Rationale

The CCIS series is devoted to the publication of proceedings of computer science conferences. Its aim is to efficiently disseminate original research results in informatics in printed and electronic form. While the focus is on publication of peer-reviewed full papers presenting mature work, inclusion of reviewed short papers reporting on work in progress is welcome, too. Besides globally relevant meetings with internationally representative program committees guaranteeing a strict peer-reviewing and paper selection process, conferences run by societies or of high regional or national relevance are also considered for publication.

Topics

The topical scope of CCIS spans the entire spectrum of informatics ranging from foundational topics in the theory of computing to information and communications science and technology and a broad variety of interdisciplinary application fields.

Information for Volume Editors and Authors

Publication in CCIS is free of charge. No royalties are paid, however, we offer registered conference participants temporary free access to the online version of the conference proceedings on SpringerLink (http://link.springer.com) by means of an http referrer from the conference website and/or a number of complimentary printed copies, as specified in the official acceptance email of the event.

CCIS proceedings can be published in time for distribution at conferences or as post-proceedings, and delivered in the form of printed books and/or electronically as USBs and/or e-content licenses for accessing proceedings at SpringerLink. Furthermore, CCIS proceedings are included in the CCIS electronic book series hosted in the SpringerLink digital library at http://link.springer.com/bookseries/7899. Conferences publishing in CCIS are allowed to use Online Conference Service (OCS) for managing the whole proceedings lifecycle (from submission and reviewing to preparing for publication) free of charge.

Publication process

The language of publication is exclusively English. Authors publishing in CCIS have to sign the Springer CCIS copyright transfer form, however, they are free to use their material published in CCIS for substantially changed, more elaborate subsequent publications elsewhere. For the preparation of the camera-ready papers/files, authors have to strictly adhere to the Springer CCIS Authors' Instructions and are strongly encouraged to use the CCIS LaTeX style files or templates.

Abstracting/Indexing

CCIS is abstracted/indexed in DBLP, Google Scholar, EI-Compendex, Mathematical Reviews, SCImago, Scopus. CCIS volumes are also submitted for the inclusion in ISI Proceedings.

How to start

To start the evaluation of your proposal for inclusion in the CCIS series, please send an e-mail to ccis@springer.com.

Linqiang Pan · Yong Wang · Jianqing Lin
Editors

Bio-Inspired Computing: Theories and Applications

18th International Conference, BIC-TA 2023
Changsha, China, December 15–17, 2023
Revised Selected Papers, Part II

Springer

Editors
Linqiang Pan (ID)
Huazhong University of Science
and Technology
Wuhan, China

Yong Wang
Central South University
Changsha, China

Jianqing Lin
Huazhong University of Science
and Technology
Wuhan, China

ISSN 1865-0929   ISSN 1865-0937 (electronic)
Communications in Computer and Information Science
ISBN 978-981-97-2274-7   ISBN 978-981-97-2275-4 (eBook)
https://doi.org/10.1007/978-981-97-2275-4

This Springer imprint is published by the registered company Springer Nature Singapore Pte Ltd.
The registered company address is: 152 Beach Road, #21-01/04 Gateway East, Singapore 189721, Singapore

Paper in this product is recyclable.

Preface

Bio-inspired computing is a field of study that abstracts computing ideas (data structures, operations with data, ways to control operations, computing models, artificial intelligence, multisource data-driven analysis, etc.) from living phenomena or biological systems such as cells, tissue, the brain, neural networks, immune systems, ant colonies, evolution, etc. The areas of bio-inspired computing include Neural Networks, Brain-Inspired Computing, Neuromorphic Computing and Architectures, Cellular Automata and Cellular Neural Networks, Evolutionary Computing, Swarm Intelligence, Fuzzy Logic and Systems, DNA and Molecular Computing, Membrane Computing, Artificial Intelligence and its Application in other disciplines such as machine learning, deep learning, image processing, computer science, cybernetics, etc. Bio-Inspired Computing: Theories and Applications (BIC-TA) is a series of conferences that aims to bring together researchers working in the main areas of bio-inspired computing, to present their recent results, exchang ideas, and cooperate in a friendly framework.

Since 2006, BIC-TA has taken place at Wuhan (2006), Zhengzhou (2007), Adelaide (2008), Beijing (2009), Liverpool and Changsha (2010), Penang (2011), Gwalior (2012), Anhui (2013), Wuhan (2014), Anhui (2015), Xi'an (2016), Harbin (2017), Beijing (2018), Zhengzhou (2019), Qingdao (2020), Taiyuan (2021), and Wuhan (2022). Following the success of the previous editions, the 18th International Conference on Bio-Inspired Computing: Theories and Applications (BIC-TA 2023) was held in Changsha, China, during December 15–17, 2023, organized by Central South University and co-organized by Xiangtan University, with the support of Operations Research Society of Hubei.

We would like to thank the keynote speakers for their excellent presentations: Weigang Chen (Tianjin University, China), Shaoliang Peng (Hunan University, China), Ke Tang (Southern University of Science and Technology, China), and Xingyi Zhang (Anhui University, China).

A special thank you is given to the general chair, Chunhua Yang, for her guidance and support of the conference. We gratefully thank Yalin Wang and Chengqing Li, for their warm welcome and inspiring speeches at the opening ceremony of the conference. We thank the local chair, Juan Zou, for her significant contribution to the organization and guidance of the conference. We thank Shouyong Jiang, Bing-chuan Wang, Zhi-zhong Liu, and Pei-qiu Huang, for their contribution in organizing the conference. We thank Zixiao Zhang for his help in collecting the final files of the papers and editing the volume and maintaining the website of BIC-TA 2023 (http://2023.bicta.org/). We also thank all the other volunteers, whose efforts ensured the smooth running of the conference.

BIC-TA 2023 received 168 submissions on various aspects of bio-inspired computing, of which 64 papers were selected for the two volumes of *Communications in Computer and Information Science*. Each paper was peer reviewed by at least three reviewers with expertise in the relevant subject area in a single-blind peer-review process. The warmest thanks should be given to the reviewers for their careful and efficient work in the reviewing process.

Special thanks are due to Springer Nature for their skilled cooperation in the timely production of these volumes.

February 2024

Linqiang Pan
Yong Wang
Jianqing Lin

Organization

Steering Committee

Xiaochun Cheng	Middlesex University London, UK
Guangzhao Cui	Zhengzhou University of Light Industry, China
Kalyanmoy Deb	Michigan State University, USA
Miki Hirabayashi	National Institute of Information and Communications Technology, Japan
Joshua Knowles	University of Manchester, UK
Thom LaBean	North Carolina State University, USA
Jiuyong Li	University of South Australia, Australia
Kenli Li	University of Hunan, China
Giancarlo Mauri	Università di Milano-Bicocca, Italy
Yongli Mi	Hong Kong University of Science and Technology, China
Atulya K. Nagar	Liverpool Hope University, UK
Linqiang Pan (Chair)	Huazhong University of Science and Technology, China
Gheorghe Păun	Romanian Academy, Romania
Mario J. Perez-Jimenez	University of Seville, Spain
K. G. Subramanian	Liverpool Hope University, UK
Robinson Thamburaj	Madras Christian College, India
Jin Xu	Peking University, China
Hao Yan	Arizona State University, USA

General Chair

Chunhua Yang	Central South University, China

Program Committee Chairs

Yong Wang	Central South University, China
Linqiang Pan	Huazhong University of Science and Technology, China

Publication Chair

Bingchuan Wang   Central South University, China

Publicity Chair

Guangwu Liu   Wuhan University of Technology, China

Local Chair

Juan Zou   Xiangtan University, China

Registration Chair

Shouyong Jiang   Central South University, China

Program Committee

Muhammad Abulaish   South Asian University, India
Andy Adamatzky   University of the West of England, UK
Guangwu Liu   Wuhan University of Technology, China
Chang Wook Ahn   Gwangju Institute of Science and Technology,
                                    South Korea
Adel Al-Jumaily   University of Technology Sydney, Australia
Bin Cao   Hebei University of Technology, China
Junfeng Chen   Hohai University, China
Wei-Neng Chen   Sun Yat-sen University, China
Shi Cheng   Shaanxi Normal University, China
Xiaochun Cheng   Middlesex University London, UK
Tsung-Che Chiang   National Taiwan Normal University, China
Sung-Bae Cho   Yonsei University, South Korea
Zhihua Cui   Taiyuan University of Science and Technology,
                                    China
Kejie Dai   Pingdingshan University, China
Ciprian Dobre   University Politehnica of Bucharest, Romania
Bei Dong   Shanxi Normal University, China
Xin Du   Fujian Normal University, China
Carlos Fernandez-Llatas   Universitat Politècnica de València, Spain

Shangce Gao	University of Toyama, Japan
Marian Gheorghe	University of Bradford, UK
Wenyin Gong	China University of Geosciences, China
Shivaprasad Gundibail	Manipal Academy of Higher Education, India
Ping Guo	Beijing Normal University, China
Yinan Guo	China University of Mining and Technology, China
Guosheng Hao	Jiangsu Normal University, China
Cheng He	Southern University of Science and Technology, China
Shan He	University of Birmingham, UK
Tzung-Pei Hong	National University of Kaohsiung, China
Pei-qiu Huang	Central South University, China
Florentin Ipate	University of Bucharest, Romania
Sunil Kumar Jha	Banaras Hindu University, India
He Jiang	Dalian University of Technology, China
Qiaoyong Jiang	Xi'an University of Technology, China
Shouyong Jiang	Central South University, China
Licheng Jiao	Xidian University, China
Liangjun Ke	Xi'an Jiaotong University, China
Ashwani Kush	Kurukshetra University, India
Hui Li	Xi'an Jiaotong University, China
Kenli Li	Hunan University, China
Lianghao Li	Huazhong University of Science and Technology, China
Yangyang Li	Xidian University, China
Zhihui Li	Zhengzhou University, China
Jing Liang	Zhengzhou University, China
Jerry Chun-Wei Lin	Western Norway University of Applied Sciences, Norway
Jianiqng Lin	Huazhong University of Science and Technology, China
Qunfeng Liu	Dongguan University of Technology, China
Xiaobo Liu	China University of Geosciences, China
Zhi-zhong Liu	Hunan University, China
Wenjian Luo	University of Science and Technology of China, China
Lianbo Ma	Northeastern University, China
Wanli Ma	University of Canberra, Australia
Xiaoliang Ma	Shenzhen University, China
Francesco Marcelloni	University of Pisa, Italy
Efrén Mezura-Montes	University of Veracruz, Mexico

Hongwei Mo	Harbin Engineering University, China
Chilukuri Mohan	Syracuse University, USA
Abdulqader Mohsen	University of Science and Technology in Yemen, Yemen
Holger Morgenstern	Albstadt-Sigmaringen University, Germany
Andres Muñoz	Universidad Católica San Antonio de Murcia, Spain
G. R. S. Murthy	Lendi Institute of Engineering and Technology, India
Akila Muthuramalingam	KPR Institute of Engineering and Technology, India
Yusuke Nojima	Osaka Prefecture University, Japan
Linqiang Pan	Huazhong University of Science and Technology, China
Andrei Paun	University of Bucharest, Romania
Gheorghe Păun	Romanian Academy, Romania
Xingguang Peng	Northwestern Polytechnical University, China
Chao Qian	University of Science and Technology of China, China
Balwinder Raj	NITTTR, India
Rawya Rizk	Port Said University, Egypt
Rajesh Sanghvi	G. H. Patel College of Engineering and Technology, India
Ronghua Shang	Xidian University, China
Zhigang Shang	Zhengzhou University, China
Ravi Shankar	Florida Atlantic University, USA
V. Ravi Sankar	GITAM University, India
Bosheng Song	Hunan University, China
Tao Song	China University of Petroleum, China
Jianyong Sun	University of Nottingham, UK
Yifei Sun	Shaanxi Normal University, China
Bing-chuan Wang	Central South University, China
Handing Wang	Xidian University, China
Yong Wang	Central South University, China
Hui Wang	Nanchang Institute of Technology, China
Hui Wang	South China Agricultural University, China
Gaige Wang	Ocean University of China, China
Sudhir Warier	IIT Bombay, India
Slawomir T. Wierzchon	Polish Academy of Sciences, Poland
Zhou Wu	Chongqing University, China
Xiuli Wu	University of Science and Technology Beijing, China
Bin Xin	Beijing Institute of Technology, China

Gang Xu	Nanchang University, China
Yingjie Yang	De Montfort University, UK
Zhile Yang	Shenzhen Institute of Advanced Technology, Chinese Academy of Sciences, China
Kunjie Yu	Zhengzhou University, China
Xiaowei Zhang	University of Science and Technology of China, China
Jie Zhang	Newcastle University, UK
Gexiang Zhang	Chengdu University of Technology, China
Defu Zhang	Xiamen University, China
Haiyu Zhang	Wuhan University of Technology, China
Peng Zhang	Beijing University of Posts and Telecommunications, China
Weiwei Zhang	Zhengzhou University of Light Industry, China
Yong Zhang	China University of Mining and Technology, China
Xinchao Zhao	Beijing University of Posts and Telecommunications, China
Yujun Zheng	Zhejiang University of Technology, China
Aimin Zhou	East China Normal University, China
Fengqun Zhou	Pingdingshan University, China
Xinjian Zhuo	Beijing University of Posts and Telecommunications, China
Shang-Ming Zhou	Swansea University, UK
Dexuan Zou	Jiangsu Normal University, China
Juan Zou	Xiangtan University, China
Xingquan Zuo	Beijing University of Posts and Telecommunications, China

Contents – Part II

Intelligent Control and Application

Contents – Part I

Membrane Computing and DNA Computing

Machine Learning and Applications

Review of Traveling Salesman Problem Solution Methods

Longrui Yang, Xiyuan Wang⬛, Zhaoqi He⬛, Sicong Wang⬛, and Jie Lin(✉)⬛

College of Electrical Engineering and Information Engineering, Lanzhou University of
Technology, Lanzhou 730050, China
449066528@qq.com

Abstract. The Traveling Salesman Problem (TSP) is a key focus in the fields of
computer science and operations research, widely applied in areas such as data collection, search and rescue, robot task allocation and scheduling, etc. This paper, by
reviewing recent literature, first introduces the definition and mathematical model
of the TSP, followed by an exposition of the concepts of classical TSP. Subsequently, an analysis of solving algorithms for the classical Traveling Salesman
Problem is conducted, categorizing them into exact algorithms, heuristic algorithms, and learning-based algorithms. The paper then provides an assessment of
the advantages and disadvantages associated with these three categories of algorithms, accompanied by an elaborate overview of the research advancements made
in recent years. Future research on TSP will focus on exploring undeveloped algorithms and integrating stable ones to address larger-scale problems, enhance solution quality, avoid local optima, and improve solution efficiency. Breakthroughs are
anticipated in the application of learning-based methods for solving the Traveling
Salesman Problem (TSP).

Keywords: Traveling Salesman Problem · Exact Algorithms · Heuristic
Algorithm · Learning-based Algorithm

1 Introduction

Traditional exact algorithms, such as Branch and Bound (BB) and Linear Programming
(LP), face significant challenges when dealing with large-scale instances of the Traveling Salesman Problem (TSP). These challenges manifest as rapidly increasing computational time and a demand for substantial computility. Consequently, for large-scale TSP
instances, the primary approaches employed today are heuristic algorithms, e.g., Ant
Colony Optimization (ACO) and Harris Hawk Optimization (HHO). For smaller-scale
Traveling Salesman Problems, the application of learning-based algorithms can enhance
solution efficiency. To further improve the efficiency of solving TSP, some research
combines heuristic algorithms with learning-based algorithms or employs heuristic algorithms for global path planning and heuristic algorithms for local path planning in solving
TSP instances. This paper aims to summarize recent developments in the TSP. Additionally, it will provide an analysis and outlook for the rapidly advancing learning-based
methods for solving the Traveling Salesman Problem, classical exact algorithms, and
heuristic algorithms.

L. Pan et al. (Eds.): BIC-TA 2023, CCIS 2062, pp. 3–16, 2024.
https://doi.org/10.1007/978-981-97-2275-4_1

2 An Overview of TSP Algorithms

As we delve into the Traveling Salesman Problem, it is essential to begin by grasping the mathematical model of the TSP and fundamental concepts associated with it. This chapter will provide an overview of the mathematical underpinnings of TSP and its related problems, followed by an exploration of how traditional TSP algorithms and cutting-edge methods are applied to tackle these challenges

2.1 TSP and Its Variants

Classical TSP can be described as follows: A salesperson begins their journey in one city to visits each city exactly once, and eventually returns to the city of departure. The goal is to minimize the total distance traveled by the salesperson while visiting all the cities. We represent this problem using a weighted graph, where V denotes the cities, and A represents the set of all arcs connecting pairs of cities. The formal formulation of this problem is as follows:

$$Z_{min} = \sum_{i \in V} \sum_{j \in V} d_{ij} x_{ij} \tag{1}$$

$$x_{ij} = \{0, 1\}(i, j \in V) \tag{2}$$

where d_{ij} represents the distance between two city points, and x_{ij} is the decision variable. Z_{min} is the shortest Hamiltonian circuit. x_{ij} equals 1 if the traveling salesman passes through cities; otherwise, it equals zero.

The TSP diagram consists of twenty city points, as shown in Fig. 1. Here, d_{ij} is the distance between two adjacent city points, and due to passing through city points i and j, the decision variable x_{ij} should be equal to 1.

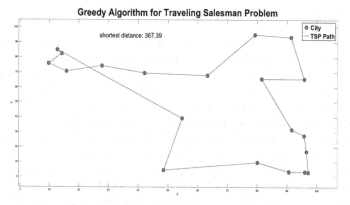

Fig. 1. TSP map obtained from 20 city points.

2.2 Traditional TSP Algorithms

The traditional methods for solving the Traveling Salesman Problem can be categorized into two types: exact algorithms and heuristic and Approximation Methods. Exact algorithms can directly compute the optimal solution, particularly for smaller instances, through methods like iteration. However, as the number of cities increases, the computational time required for exact solutions significantly grows, demanding high comp-utility. In such cases, the direct solving approach becomes impractical, and approximation algorithms often yield better results. Currently, exact algorithms primarily employ direct solving methods, while approximation algorithms mainly utilize heuristic algorithms to address the Traveling Salesman Problem. Additionally, learning-based solvi-ng methods can achieve exact solutions for small-scale TSP instances and can be combined with heuristic algorithms for approximate solving. A comparative analysis of the strengths and weaknesses of these algorithms is presented in Fig. 2.

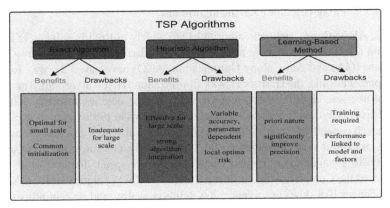

Fig. 2. Analysis of advantages and disadvantages of various algorithms

Exact Algorithms. In the realm of exact solution algorithms, Dynamic Programming (DP), Branch and Bound (BB), and Linear Programming (LP) are commonly used approaches. Dynamic Programming, introduced by R. Bellman and others in the 1950s [1], operates by iteratively finding local optimal solutions and then determining the global optimum. This algorithm reduces redundant calculations and preserves previous solution advantages. However, DP lacks a standardized model, and its computational demands increase significantly with problem size. To enable DP to handle larger-scale problems, Xu et al. (2020) [2] proposed a general framework that combines deep neural networks with dynamic programming to enhance DP's scalability for solving larger city point instances. Branch and Bound, initially presented by M. Karp and others in the 1960s [3], is fundamentally an enumeration process. It seeks to reduce ineffective computations compared to brute-force enumeration. However, this method has algorithmic space constraints. To mitigate these constraints, Zhang et al. (2023) [4] improved existing branch and bound algorithms. Their approach reduces space requirements by discovering and proving the optimality of solutions through partial sequence searches.

Linear Programming offers flexibility in practical applications but can be challenging to set up due to the complexity of constraint conditions, including potentially redundant constraints. Dong Chuanbo (2019) [5] introduced a relaxation algorithm to address the problem of eliminating subtour constraints in linear programming. Although exact solution algorithms have limitations when dealing with large-scale problems, combining them with heuristic algorithms makes it possible to tackle such challenges. Weise (2019) [6] proposed a new crossover operator for evolutionary algorithms (EAs) in the context of the Traveling Salesman Problem. This operator utilizes branch and bound to identify the optimal combinations of directed edges in parent solutions. When compared with ten other operators on instances with 110 cities, this method exhibited significant performance improvements. Branch and Bound can thus play a role in evolutionary algorithms by enhancing the efficiency of heuristic algorithms through the search for directed edges. In summary, recent research in standalone exact solution algorithms has seen limited breakthroughs. These algorithms often face limitations when dealing with large-scale TSP instances. Most approaches related to exact algorithms combine elements of learning-based or heuristic algorithms.

Heuristic Algorithms. Heuristic algorithms are widely utilized in solving TSP problems, with a focus on mimicking natural processes. Early algorithms such as Genetic Algorithms (GA) and Ant Colony Optimization (ACO) have paved the way for more recent approaches like the Harris's Hawks Optimization (HHO) algorithm. The performance of heuristic algorithms continues to improve with ongoing research. John Holland and his colleagues [7] introduced Genetic Algorithms in 1985. GA possesses strong global search capabilities, but it often struggles with local search and may prematurely converge or exhibit slow convergence rates. Additionally, GA can suffer from solution instability. To enhance solution quality and convergence speed, Toathom et al. (2022) [8] proposed the Complete Subtour Order Crossover (CSOX) algorithm for GA. However, challenges like slow convergence, low solution quality, and susceptibility to local optima persist. Addressing these concerns, Zhang et al. (2022) [9] proposed a Genetic Algorithm with Jumping Genes and Heuristic Operators to bolster GA's performance. Xu et al. (2022) [10] introduced a Bio-Inspired Heuristic Genetic Algorithm to mitigate instability issues associated with traditional GA. Ant Colony Optimization (ACO)[11], introduced by Dorigo et al. in 1992, excels at finding global optimal solutions with few parameters. However, ACO's convergence speed is influenced by population diversity and may fall into local optima, resulting in reduced precision. Yang et al. (2020) [12] introduced a novel game-based ACO (NACO) algorithm that combines ant colony systems and maximum-minimum ant systems with an entropy-based learning strategy. By adaptively increasing diversity, NACO enhances the precision of optimal solutions. Stodola et al. (2020) [13] combined Ant Colony Optimization (ACO) with simulated annealing to tackle dynamic combinatorial optimization problems, boosting convergence speed. To mitigate issues related to local optima, Skinderowicz et al. (2022) [14] proposed Focused ACO (FACO), which controls the difference in quantity between newly constructed and previously selected solutions, concentrating the search process. This approach effectively integrates specific problem-specific local optima while preserving the quality of existing solutions. Li (2023) [15] introduced a novel ant colony algorithm to address ACO's slow convergence to local optima and lack of computational precision. Furthermore, combining ACO with heuristic algorithms significantly enhances its local search capabilities. Mathurin Soh et al. (2022) [16] combined multi-population ant colony optimization with the Lin-Kernighan heuristic algorithm, resulting in substantial improvements in convergence speed and solution accuracy. In 2019, Heidari et al. [17] introduced the Harris's Hawks Optimization (HHO) algorithm, which excels in global search capabilities and high-dimensional convergence. However, HHO is prone to local optima and early convergence. Strengthening HHO's local search capabilities can enhance its overall performance. To improve population diversity and convergence speed while preventing local optima, Tang Andi et al. (2021) [18] proposed a Chaos-Elite Harris's Hawks Optimization (HHO) algorithm with random walks. Gharehchopogh et al. (2021) [19] presented a method that generates travel routes using random key encoding while preserving HHO's performance. This approach employs the LK local search mechanism to enhance search capabilities and uses the Metropolis acceptance strategy to avoid local optima. Hao Chen et al. (2020) [20] introduced the CMDHHO algorithm, significantly reducing HHO's convergence time. G. Hussien and Amin (2021) [21] proposed an improved version of HHO (IHHO), which significantly reduces the risk of falling into local optima and enhances the algorithm's convergence speed when applied to complex optimization tasks. In addition to the above algorithms, numerous other heuristic algorithms are available, some of which are listed in Table 1.

2.3 Learning-Based Algorithms

Since 1985 when J.J. Hopfield and D.W. Tank [38] employed Hopfield neural networks to tackle the TSP instances, the development of learning-based methods for solving TSP has gradually progressed. However, the earliest neural network approaches for solving TSP faced challenges such as converging to local optima and the difficulty of setting node penalties and weights. In recent years, methods based on graph neural networks and deep reinforcement learning have been applied to address the TSP. These approaches have shown promising results, particularly for small-scale TSP instances. Nevertheless, for larger-scale TSP instances, they are often combined with heuristic algorithms. Bengio et al. (2021) [39] suggested that using machine learning to solve combinatorial optimization problems may offer certain advantages over heuristic algorithms in decision-making processes. Common learning-based TSP solving methods are illustrated in Fig. 3.

Learning Heuristic Methods. In this section, we delve into the application of learning heuristic methods in solving the Traveling Salesman Problem (TSP). We begin by introducing reinforcement learning (RL) methods, which offer several common approaches to tackle TSP such as Q-Learning, Deep Q-Networks (DQN), etc. Transitioning to graph neural networks (GNNs), we discuss their role in solving TSP and strategies for improving GNN-based methods' speed and scalability. Finally, we summarize recent research developments, providing readers with a comprehensive overview of the latest trends and applications of RL and GNN methods in addressing the TSP.

Table 1. Partial heuristic algorithm

Time	The author	Algorithm
2006	Karaboga et, al.	Artificial Bee Colony Algorithm [22]
2008	Xin-She Yang	Firefly Algorithm [23]
2009	Xin-She Yang & Suash Deb	Cuckoo Search Algorithm [24]
2011	R.V. Rao	Teaching-Learning-Based Optimization Algorithm [25]
2011	Rui Tang et, al.	Wolf Pack Algorithm [26]
2014	Hojjat Emami & Farnaz Derakhshan	Election Algorithm [27]
2014	Seyedali Mirjalili	Grey Wolf Algorithm [28]
2016	Seyedali Mirjalili & Andrew Lewis	Whale Algorithm [29]
2016	Alireza Askarzadeh	Crow Search Algorithm [30]
2017	Seyedali Mirjalili et, al.	Tunicate Optimization Algorithm [31]
2018	Meeta Kumar et, al.	Social Evolution and Learning Optimization Algorithm [32]
2018	Shahrzad Saremi	Locust Optimization Algorithm [33]
2019	Sankalap Arora & Satvir Singh	Butterfly Optimization Algorithm [34]
2019	Mohit Jain et, al.	Squirrel Search Algorithm [35]
2019	Heidari et, al.	Harris's Hawk Algorithm
2020	Amin Foroughi Nematollahi	Golden Section Method [36]
2021	Jiankai Xue & Bo Shen	Sparrow Search Algorithm [37]

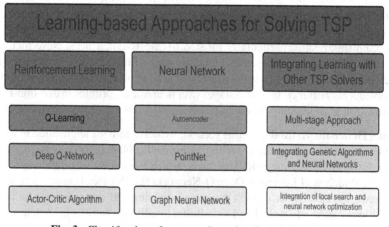

Fig. 3. Classification of common Learning-Based Algorithms

Reinforcement Learning (RL) offers several common methods for solving the Traveling Salesman Problem (TSP). These methods include modeling based on Markov Decision Processes (MDP), solving through Policy Gradient (PG) techniques, and using DQN for TSP. Figure 4 illustrates the Actor-Critic algorithm, a reinforcement learning technique that combines value functions with policy-based algorithms. Unlike action probability-based methods that update values only at the end of episodes, this algorithm updates values at each step. This approach reduces the variance in policy gradients, enhancing solution efficiency. In Fig. 4, the actor network progressively generates node selections, while the critic network evaluates the overall quality of the generated path. The reward signal is typically based on the length of the generated path, implying that the actor network's objective is to produce a solution that minimizes the path length. Shorter paths receive positive rewards, whereas longer paths incur negative rewards, guiding the actor network to learn and generate higher-quality paths. The entire framework involves the collaborative training of actor and critic networks, aiming to ultimately obtain a more effective algorithm for solving TSP instances.

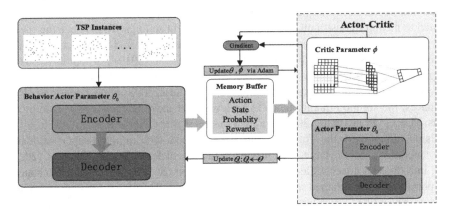

Fig. 4. Illustration of Actor-Critic algorithm

Reinforcement Learning (RL) offers certain advantages when compared to other solving algorithms for the Traveling Salesman Problem (TSP). Through learning and training, RL can significantly reduce solution times, especially for small-scale problems, where it outperforms exact solving methods in terms of speed. Kim et al. (2022) [40] improved the performance of deep reinforcement learning in solving the Traveling Salesman Problem by introducing the training method Symmetricity for Neural Combinatorial Optimization (SYM-NCO). Compared to traditional iterative solving, this approach dramatically enhances solution speed. RL can greatly enhance a system's generalization capabilities and reduce the requirements for initial solutions. However, it faces challenges related to accuracy and convergence speed when solving TSP. It struggles with solving large-scale TSP instances. To address this, Ouyang et al. (2021) [41] introduced a novel approach called MAGIC (MLP for M, Attention for A, GNN for G, Inter-leaved local search for I, and Curriculum Learning for C) to improve the performance of deep reinforcement learning. Experimental results demonstrate that MAGIC outperforms other deep reinforcement learning-based methods in terms of both performance and generalizability for solving large-scale city-point problems. It also excels in computational time compared to certain TSP heuristics. Furthermore, Y Xu et al. (2022) [42] introduced a scheduled data utilization strategy to stabilize the reinforcement learning process across different problem sizes. Experimental results demonstrate that this model achieved improved generalization to larger TSP instances not seen before in both zero-shot and few-shot scenarios. Addressing the issue of enhancing generalization opens up the potential for learning-based methods to tackle large-scale problems.

Graph Neural Networks (GNNs) find extensive application in solving TSP due to their ability to leverage node and edge features. Common methods for solving TSP using GNNs include Graph Convolution Networks (GCN) and Graph Attention Networks (GAT). GNN-based TSP solvers offer speed and simplicity, requiring relatively little training data to learn better approximations for solutions. Fei et al. (2022) [43] proposed an enhanced ant colony optimization algorithm based on graph convolutional networks. This algorithm improves the efficiency of solving the problem and enhances the capability to escape local optima. It is evident that solving the TSP using graph convolutional neural networks has considerable potential. However, they face challenges related to memory constraints and the impact of weight settings. To enhance the speed and scalability of GNN-based solvers, researchers have explored various approaches. In terms of both speed and solution quality, K. Joshi et al. (2019) [44] employed deep graph convolutional networks to construct TSP graphs and used Beam search to output paths in a non-autoregressive manner. Nevertheless, for solving large-scale TSP instances, the solution quality tends to be lower. For tackling even larger-scale problems, Marcelo Prates et al. (2019) [45] demonstrated that GNNs can learn how to solve decision variables for the Traveling Salesman Problem with minimal supervision and can generalize to handle large-scale instances. In a recent development, Kim et al. (2022) [46] introduced a novel Scale-Conditional Adaptation (SCA) scheme, which enhances the portability of pre-trained solvers to larger-scale tasks. This scheme assists pretrained models in adapting to even larger-scale problems effectively. Utilizing deep graph neural networks to construct efficient TSP graphs has shown significant improvements in solving efficiency. Martin J. A. Schuetz et al. (2022) [47] proposed a scalable GNN-based solver that efficiently addresses TSP instances at the million-city scale. Recent research in this domain is summarized in Table 2 below:

Table 2. Partly research on Learning-Based TSP in Recent Five Years

Time	The author	Solving TSP with learning-based methods
2018	Kool et, al.	Transformer and Attention Mechanism [48]
2018	Deudon et, al.	Combining Transformer Architecture with LKH3 [49]
2019	Joshi et, al.	Combining Neural Network PN with GCN
2019	Prates et, al.	Typed Graph Networks Model based on Graph Neural Networks
2020	Cappart et, al.	Combining Transformer Architecture with GAT (Graph Attention Network) [50]
2021	Bresson et, al.	Transformer Architecture and Reinforcement Learning Training [51]
2021	Omar Gutiérrez et, al.	Enhancing Generalization using Out-of-Distribution Application of Graph Neural Networks [52]
2021	Jiongzhi Zheng et, al.	Combining Q-Learning, Sarsa, and Monte Carlo with LKH [53]
2021	Wenbin Ouyang et, al.	Training Models in Deep Reinforcement Learning using Curriculum Learning [41]
2021	Zhang-Hua Fu et, al.	Using Deep Reinforcement Learning with Heatmaps [54]
2022	Chenchen Fu et, al.	Pointer Networks based on Graph Attention [55]
2023	Haoran Ma et, al.	Complementary GNN Methods for Multiple Models with Strong Generalization [56]
2023	Elīza Gaile et, al.	Graph Neural Networks with Loss Functions [57]
2023	Nasrin Sultana et, al.	Combining Convolutional Neural Networks with Long Short-Term Memory [58]
2023	Yan Jin et, al.	End-to-End Deep Reinforcement Learning based on Multi-Pointer Transformers [59]

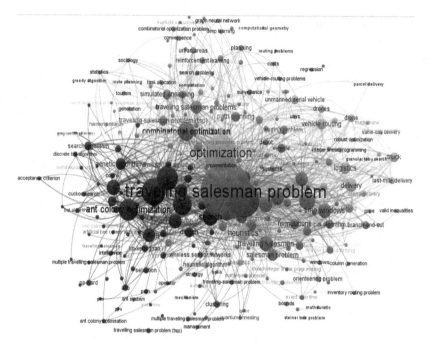

Fig. 5. Collinear analysis by Vosviewer

3 Conclusion and Future Prospects

A total of 2,000 SCI papers from 2018 to 2023 were collected using Web of Science, and a TSP keyword co-occurrence analysis was conducted using VOSviewer, as shown in Fig. 5. This analysis reveals that exact algorithms have found partial applications in solving TSP-related problems. However, their scalability remains a challenge, and they are often combined with heuristic algorithms. In recent years, the primary approach for solving TSP problems has been heuristic algorithms, with a focus on nature-inspired algorithms. Among these, Genetic Algorithms (GA) and Ant Colony Optimization (ACO) have been widely used. While there are various heuristic algorithms, many do not yield satisfactory results, which limits their widespread adoption. Therefore, several metaheuristic algorithms incorporate the LK local search algorithm to enhance their performance.

Learning-based methods have not yet become the preferred approach for solving TSP. However, they differ distinctly from previous algorithms and offer significant advantages in terms of solving time. In the future, learning-based methods may become the mainstream approach, but they need further optimization to address issues such as convergence and stability during model training.

Derived problems from TSP, such as vehicle routing and orienteering problems, have practical deployments, and they are expected to consider additional factors in the future.

In future research, a synergistic blend of algorithms that have not been extensively studied or algorithms with stable performance will be used to tackle larger-scale problems, with an emphasis on improving solution quality, avoiding local optima, and enhancing solving efficiency. More algorithms for solving TSP problems will be created, leading to significant efficiency improvements. Future TSP will also consider factors beyond distance to meet real-world application requirements. Solving TSP problems using learning-based algorithms, coupled with enhanced graph neural networks, may enable the efficient solution of large-scale TSP problems, and the development of new models can boost solution efficiency. Breakthroughs in learning-based TSP solving methods are expected in the future. With the increasing computational power and the emergence of new solvers, significant improvements in TSP instances solving speed, scale, and quality are anticipated. The application of superior solvers will bring greater convenience to future life.

References

1. Bellman, R.: Dynamic programming. Science **153**(3731), 34–37 (1966)
2. Xu, S., Panwar, S.S., Kodialam, M., Lakshman, T.V.: Deep neural network approximated dynamic programming for combinatorial optimization. In: Proceedings of the AAAI Conference on Artificial Intelligence, vol. 34, pp. 1684–1691 (2020)
3. Lawler, E.L., Wood, D.E.: Branch-and-bound methods: a survey. Oper. Res. **14**(4), 699–719 (1966)
4. Zhang, W., Sauppe, J.J., Jacobson, S.H.: Results for the close-enough traveling salesman problem with a branch-and-bound algorithm. Comput. Optim. Appl. **85**(2), 369–407 (2023)
5. Donog, C.: A relaxation algorithm for solving the traveling salesman problem. Shandong Sci. **32**(4), 74–79 (2019)
6. Weise, T., Jiang, Y., Qi, Q., Liu, W.: A branch-and-bound-based crossover operator for the traveling salesman problem. Int. J. Cogn. Inform. Nat. Intell. (IJCINI) **13**(3), 1–18 (2019)
7. Holland, J.H.: Genetic algorithms. Sci. Am. **267**(1), 66–73 (1992)
8. Toathom, T., Champrasert, P.: The complete subtour order crossover in genetic algorithms for traveling salesman problem solving. In: 2022 37th International Technical Conference on Circuits/Systems, Computers and Communications (ITC-CSCC), pp. 904–907. IEEE (2022)
9. Zhang, P., Wang, J., Tian, Z., Sun, S., Li, J., Yang, J.: A genetic algorithm with jumping gene and heuristic operators for traveling salesman problem. Appl. Soft Comput. **127**, 109339 (2022)
10. Xu, J., Han, F., Liu, Q., Xue, X.: Bioinformation heuristic genetic algorithm for solving TSP. J. Syst. Simul. **34**(8), 1811–1819 (2022)
11. Dorigo, M.: Optimization, learning and natural algorithms. Ph.D. thesis, Politecnico di Milano (1992)
12. Yang, K., You, X., Liu, S., Pan, H.: A novel ant colony optimization based on game for traveling salesman problem. Appl. Intell. **50**, 4529–4542 (2020)
13. Stodola, P., Michenka, K., Nohel, J., Rybanský, M.: Hybrid algorithm based on ant colony optimization and simulated annealing applied to the dynamic traveling salesman problem. Entropy **22**(8), 884 (2020)
14. Skinderowicz, R.: Improving ant colony optimization efficiency for solving large tsp instances. Appl. Soft Comput. **120**, 108653 (2022)
15. Li, W., Wang, C., Huang, Y., Cheung, Y.M.: Heuristic smoothing ant colony optimization with differential information for the traveling salesman problem. Appl. Soft Comput. **133**, 109943 (2023)

16. Soh, M., Tsofack, B.N., Djamegni, C.T.: A hybrid algorithm based on multi-colony ant optimization and lin-kernighan for solving the traveling salesman problem. Rev. Afr. Recherche Inform. Math. Appl. **35** (2022)

17. Heidari, A.A., Mirjalili, S., Faris, H., Aljarah, I., Mafarja, M., Chen, H.: Harris hawks optimization: algorithm and applications. Futur. Gener. Comput. Syst. **97**, 849–872 (2019)

18. Tang, A., Han, T., Xu, D., Xie, L.: Chaotic elite Harris' hawk optimization algorithm. Computer Applications **41**(8), 2265–2272 (2021)

19. Gharehchopogh, F.S., Abdollahzadeh, B.: An efficient Harris hawk optimization algorithm for solving the travelling salesman problem. Clust. Comput. **25**(3), 1981–2005 (2022)

20. Chen, H., Heidari, A.A., Chen, H., Wang, M., Pan, Z., Gandomi, A.H.: Multi-population differential evolution-assisted Harris hawks optimization: framework and case studies. Futur. Gener. Comput. Syst. **111**, 175–198 (2020)

21. Hussien, A.G., Amin, M.: A self-adaptive harris hawks optimization algorithm with opposition-based learning and chaotic local search strategy for global optimization and feature selection. Int. J. Mach. Learn. Cybern. 1–28 (2022)

22. Basturk, B.: An artificial bee colony (ABC) algorithm for numeric function optimization. In: IEEE Swarm Intelligence Symposium, Indianapolis, USA, vol. 2006, p. 12 (2006)

23. Yang, X.S.: Firefly algorithm, stochastic test functions and design optimisation. Int. J. Bio-Inspir. Comput. **2**(2), 78–84 (2010)

24. Yang, X. S., Deb, S.: Cuckoo search via Lévy flights. In: 2009 World Congress on Nature & Biologically Inspired Computing (NaBIC), pp. 210–214. IEEE (2009)

25. Rao, R.V., Savsani, V.J., Vakharia, D.P.: Teaching–learning-based optimization: a novel method for constrained mechanical design optimization problems. Comput. Aided Des. **43**(3), 303–315 (2011)

26. Tang, R., Fong, S., Yang, X. S., Deb, S.: Wolf search algorithm with ephemeral memory. In: Seventh International Conference on Digital Information Management (ICDIM 2012), pp. 165–172. IEEE (2012)

27. Emami, H., Derakhshan, F.: Election algorithm: a new socio-politically inspired strategy. AI Commun. **28**(3), 591–603 (2015)

28. Mirjalili, S., Mirjalili, S.M., Lewis, A.: Grey wolf optimizer. Adv. Eng. Softw. **69**, 46–61 (2014)

29. Mirjalili, S., Lewis, A.: The whale optimization algorithm. Adv. Eng. Softw. **95**, 51–67 (2016)

30. Askarzadeh, A.: A novel metaheuristic method for solving constrained engineering optimization problems: crow search algorithm. Comput. Struct. **169**, 1–12 (2016)

31. Mirjalili, S., Gandomi, A.H., Mirjalili, S.Z., Saremi, S., Faris, H., Mirjalili, S.M.: Salp swarm algorithm: a bio-inspired optimizer for engineering design problems. Adv. Eng. Softw. **114**, 163–239 (2017)

32. Kumar, M., Kulkarni, A.J., Satapathy, S.C.: Socio evolution & learning optimization algorithm: a socio-inspired optimization methodology. Futur. Gener. Comput. Syst. **81**, 252–272 (2018)

33. Saremi, S., Mirjalili, S., Lewis, A.: Grasshopper optimization algorithm: theory and application. Adv. Eng. Softw. **105**, 30–47 (2017)

34. Arora, S., Singh, S.: Butterfly optimization algorithm: a novel approach for global optimization. Soft. Comput. **23**, 715–734 (2019)

35. Jain, M., Singh, V., Rani, A.: A novel nature-inspired algorithm for optimization: squirrel search algorithm. Swarm Evol. Comput. **44**, 148–175 (2019)

36. Nematollahi, A.F., Rahiminejad, A., Vahidi, B.: A novel meta-heuristic optimization method based on golden ratio in nature. Soft. Comput. **24**, 1117–1151 (2020)

37. Xue, J., Shen, B.: A novel swarm intelligence optimization approach: sparrow search algorithm. Syst. Sci. Control Eng. **8**(1), 22–34 (2020)

38. Hopfield, J.J., Tank, D.W.: "Neural" computation of decisions in optimization problems. Biol. Cybern. **52**(3), 141–152 (1985)
39. Yoshua, B., Andrea, L., Antoine, P.: Machine learning for combinatorial optimization: a methodological tour d'horizon. Eur. J. Oper. Res. **290**(2), 405–421 (2021)
40. Kim, M., Park, J., Park, J.: Sym-nco: leveraging symmetricity for neural combinatorial optimization. arXiv preprint arXiv:2205.13209 (2022)
41. Ouyang, W., Wang, Y., Weng, P., Han, S.: Generalization in deep RL for TSP problems via equivariance and local search. arXiv preprint arXiv:2110.03595 (2021)
42. Xu, Y., Fang, M., Chen, L., Du, Y., Xu, G., Zhang, C.: Shared dynamics learning for large-scale traveling salesman problem. Adv. Eng. Inform. **56**, 102005 (2023)
43. Fei, T., Wu, X., Zhang, L., Zhang, Y., Chen, L.: Research on improved ant colony optimization for the traveling salesman problem. Math. Biosci. Eng. **19**(8), 8152–8186 (2022)
44. Joshi, C.K., Laurent, T., Bresson, X.: On learning paradigms for the traveling salesman problem. arXiv preprint arXiv:1910.07210 (2019)
45. Prates, M., Avelar, P.H., Lemos, H., Lamb, L.C., Vardi, M.Y.: Learning to solve NP-complete problems: a graph neural network for decision TSP. In: Proceedings of the AAAI Conference on Artificial Intelligence, vol. 33, no. 01, pp. 4731–4738 (2019)
46. Kim, M., Jiwoo, S.O.N., Kim, H., Park, J.: Scale-conditioned adaptation for large scale combinatorial optimization. In: NeurIPS 2022 Workshop on Distribution Shifts: Connecting Methods and Applications (2022)
47. Schuetz, M.J., Brubaker, J.K., Katzgraber, H.G.: Combinatorial optimization with physics-inspired graph neural networks. Nat. Mach. Intell. **4**(4), 367–377 (2022)
48. Kool, W., Van Hoof, H., Welling, M.: Attention, learn to solve routing problems!. arXiv preprint arXiv:1803.08475 (2018)
49. Deudon, M., Cournut, P., Lacoste, A., Adulyasak, Y., Rousseau, L.-M.: Learning heuristics for the TSP by policy gradient. In: van Hoeve, W.-J. (ed.) CPAIOR 2018. LNCS, vol. 10848, pp. 170–181. Springer, Cham (2018). https://doi.org/10.1007/978-3-319-93031-2_12
50. Cappart, Q., Moisan, T., Rousseau, L.M., et al.: Combining reinforcement learning and constraint programming for combinatorial optimization. arXiv:2006.01610 (2018)
51. Bresson, X., Laurent, T.: The transformer network for the traveling salesman problem. arXiv preprint arXiv:2103.03012 (2021)
52. Gutiérrez, O., Zamora, E., Menchaca, R.: Graph representation for learning the traveling salesman problem. In: Roman-Rangel, E., Kuri-Morales, Á.F., Martínez-Trinidad, J.F., Carrasco-Ochoa, J.A., Olvera-López, José Arturo. (eds.) MCPR 2021. LNCS, vol. 12725, pp. 153–162. Springer, Cham (2021). https://doi.org/10.1007/978-3-030-77004-4_15
53. Zheng, J., He, K., Zhou, J., Jin, Y., Li, C.M.: Combining reinforcement learning with Lin-Kernighan-Helsgaun algorithm for the traveling salesman problem. In: Proceedings of the AAAI Conference on Artificial Intelligence, vol. 35, vol. 14, pp. 12445–12452 (2021)
54. Fu, Z. H., Qiu, K. B., Zha, H.: Generalize a small pre-trained model to arbitrarily large TSP instances. In: Proceedings of the AAAI Conference on Artificial Intelligence, vol. 35, no. 8, pp. 7474–7482 (2021)
55. Fu, C., et al.: A learning approach for multi-agent travelling problem with dynamic service requirement in mobile IoT. Comput. Electr. Eng. **104**, 108397 (2022)
56. Ma, H., Tu, S., Xu, L.: IA-CL: a deep bidirectional competitive learning method for traveling salesman problem. In: Tanveer, M., Agarwal, S., Ozawa, S., Ekbal, A., Jatowt, A. (eds.) ICONIP 2022, Part I. LNCS, vol. 13623, pp. 525–536. Springer, Cham (2023). https://doi.org/10.1007/978-3-031-30105-6_44
57. Gaile, E., Draguns, A., Ozoliņš, E., Freivalds, K.: Unsupervised training for neural TSP solver. In: Simos, D.E., Rasskazova, V.A., Archetti, F., Kotsireas, I.S., Pardalos, P.M. (eds.) LION 2022. LNCS, vol. 13621, pp. 334–346. Springer, Cham (2023). https://doi.org/10.1007/978-3-031-24866-5_25

58. Sultana, N., Chan, J., Sarwar, T., Qin, A.K.: Learning to optimise general TSP instances. Int. J. Mach. Learn. Cybern. **13**(8), 2213–2228 (2022)
59. Jin, Y., et al.: PointerFormer: deep reinforced multi-pointer transformer for the traveling salesman problem. arXiv preprint arXiv:2304.09407 (2023)
60. Wang, Y., Chen, Z., Yang, X., Wu, Z.: Deep reinforcement learning combined with graph attention model to solve TSP. J. Nanjing Univ. (Nat. Sci.) **58**(3), 420–429 (2022)
61. Zhang, S., Guo, G.: A review of the multi-traveling salesman model and its applications. Comput. Sci. Explor. **16**(7), 1516 (2022)
62. Dong, S., Wang, P., Abbas, K.: A survey on deep learning and its applications. Comput. Sci. Rev. **40**, 100379 (2021)
63. Wang, Y., Chen, Z., Wu, Z., Gao, Y.: Review of reinforcement learning for combinatorial optimization problem. J. Front. Comput. Sci. Technol. **16**(2), 261–279 (2022)

MAD-SGS: Multivariate Anomaly Detection with Multi-scale Self-learned Graph Structures

Junnan Tang, Dan Li[✉], and Zibin Zheng

School of Software Engineering, Sun Yat-Sen University, Zhuhai 528478, China
tangjn5@mail2.sysu.edu.cn, {lidan263,zhzibin}@mail.sysu.edu.cn

Abstract. Cyber-Physical Systems (CPSs) integrate sensing, computation, cybernetics, and networking to control a hybrid physical system consisting of different functional subsystems. Accurate and efficient
anomaly detection for Multivariate Time Series (MTS) with rich temporal and spatial information generated by highly intertwined sensors and
actuators in CPS to reduce the negative influence caused by abnormal-
ities. Since existing methods with temporal modeling cannot effectively
recognize anomalies without sufficiently utilizing spatial correlations conveyed by MTS data, Graph Convolution Networks (GCNs) have recently
been exploited to extract spatial correlations in MTS data. However,
most previous works that utilized spatial information focused on learning
static long-term graph structures and did not explore the potential variations of short-term spatial correlations. This paper proposed a Multivariate Anomaly Detection framework with multi-scale Self-learned Graph
Structures (MAD-SGS) based on the Variational Autoencoder (VAE)
architecture. Specifically, the Long Short-Term Memory (LSTM) was
applied to extract and exploit the temporal information, and the Graph
Convolution Network (GCN) was employed to exploit spatial correlations at different scales (the long-term static correlation and short-term
dynamic correlation). Besides, we utilized a self-learning approach for
long-term graph static structure learning and employed feature similarity
to learn short-term dynamic graph structures. The proposed MAD-SGS
framework was tested on four datasets collected from three real-world
CPSs: the Secure Water Treatment (SWaT), the Water Distribution
(WADI), and the BATtle of Attack Detection Algorithm (BATADAL)
datasets. Experimental results indicated that the proposed MAD-SGS
outperformed the state-of-the-art methods.

Keywords: Cyber-Physical Systems · Anomaly Detection · Graph
Convolution Networks · Spatial-Temporal Mining

1 Introduction

Today's Cyber-Physical Systems (CPSs), such as smart buildings, factories,
power plants, water distribution systems, and automatic driving systems, are

L. Pan et al. (Eds.): BIC-TA 2023, CCIS 2062, pp. 17–31, 2024.
https://doi.org/10.1007/978-981-97-2275-4_2

large, complex, and affixed with highly intertwined sensors, actuators, and controllers that generate substantial amounts of Multivariate Time Series (MTS) data [1,7]. Since CPSs are usually designed for mission-critical tasks, they are vulnerable to worn-out equipment, malfunctioned controllers, and cyber-attacks [23]. Leaving disregarded, CPS anomalous behaviors would lead to serious economic losses or safety issues. Thus, It is important to closely monitor the CPS behaviors for abnormal events through anomaly detection using the MTS data generated by the systems.

Anomalies in CPSs are commonly defined as points at certain timestamps where the system's behavior differs obviously from the normal status, and the proportion of abnormal data is usually low due to infrequent occurrences [10]. The work aims at addressing the anomalies caused by cyber-attacks, which pose a greater threat to CPSs with the advancement of internet technologies. Statistically, MTS data encompasses multiple interdependent time series, necessitating the simultaneous mining of spatial-temporal information. Physically, the spatial information of different sub-systems and functional parts reflect the interdependencies among CPS nodes. Hence, accurate mining of spatial-temporal information poses a challenge in anomaly detection in CPSs.

Previously, Machine Learning (ML)-based methods applied to CPS anomaly detection treated MTS data as a collection of individual vectors resulting in the neglect of the spatial-temporal information [11,12]. To capture temporal information, Convolutional Neural Networks (CNNs) and Recurrent Neural Networks (RNNs), such as Temporal Convolution Networks (TCN) and Long Short-Term Memory (LSTM), have been integrated into deep neural networks. He and Zhao [8] trained the TCN on normal sequences and used it to predict trends in several time steps. The Encoder-Decoder scheme for Anomaly Detection (EncDec-AD) proposed in [15] implemented LSTM to the encoder-decoder architecture to reconstruct the MTS sequences. These studies did not specifically mention extracting spatial information. With the advancement of Graph Neural Networks (GNN), the Spatial-temporal Graph Neural Network (STGNN) model has been employed to extract spatial-temporal information from MTS data. [3] treated nodes of traffic trajectories (another type of MTS data) as points on an image and utilized CNN to process images at different locations to capture the spatial correlations. In CPSs, the spatial information conveyed by sensors/actuators only relates to the system's working status, which is not as intuitive as the temporal correlation reflected by timestamps. As a result, CNN and RNN are unsuitable for handling hidden spatial correlations in MTS. Given the excellent performance of Graph Neural Networks (GNNs) in capturing spatial properties in networks, one promising way to capture and deal with spatial information is to employ GNNs. MTAD-GAT [24] set two graph attention layers with complete graphs to capture the relationships between multiple features and the temporal information between different timestamps. Many works have attempted the approach of STGNNs and achieved promising results. However, they did not distinguish the multi-scaled spatial correlations varying alone time.

Although the physical connection of CPS can be regarded as a naive spatial correlation, it cannot accurately reflect the implicit correlations among sensors/actuators. Therefore, finding underlying spatial correlation and combining it with temporal dependence in MTS anomaly detection tasks for CPS is challenging. Recently, graph structures mined from MTS data have served as essential tools for anomaly detection. As a forerunner, MAD-SGCN [19] adopted GCN and LSTM as the core models of the encoders and the decoder in the variational autoencoder framework to capture the spatial-temporal information and the adjustment matrix, i.e., the structure of the graph, is generated from self-supervised learning with sparsity and uncertainty constraints automatically. Authors of [22] considered a static graph that would not change after enough training iterations based on the assumption that the relationship among nodes is relatively stable during long-term running. However, it is intuitive that although the relative physical positions of sensor/actuator nodes are unchanged, the working status of CPS would always change over time, which indicates that the potential correlation among nodes is indeed dynamic rather than static. Thus, carefully considering stable long-term static graphs and variable short-term dynamic graphs could provide rich spatial information to guide anomaly detection. To the best of the authors' knowledge, few works have considered spatial information from the short-term period for anomaly detection tasks.

To employ a combination of stable long-term graphs and variable short-term graphs for CPSs' anomalies detection task, in this paper, we proposed a Multivariate time series Anomaly Detection framework with multi-scale Self-learned Graph Structures (MAD-SGS) based on sequence reconstruction. We deployed Graph Convolution Networks (GCN) and LSTM as the encoder and decoder to extract and exploit the spatial and temporal information in the MTS sequence, respectively. In addition, we captured the spatial correlation in the system using similarity measures for short-term graph structures and applied a self-learning method for long-term graph structures. A self-attention mechanism is set to measure the importance of different nodes.

The contributions of this paper are summarized as follows:

1. We proposed an unsupervised anomaly detection framework for multivariate time series using GCN and LSTM to capture spatial-temporal information effectively.
2. We adopted a self-learning method to construct the long-term graph structure and the nodes' behavior similarity and propagation to form the short-term graph structure.
3. Extensive experiments on three large raw public CPS datasets demonstrated the effectiveness of the proposed method. An ablation study is also conducted to verify the contribution of each sub-component of the proposed method.

The remaining part of this paper is organized in the following order. Section 2 introduces the related work of this article from four progressive aspects. The model is described in detail in Sect. 3. To verify the model's performance, experiments on three raw datasets from iTrust are arranged in Sect. 4, and the results are analyzed. The last Sect. 5 summarizes the results and prospects.

2   Related Work

2.1   Machine Learning and Traditional Deep Learning Method

The traditional unsupervised method treats each timestamp individually and applies outlier detection algorithms to find the points which is the minority of the data. As a kernel-based method, One-class SVM (OCSVM) [20] tries to find a supporting hyperplane in a kernel feature space so that all normal points would be separated from the origin, thus anomalies' distance to the separating hyperplane could be different. Based on density metric, Breunig et al. [4] presented a Local Outlier Factor (LOF), in which each point would calculate an outlier factor and get a LOF score to find outliers as anomaly points. Defining anomalies as easily isolated outliers, Liu et al. [14] designed iForest (Isolation Forest) in which anomalies would be isolated closer to the tree's root. Due to the feature extraction ability of the deep learning method, unsupervised deep learning models have indicated encouraging performance in recent years. Deep AutoEncoding Gaussian Mixture Model (DAGMM) [26] combines a deep autoencoder that targets to learn a low-dimensional representation and a Gaussian Mixture Model (GMM) that utilizes the reduced representation and the reconstruction error as input and evaluates the energy for each sample where higher energy demonstrates more anomalies. These unsupervised machine learning and deep learning methods cannot capture temporal and spatial information in multivariate time series data, resulting in limited performance.

2.2   Deep Learning Method with Temporal Modeling

Convolutional Neural Networks (CNN) and Recurrent Neural Networks have utilized temporal information between timestamps. Li et al. [10] propound a GAN-based model Multivariate Anomaly Detection with GAN (MAD-GAN), which employs LSTM-RNN as the base model of the generator and discriminator to extract the temporal information from normal multivariate sequences. The anomaly score considers the discrimination loss and the residual loss together. Audibert et al. [2] demonstrated UnSupervised Anomaly Detection for multivariate time series (USAD); two autoencoders with the same encoder are trained to reconstruct the normal sequence well. After that, adversarial training is performed between these two autoencoders to maximize the reconstruction error of the input data reconstructed by the other autoencoder. These CNN-based and GNN-based works have effectively learned the temporal information in multivariate time series. However, spatial correlation between the nodes in multivariate time series could not be captured without spatial modeling.

2.3   Deep Learning Method with Spatial-Temporal Modeling

Spatial information is practical for multivariate time series anomaly detection and temporal information. Huang et al. [9] proposed Spatio-Temporal Pattern Network (STPN) for a traffic system based on the concepts of symbolic dynamics

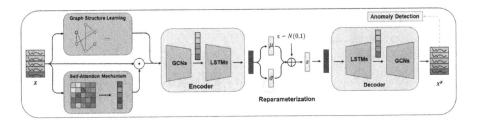

Fig. 1. An illustration of the proposed MAD-SGS framework

filtering. Custom domain knowledge-based partitioning is applied to transform the continuous time-series data into a 6-symbol sequence. Then, the mutual information (MI) is used to measure the correlations between the sensors, and the structural similarity (SSIM) indices are utilized to identify the anomalous days by measuring the similarity between each day to the other days. Miele et al. [17] propose an original unsupervised deep anomaly detection framework with a neural architecture combining AEs and Graph Convolutional Networks (GCNs) at its core. The encoder and decoder are based on GCNs, and the graph structure comprises the mutual information (MI) between them. During detection, by defining local and global indicators based on the model reconstruction errors, the framework triggers warnings after applying a four-stage threshold method to minimize false alarms during normal operating conditions. The above works try to combine the graph neural network to use the spatial information in the multivariate time series. Recently, some traffic forecasting works have attempted to deploy a dynamic graph learning method to capture and utilize more spatial information from multivariate time series. Li et al. [13] propose a static and dynamic graph learning network (SDGS) for multivariate time series forecasting, in which two graphs are learned for static and dynamic graphs, respectively. The static graph is developed to capture the fixed long-term pattern via node embedding, and dynamic graphs are time-varying matrices based on changing node-level features to model dynamic dependencies over the short term. To track the spatial dependencies among traffic data, Diao et al. [6] propose a dynamic spatial-temporal GCNN for accurate traffic forecasting whose core is the finding of the change of Laplacian matrix with a dynamic Laplacian matrix estimator. The real-time traffic data are decomposed into a stable global component based on long-term temporal-spatial traffic relationships and a local component based on tensor decomposition.

3   Methodology

3.1   Architecture

Figure 1 shows the overall architecture of our proposed model. MAD-SGS reconstructed the input sequence based on an encoder-decoder architecture, where different parameterized GCN and LSTM are applied in the encoder and decoder.

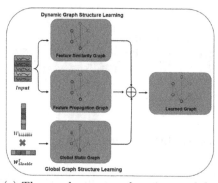
(a) The graph structure learning module

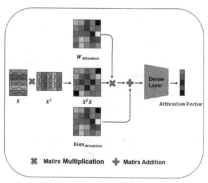
(b) The self-attention mechanism

Fig. 2. An illustration of the proposed MAD-SGS submodule

The reparameterization trick is utilized for optimization. Specifically, the input sequence is multiplied with attention from the "self-attention mechanism" and fed into the encoder along with the matrix obtained from the "graph structure learning module" and encoded by GCNs and LSTMs. Then the sequence was used to generate μ and σ. These two variables are combined with a standard Gaussian distribution for the reparameterization trick. The latent variable z will proceed into the decoder, generating the final reconstructed sequence.

The graph structure learning module in Fig. 2(a) consists of a dynamic graph structure learning module and a static graph structure learning module. The dynamic part captures the spatial information in the input sequence from two aspects and generates the feature similarity and propagation graphs. The static part multiplies a weight vector to obtain a global graph. Then three graphs are summed to get the learned graph.

As shown in Fig. 2(b), a self-attention mechanism enhances learning capability. After element-wise multiplication, the input sequence is multiplied by a weight matrix and added to another weighted matrix. Then a dense layer compresses it to gain the attention of nodes.

3.2   Graph Structure Learning

We utilize the Graph Convolution Network (GCN) to extract the spatial information of CPSs' MTS. GCN has accomplished decent success in many graph data mining tasks due to its ability to exploit arbitrary graph structures and manifolds. Especially, given a graph $G = (V, E)$ in which V is vertices set, and E is edges set, its structure can be represented by a $N_v \times N_v$ adjacency matrix A defined as

$$A_{ij} = \begin{cases} w(i,j), & e_{ij} \in E \\ 0, & otherwise \end{cases} \tag{1}$$

The following formula shows the propagation rule of the GCN layers.

$$H^{l+1} = f(H^l, A) = \sigma(D^{-\frac{1}{2}}\tilde{A}D^{-\frac{1}{2}}H^lW^l), \tag{2}$$

where l is the layer index. Considering the $l + 1^{th}$ layer, H^l denotes the feature from the lth layer, with $H^0 = X$ corresponding to the input of the GCN. W^l is a training matrix, and σ is the activation function, usually ReLU or Sigmoid.

The long-term graph structure is unaffected by nodes' behavior, making it static and capable of reflecting global spatial information. The short-term graph structure is dynamic and capable of capturing local spatial information as it is computed based on the current behavior of nodes. Since the adjacency matrix A is not explicitly given in CPSs' dataset, we apply a self-learning method to learn the steady long-term graph structure and employ a feature similarity measure method to learn the changing short-term graph structure.

Long-Term Graph Structure. The long-term stable operational state of the system can be represented as a static graph. This graph should be symmetric, implicit, and not influenced by the states of the nodes. Therefore, we apply a matrix A_g obtained by multiplying a randomly initialized weight matrix W_{AG} by itself, and hand it over to the model for self-learning to capture the long-term operational state of the system.

$$A_g = \sigma(W_{AG}^T W_{AG}) \tag{3}$$

where σ is a sigmoid activate function to make the element in the adjustment matrix belong to $(0, 1)$

Short-Term Graph Structure. Taking the readings as features of the nodes, inspired by [25], we calculate two graph structures by feature similarity measure and combine them to form the short-term graph. These two graph structures are based on the following two assumptions respectively: (1) Feature similarity graph. If two nodes show similar trends in changes in the readings at the same time, there is a strong mutual influence between them. (2) Feature propagation graph. In two nodes with a strong mutual influence, if one is influenced by a third node through a physical connection, there is a higher likelihood that the other node will also be influenced by the third one, i.e., the influence of the third node propagates. Similarly, if one node influences a third node through a physical connection, there is a higher likelihood that the other node will influence the third node through its influence on the other, i.e., the influence between them propagates to the third one.

The feature similarity graph G^S represents the similarity of the features of two sensors/actuators, thus reflecting their degree of correlation. The edges of G^S are obtained by

$$G^S(i, j) = \begin{cases} s^S(f_i, f_j) & s^S(f_i, f_j) \geq \epsilon^S \\ 0 & otherwise \end{cases} \tag{4a}$$

$$s^S(f_i, f_j) = \frac{1}{K} \sum_{k=1}^{K} \cos(w_{k,i}^S \odot f_i, w_{k,j}^S \odot f_j) \tag{4b}$$

where $\epsilon^S \in [0, 1]$ is the threshold guaranteeing sparsity of G^S, $s^S(f_i, f_j)$ calculates the similarity and $W_{k,i}^S = \{w_{k,i}^S\}$ is a learning parameter matrix to enhance randomness.

The feature propagation graph consists of two parts: calculating the similarity of head nodes to obtain possible tail nodes G^{PH}, and calculating the similarity of tail nodes to obtain possible head nodes G^{PT}.

The similarity of the calculated head node G^P is designed the same as G^S in (4a). Similar to ϵ^S and $s^S(f_i, f_j)$ in (4b), ϵ^P and $s^P(f_i, f_j)$ with different learning matrix W^P control sparsity of G^P. With these, we can obtain the possible influenced tail nodes as follows,

$$G^{PH} = G^P A_o \tag{5}$$

in which A_o is a physical connection graph that doesn't change throughout the entire system run time. The G^{PT} is calculated from $G^{PT} = A_o G^P$, The short-term graph A_{cur} is represented by fusing these graphs through a channel attention layer as The short-term graph synthesizes the above three aspects based on node reading features. Therefore, a weight combination is applied to fuse the three graphs and generate the short-term dynamic graph A_s.

$$A_s = \sigma(w_S, w_{PH}, w_{PT}) \cdot (G^S, G^{PH}, G^{PT}) \tag{6}$$

where $w_S + w_{PH} + w_{PT} = 1$ and σ is a softmax activate function.

The final adjustment matrix A utilized to capture spatial information is defined as follows,

$$A = \sigma(w_g, w_s) \cdot (A_g, A_s) \tag{7}$$

where $w_g + w_s = 1$ and σ is a softmax activate function. The short-term graph A_s captures the mutual interactions among nodes over a short period, while the long-term graph A_g captures the relationships among nodes during stable system operation. Through this combined operation, The spatial matrix A can simultaneously consider both effects.

3.3   Sequence Reconstruction Module

Models for anomaly detection tasks typically adopt prediction or sequence reconstruction methods to learn normal data patterns. We establish a sequence reconstruction model based on the encoder-decoder paradigm. We leverage the reparameterization trick in VAE architecture to avoid overfitting and ensure that the latent space has good properties.

Besides, the Graph Convolutional Network (GCN) and LSTM have been used as autoencoders [15,17,19]. Hence GCNs and LSTMs are set respectively to encode the spatial and temporal information of the multivariate sequences. The graph structure A for GCNs comes from the graph structure learning module. Specifically, the multivariate sequence X_n is first fed into the GCN for spatial information. The LSTM further encodes the generated feature X_n^{GE} to extract the temporal information. The latent variable is sampled to generate the decoding representation through the reparameterization trick. The decoding part also consists of LSTM and GCN and generates X_n^{GD}, and the loss function of reconstruction is defined as

$$L_{rec} = \frac{1}{N} \sum_{n=1}^{N} (MSE(X_n, X_n^R) + MSE(X_n^{GE}, X_n^{GD})) \tag{8}$$

where MSE denotes the Mean Squared Error, X_n means input sequences, and X_n^R means reconstructed sequence. In this way, L_{rec} can both consider the performance of capturing the spatial and temporal information. In addition, a KL divergence between the prior distribution $p(z)$ and posterior distribution $q(z|X_n)$ pushes the latter towards the former and is written as

$$L_{KLD} = \frac{1}{2} \sum_{d=1}^{D_z} (-\log \sigma_d^2 + \mu_d^2 + \sigma_d^2 - 1) \tag{9}$$

where D_z represents the dimension of the latent variable z.

We employ a self-attention mechanism to separate the importance of different nodes and enhance the learning ability

$$X = X \odot \sigma(W_{Attention}(X^T X) + bias_{Attention}), \tag{10}$$

in which σ is softmax and $\odot$ denotes an element-wise multiplication.

After all, the loss function of the model is calculated by

$$L = \alpha(||A||) + L_{rec} + L_{KLD} \tag{11}$$

where $||\cdot||$ denotes the matrix p-norm (usually 2-norm) and α is a hyperparameter to balance the sequence reconstruction loss and graph construction loss.

3.4   Anomaly Detection Method

The model is well trained without anomalies resulting in low reconstruction error of the normal data. If the reconstruction error at timestamp t in test data is higher than a certain threshold s, we consider it is anomalous.

The reconstruction error at each timestamp t is formulated as

$$Error_t = \frac{1}{N_t} \sum_{t=1}^{N_t} MSE(x_{m,t}, x_{m,t}^R) \tag{12}$$

where $x_{m,t}$ and $x_{m,t}^R$ are the data from test sequences X_m and reconstructed sequences X_m^R at timestamp t, with $N_t = |\{(x_{m,t}, x_{m,t}^R)|x_t \in X_m, m \in (1, M)\}|$ is the number of the sequences containing timestamp t. The reconstruction errors of x_t for all sequences containing it are uniformly accumulated.

Table 1. Dataset Information

Dataset	Nodes	Training	Testing	Anomaly (%)
SWaT	24(S)+27(A)	496800	449918	12.13
SWaT A4&A5	24(S)+53(A)	8097	6600	30.00
BATADAL	21(S)+12(A)	8761	4177	5.24

S: sensors in the dataset. A: actuators in the dataset

4   Experiments

4.1   Dataset

To verify the effectiveness of the proposed MAD-SGS framework, we conducted experiments on three raw CPSs datasets with anomalies caused by cyber-attacks from two publicly available CPSs collected on the iTrust testbed. Nodes in these CPSs are sensors and actuators. Sensors convert a physical parameter into an electronic output, i.e., an electronic value. Actuators convert a signal into a physical output, i.e., turning the pump off or on [16]. Details of each dataset are summarized in Table 1. Note that, unlike some previous works that reprocessed the aforementioned datasets [18], we directly fed raw data to our model without additional data pre-processing.

SWaT. The Secure Water Treatment (SWaT) dataset [16] is collected from the SWaT Testbed by running the system continuously for 11 days, of which the first 7 days were under normal conditions, and the last 4 days were subjected to 36 simulated attacks. The SWaT dataset consists of six mechanical processes: raw water storage (P1), pre-treatment (P2), ultrafiltration (P3), ultraviolet dechlorinator (P4), reverse osmosis (P5), and treated water storage (P6). Variables include readings collected per second coming from 51 sensors and actuators.

SWaT A4&A5. This dataset is also collected from the SWaT testbed by running the system without any attacks from 12:35 PM to 2:50 PM, and then 6 attacks were carried out between 3:08 PM to 4:16 PM. As an extended version of the SWaT dataset, SWaT A4&A5 contains 77 variables, including 71 recorded by sensors and actuators and 6 process states.

BATADAL. The BATtle of Attack Detection Algorithms (BATADAL) system [21] is another water distribution system whose data is collected by the C-Town Public Utility (CPU), the main water distribution system operator of C-Town. The BATADAL testbed consists of 9 programmable logic controllers (PLCs) that continuously monitor and control the sensors and actuators in each sub-process. The normal part of the BATDAL dataset is generated from a one-year-long simulation under a normal state. In contrast, the abnormal part spans 6 months with 7 simulated attacks inserted into the system. Variables include readings collected per hour coming from 43 sensors and actuators.

4.2   Baselines

To verify the superiority of the proposed MAD-SGS, we compared it with the following baselines: OCSVM [20], which is a kernel-based method that aims at finding a supporting hyperplane to separate all normal points from the origin; iForest [14], which utilizes random isolation trees to isolate the data points; USAD [2], which is an unsupervised reconstruction-based model with temporal modeling using two autoencoders using the same encoder against each other to reconstruct the sequence; MAD-GAN [10] which is an unsupervised reconstruction-based model with temporal modeling that applies LSTM-RNN in the GAN's generator and discriminator to learn the temporal information; MTAD-GAT [24], which is a deep learning method with temporal and spatial modeling using two Graph Attention (GAT) layers with a complete graph; GDN [5], which utilizes the spatial-based GAT to form a forecasting model to predict the future value of each node; MAD-SGCN [19], which is a reconstruction-based method with GCNs and LSTMs as its encoder and decoder.

4.3   Experimental Set-Up

For each dataset, the optimal window size w and stride s $(s < w)$ of the subsequences fed to models were determined by grid search. Since the datasets were collected differently, different search ranges of w and s are concerned with each dataset separately. The search range for SWaT (SWaT A4&A5) is [60, 90, 120, 150] and [20, 30, 60, 80, 90] since they were recorded once per second. For BATADAL, w and s were searched within ranges [12, 24, 36, 48, 60] and [4, 8, 12, 24], respectively, since it was collected once per hour. At last, these two hyperparameters were set as (120, 80) for SWaT, (90, 30) for SWaT A4&A5, and (60, 12) for BATADAL.

Besides, feature similarity graph threshold ϵ^S and feature propagation graph threshold ϵ^H were also determined by grid search. As they are the standards to measure the similarity between two nodes, the value should not be too low to reduce the influence of noise and should not be too high to avoid overlooking information. Thus, the search range is [0.6, 0.7, 0.8, 0.9]. Finally, they were set as 0.8.

For the encoder, the input size and hidden size of GCNs were $(w, 32)$ and $(32, 16)$, while that of LSTMs were $(s, 32)$, $(32, 16)$. Then two compression layers from 256 to D_z were utilized for encoding μ and σ of the hidden variables z. With a re-parameterization trick, an expansion layer from D_z to 256 was used for decoding z. Then, the VAE output was reverted to 16×16. Then after the opposite operations by LSTM and GCN-based decoders, the sequence shape altered, $16 \times 16 \rightarrow N_v \times 16 \rightarrow N_v \times w$. To enhance the learning ability of the model, a BatchNorm1D layer is employed to regularize the input, and the encoded and decoded data of each layer passes through the dropout layer at a rate of 0.2. The maximum value of epoch and patient for training are 500 and 50. The learning rate of the Adam optimizer is 10^{-2}, and the weight decay is $5e - 4$.

Table 2. Point-wise Anomaly Detection Results

Method	SWaT			SWaT A4&A5			BATADAL		
Metric	Pre	Rec	F1	Pre	Rec	F1	Pre	Rec	F1
OCSVM	0.1882	**0.8396**	0.3074	0.2962	**0.9948**	0.4565	0.7560	0.2959	0.4254
iForest	0.9612	0.5992	0.7382	0.2894	<u>0.9943</u>	0.4482	0.2060	0.4313	0.2750
USAD	0.9880	0.6496	0.7838	0.3957	0.6314	0.4865	0.7628	0.4715	0.5828
MTAD-GAT	0.9805	0.6647	0.7923	0.3714	0.6899	0.4828	0.5081	0.3646	0.4177
GDN	0.9463	0.6654	0.7806	0.2916	0.9900	0.4504	**0.8450**	0.3504	0.4947
MAD-GAN	<u>0.9897</u>	0.6374	0.7757	0.3267	0.7739	0.4594	0.7143	0.5505	0.6218
MAD-SGCN	0.9857	0.6879	<u>0.8103</u>	<u>0.3753</u>	0.8720	<u>0.5136</u>	0.5212	**0.8995**	<u>0.6600</u>
Ours	**0.9916**	<u>0.6907</u>	**0.8143**	**0.5389**	0.7591	**0.6303**	<u>0.8361</u>	<u>0.7338</u>	**0.7816**

4.4 Results

Table 2 summarised the numerical anomaly detection results. The best performance was marked in bold, and the second-best results were underlined. It can
be observed that MAD-SGS performed better than the baselines in terms of
point-wise anomaly detection. Specifically, the F1-score on the three datasets
are 0.40%, 11.67%, and 12.16% higher than the best performance in the baseline, respectively. This demonstrates the effectiveness of the proposed model in
the task of MTS anomaly detection.

Further observation shows that despite decent precision and F1-score, the
performance improvement by MAD-SGS compared with baselines for SWaT is
not as significant as that for the BATADAL. Specifically, compared to the baselines (USAD and MAD-GAN) with temporal modeling, MAD-SGS presented an
improvement of around 3% in the F1-score for the SWaT dataset, while it was
16–20% for the other dataset.

Additionally, compared with MAD-SGCN (the spatial-temporal modeling
SOTA method), MAD-SGS improved the point-wise anomaly detection only
by 0.40% of the F1-score for the SWaT dataset. The fact that actuators only
measure discrete status made learning their spatial information with other nodes
challenging. With fewer actuators (shown in Table 1), it was easier for the spatial-
temporal modeling model to handle spatial information from the SWaT dataset
to construct graph structures. Thus both MAD-SGCN and MAD-SGS achieved
similar good performance on the SWaT dataset.

Similarly, by inspecting the point-wise anomaly detection results on SWaT
and SWaT A4&A5, the performances by SOTA and MAD-SGS were generally
better on SWaT A4&A5. This is also because SWaT A4&A5 contains more
actuators than SWaT, making its spatial information more challenging to mine
for learning the graph structure.

Table 3. Ablation Study on SWaT

Setting	Pre	Rec	F1
Ours	0.9916	**0.6907**	**0.8143**
w/o Graph Structure Learning	0.9942	0.6280	0.7698
w/o Long-term Graph	0.9962	0.6278	0.7702
w/o Short-term Graph	0.9918	0.6362	0.7752
w/o Attention Mechanism	**0.9969**	0.6083	0.7556

4.5 Ablation Study

To assess the effect of the individual sub-modules and evaluate the design choices of the model, an ablation study was applied to SWaT, which performed comparative experiments by deleting or replacing some sub-modules in the model. Results are given in Table 3.

To verify the importance of the graph structure learning module, we used a fully connected graph instead. The model's performance has declined significantly from 0.8143 to 0.7689. The fully connected matrix contained redundant information and impaired GCN to learn spatial information effectively.

The absence of the long-term static graph prevented the model from learning long-term dependencies information between nodes, leading to a decline in the F1 score to 0.7702.

We also removed the short-term graph to observe the performance contributed by the dynamic graph structure on the overall sequence reconstruction. The experimental results indicate that without the guidance of a dynamic graph, the model's accuracy is reduced without the ability to learn the dynamic interrelationships among nodes during attacks.

The verification experiment for the role of the attention mechanism is carried out by directly feeding the sequence to the encoder. The results indicate that the absence of a self-attention mechanism led to significant changes in the model's anomaly detection performance since the self-attention mechanism helped to introduce randomness into the model, enhancing its robustness. Additionally, it enabled the model to focus on important information and reduced the chances of overlooking critical information due to equal learning.

5 Conclusions

In this paper, we have introduced a new unsupervised anomaly detection method, MAD-SGS, for detecting anomalies caused by cyber-attacks in CPSs. We utilized GCNs and LSTMs to capture the spatial and temporal information in multivariate time series (MTS) data generated from CPSs. Specifically, we designed a graph structure module to learn nodes' long-term dependence and short-term variations. The point-wise anomaly detection experiments on four real datasets from iTrust confirmed the effectiveness of MAD-SGS for anomaly detection.

Acknowledgment. This research was supported by the Guangdong Basic and Applied Basic Research Foundation (2022A1515011713).

References

1. Ahmed, C.M., Zhou, J., Mathur, A.P.: Noise matters: using sensor and process noise fingerprint to detect stealthy cyber attacks and authenticate sensors in cps. In: Proceedings of the 34th Annual Computer Security Applications Conference, pp. 566–581 (2018)
2. Audibert, J., Michiardi, P., Guyard, F., Marti, S., Zuluaga, M.A.: USAD: unsupervised anomaly detection on multivariate time series. In: Proceedings of the 26th ACM SIGKDD International Conference on Knowledge Discovery & Data Mining, pp. 3395–3404 (2020)
3. Bogaerts, T., Masegosa, A.D., Angarita-Zapata, J.S., Onieva, E., Hellinckx, P.: A graph CNN-LSTM neural network for short and long-term traffic forecasting based on trajectory data. Transp. Res. Part C: Emerg. Technol. **112**, 62–77 (2020)
4. Breunig, M.M., Kriegel, H.P., Ng, R.T., Sander, J.: LOF: identifying density-based local outliers. In: Proceedings of the 2000 ACM SIGMOD International Conference on Management of Data, pp. 93–104 (2000)
5. Deng, A., Hooi, B.: Graph neural network-based anomaly detection in multivariate time series. In: Proceedings of the AAAI Conference on Artificial Intelligence, vol. 35, pp. 4027–4035 (2021)
6. Diao, Z., Wang, X., Zhang, D., Liu, Y., Xie, K., He, S.: Dynamic spatial-temporal graph convolutional neural networks for traffic forecasting. In: Proceedings of the AAAI Conference on Artificial Intelligence, vol. 33, pp. 890–897 (2019)
7. Garg, A., Zhang, W., Samaran, J., Savitha, R., Foo, C.S.: An evaluation of anomaly detection and diagnosis in multivariate time series. IEEE Trans. Neural Netw. Learn. Syst. **33**(6), 2508–2517 (2021)
8. He, Y., Zhao, J.: Temporal convolutional networks for anomaly detection in time series. J. Phys: Conf. Ser. **1213**(4), 42050 (2019)
9. Huang, T., Liu, C., Sharma, A., Sarkar, S.: Traffic system anomaly detection using spatiotemporal pattern networks. Int. J. Prognost. Health Manage. **9**(1) (2018)
10. Li, D., Chen, D., Jin, B., Shi, L., Goh, J., Ng, S.-K.: MAD-GAN: multivariate anomaly detection for time series data with generative adversarial networks. In: Tetko, I.V., Kůrková, V., Karpov, P., Theis, F. (eds.) ICANN 2019, Part IV. LNCS, vol. 11730, pp. 703–716. Springer, Cham (2019). https://doi.org/10.1007/978-3-030-30490-4_56
11. Li, D., Hu, G., Spanos, C.J.: A data-driven strategy for detection and diagnosis of building chiller faults using linear discriminant analysis. Energy Build. **128**, 519–529 (2016)
12. Li, D., Zhou, Y., Hu, G., Spanos, C.J.: Fault detection and diagnosis for building cooling system with a tree-structured learning method. Energy Build. **127**, 540–551 (2016)
13. Li, Z.L., Zhang, G.W., Yu, J., Xu, L.Y.: Dynamic graph structure learning for multivariate time series forecasting. Pattern Recogn. **138**, 109423 (2023)
14. Liu, F.T., Ting, K.M., Zhou, Z.H.: Isolation forest. In: 2008 Eighth IEEE International Conference on Data Mining, pp. 413–422. IEEE (2008)
15. Malhotra, P., Ramakrishnan, A., Anand, G., Vig, L., Agarwal, P., Shroff, G.: LSTM-based encoder-decoder for multi-sensor anomaly detection. arXiv preprint arXiv:1607.00148 (2016)

16. Mathur, A.P., Tippenhauer, N.O.: SWaT: a water treatment testbed for research and training on ICS security. In: 2016 International Workshop on Cyber-Physical Systems for Smart Water Networks (CySWater), pp. 31–36. IEEE (2016)
17. Miele, E.S., Bonacina, F., Corsini, A.: Deep anomaly detection in horizontal axis wind turbines using graph convolutional autoencoders for multivariate time series. Energy AI **8**, 100145 (2022)
18. Perales Gómez, Á.L., Fernández Maimó, L., Huertas Celdrán, A., García Clemente, F.J.: MADICS: a methodology for anomaly detection in industrial control systems. Symmetry **12**(10), 1583 (2020)
19. Qi, P., Li, D., Ng, S.K.: MAD-SGCN: multivariate anomaly detection with self-learning graph convolutional networks. In: 2022 IEEE 38th International Conference on Data Engineering (ICDE), pp. 1232–1244. IEEE (2022)
20. Scholkopf, B., Platt, J.C., Shawe-Taylor, J., Smola, A.J., Williamson, R.C.: Estimating the support of a high-dimensional distribution. Neural Comput. **13**(7), 1443–1471 (2001)
21. Taormina, R., et al.: Battle of the attack detection algorithms: disclosing cyber attacks on water distribution networks. J. Water Resour. Plan. Manag. **144**(8), 04018048 (2018)
22. Yang, J., Yue, Z.: Learning hierarchical spatial-temporal graph representations for robust multivariate industrial anomaly detection. IEEE Trans. Industr. Inform. (2022)
23. Zhang, J., Pan, L., Han, Q.L., Chen, C., Wen, S., Xiang, Y.: Deep learning based attack detection for cyber-physical system cybersecurity: a survey. IEEE/CAA J. Autom. Sinica **9**(3), 377–391 (2021)
24. Zhao, H., et al.: Multivariate time-series anomaly detection via graph attention network. In: 2020 IEEE International Conference on Data Mining (ICDM), pp. 841–850. IEEE (2020)
25. Zhao, J., Wang, X., Shi, C., Hu, B., Song, G., Ye, Y.: Heterogeneous graph structure learning for graph neural networks. In: Proceedings of the AAAI Conference on Artificial Intelligence, vol. 35, pp. 4697–4705 (2021)
26. Zong, B., et al.: Deep autoencoding gaussian mixture model for unsupervised anomaly detection. In: International Conference on Learning Representations (2018)

A Two-Level Game-Theoretic Approach for Joint Pricing and Resource Allocation in Multi-user Mobile Edge Computing

Erqian Ge[1], Hao Tian[1], Wanyue Hu[1], and Fei Li[1,2(✉)]

[1] School of Electrical and Information Engineering, Anhui University of Technology, Maanshan 243032, China
`lanceleeneu@126.com`
[2] Anhui Province Engineering Laboratory of Intelligent Demolition Equipment, Anhui University of Technology, Maanshan 243032, China

Abstract. Mobile Edge Computing (MEC) offers an efficient model to extend cloud computing capabilities to IoT devices and mobile users. This research paper focuses on multi-user MEC systems. It presents a two-level approach, with the upper level optimizing the allocation and pricing of computing resource by the Edge Server Provider (ESP), while the lower level handles offloading strategy, channel selection for the Device Manager (DM). However, the problem's complexity, involving mixed variables, classifies it as NP-hard. To address this challenge, we formulate the computational offloading problem among DMs as a multi-user computational offloading game. Our analysis confirms the existence of Nash equilibriums and finite improvement properties. Based on this, we propose a two-tier distributed computational offloading algorithm that achieves Nash equilibrium. Experiments demonstrate its effectiveness in optimizing resource allocation and maximizing ESP and DM profits.

Keywords: Mobile edge computing · Bilevel optimization · Game theory

1 Introduction

The rapid development of mobile internet and the widespread adoption of smart devices have brought forth a myriad of services [1]. This includes a surge in real-time applications and services, such as augmented reality [2], virtual reality [3], and autonomous driving [4]. Despite the recent increase in computational power of mobile devices, their computing and battery resources remain relatively limited, making it difficult to meet the demands of high-energy-consuming complex tasks [5–7]. The traditional cloud computing model centralizes data and application processing in remote cloud data centers [8]. However, this centralized approach has limitations in certain application scenarios. Cloud servers are often far from users, resulting in significant communication delays, which pose

L. Pan et al. (Eds.): BIC-TA 2023, CCIS 2062, pp. 32–49, 2024.
https://doi.org/10.1007/978-981-97-2275-4_3

substantial challenges for real-time applications, interactive games, video streaming, and other latency-sensitive applications. To address this issue, Mobile Edge Computing (MEC) has emerged. MEC extends computing resources to the network's edge, bringing them closer to end-users and device locations [9]. This means that computational tasks are distributed to edge nodes situated closer to users, such as mobile communication base stations, routers, and edge servers. By doing so, MEC achieves lower transmission latency and higher bandwidth, providing support for latency-sensitive applications [10]. However, mobile cloud computing faces challenges in terms of limited resource capacity [11]. Therefore, the development of an efficient resource allocation mechanism is crucial for fully unleashing the potential of mobile cloud computing. It can optimize the utilization of resources to the maximum extent, enhance the overall system performance, reduce communication latency, lower energy consumption, and improve the throughput of the entire network, among other functions [12–14].

Several studies have proposed for resource allocation in MEC [15,16]. For instance, Wang et al. [17] jointly optimize the computational speed, offloading rate and transmission power of Smart Mobile Devices, expressing them as two optimization objectives: minimizing energy consumption and minimizing application execution latency. They employ variable substitution techniques and single-variable search methods to address these two optimization problems. Dinh et al. [18] proposed an optimization framework that extends offloading tasks from a single mobile device to multiple edge devices. This framework jointly optimizes task allocation decisions and the central processing unit frequency of mobile devices, aiming to minimize both the total task execution latency and mobile device energy consumption. Wu et al. [19] consider a multi-user, multi-server Mobile Edge Computing (MEC) network with time-varying fading channels. They propose three distributed methods based on Q-learning to learn effective offloading strategies. Du et al. [20] considered a Mobile Edge Computing system that supports unmanned aerial vehicles (UAVs). They transformed and relaxed the optimization problem into a convex problem and applied pipeline job-shop scheduling techniques to address it.

Nonetheless, the majority of research within the MEC domain, as observed in studies like [20–23], primarily focuses on maximizing the overall system performance while neglecting the interests of individual entities. In complex MEC scenarios, the Edge Server Providers (ESPs) and Device Managers (DMs) often represent distinct entities with divergent objectives. To address this intricate landscape, several studies have proposed the utilization of bilevel optimization techniques. For instance, Zhou et al. in [24] introduced a bilevel optimization algorithm designed for resource allocation and task offloading in cooperative MEC scenarios. The upper level is tailored to enhance system efficiency by optimizing computing offloading strategies, while the lower level ensures fairness in task allocation by optimizing server resources. In a similar vein, Zhang et al. as documented in [25], presented a bilevel optimization algorithm with a primary focus on balancing user computational demands and energy conservation to optimize overall system performance. Furthermore, in the work by

Huang et al. [26], a bilevel optimization algorithm was proposed with the aim of optimizing computing resource costs and energy consumption at the upper level, while concurrently devising offloading strategies at the lower level. However, when solving multi-objective, multi-user, bilevel optimization problems, this method may not be suitable for this situation.

This paper delves into the complexities of multi-user resource allocation in a Mobile Edge Computing system under complex environments. Specifically, it explores scenarios in which Edge Server Provider (ESP) offer computing resources to Device Managers (DMs) with the aim of generating profits while concurrently minimizing task execution times for the lower-level DMs. Each DM autonomously selects their offloading strategy, driven by the goal of maximizing their individual profits. Notably, this study departs from previous research by addressing a unique scenario where the optimization objectives of the upper level are no longer singular, where each IoT device is affiliated with a distinct DM, and each DM independently determines their task execution mode. The primary contributions of this paper are summarized as follows:

1. We analyze the interaction between ESPs and DMs and formulate the computing resource allocation and pricing optimization problem as a bilevel optimization problem (BOP).The ESPs optimize the prices of the provided computing resources as well as the computing resources allocated to each device at the upper level, while the offloading strategies of the DMs are modeled as a non-cooperative game at the lower level.
2. We introduce a two-tier optimization algorithm called Distributed Bilevel Optimization Algorithm Based on Nash Equilibrium (DBGOA) to solve this BOP. we prove the existence of Nash equilibrium in the lower-level optimization problem and obtain the Nash equilibrium for a given price of computing resources and allocation of computing resources. In addition, we use NSGA-II to solve the upper level optimization problem.
3. The experiment evaluations conducted across multiple instances robustly validate the performance of the proposed algorithm. Through a systematic comparison with other algorithms, the strengths and weaknesses of the algorithm are elucidated, along with its relative efficacy in different scenarios. This provides valuable insights for the broader field of optimization algorithms.

2   System Model

In the context of mobile edge computing, we examine a system comprising an edge server and n IoT devices, visually represented in Fig. 1. The Edge Server Provider (ESP) is the manager of the MEC server, which in turn offers computing resources to the various IoT devices. Each IoT device is associated with a distinct Device Manager (DM). Consider that every IoT device possesses a pending task, and these tasks are represented by the set $\mathcal{N} = \{1, 2, \cdots, n\}$. Each individual IoT device's task can be described by the triplet $\sigma_i = (D_i, C_i)$, where D_i signifies the data size of σ_i, C_i indicates the computing resources required for σ_i.

1. If σ_i is executed locally, $o_i = -1$ (i.e., the local mode);
2. If σ_i is executed on the edge server, $o_i = l$ (i.e., the MEC mode);
3. If σ_i is not executed, $o_i = 0$ (i.e., the non-execution mode).

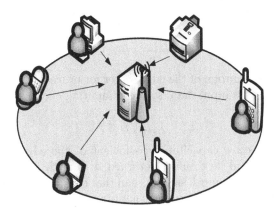

Fig. 1. The optimization of task average completion time for DM.

2.1   MEC Mode

When σ_i is executed in the MEC mode, The completion of σ_i requires three steps. First, the ith IoT device needs to offload σ_i to the edge server through a wireless channel; then, the edge server execute σ_i; and finally, the edge server offloads the result of σ_i to the IoT device.

Assume that there are two channels in the system, denoted as $l \in \{1,2\}$. $I_{\{A\}}$ denotes the channel selection, where $I_{\{o_i=l\}} = 1$ if σ_i transmits over channel l, otherwise $I_{\{o_i=l\}} = 0$. But the offload rates of tasks under the same channel affect each other. Thus, the uplink rate of σ_i executing through channel l in MEC mode can be expressed as

$$R_i^l = W log_2 \left(1 + \frac{P_i h_i}{\omega + \sum_{n \neq i} I_{\{l=k\}} P_n h_n} \right) \qquad (1)$$

where, W denotes the channel bandwidth, ω signifies the background noise power, and P_i represents the transmission power of the ith IoT device. Note that P_i can be predetermined based on the computing capability and location of the ith IoT device.

Then, we can obtain the transmission time and transmission power of σ_i offload to the edge server based on the uplink rate.

$$T_i^{t,l} = \frac{D_i}{R_i^l}, \forall i \in \mathcal{N} \qquad (2)$$

and

$$E_i^{t,l} = \frac{P_i D_i}{R_i^l}, \forall i \in \mathcal{N}. \tag{3}$$

When σ_i is in the edge server, the edge server needs to allocate computing resources for the completion of σ_i, let $\mathbf{r} = [r_1, r_2, \ldots, r_n]$ denote the computing resources allocated to each task. Then, the computing time of σ_i on the edge server is expressed as

$$T_i^c = \frac{C_i}{r_i}, \forall i \in \mathcal{N}. \tag{4}$$

And the energy consumption of the edge server for processing the task is usually related to the size of its input data, so the computing energy consumption of σ_i on the edge server is

$$E_i^c = k_1 D_i, \forall i \in \mathcal{N}. \tag{5}$$

where k_1 is the effective capacitance coefficient of the edge server. The edge server completes σ_i and its result is returned to the ith IoT device. Since the size of the result is usually much smaller than that of input data for the offloading task, therefore the time and energy consumption for receiving the result is not considered in this paper.

2.2   Local Mode

When σ_i is executed in the local mode, the task utilizes solely the computing resources of the ith IoT device, without involving the edge server's resources. The energy consumption and time required for the ith IoT device to execute σ_i locally are expressed as follows:

$$T_i^l = \frac{C_i}{r_i^l}, \forall i \in \mathcal{N} \tag{6}$$

and

$$E_i^l = k_0 \left(r_i^l\right)^3 T_i^l, \forall i \in \mathcal{N} \tag{7}$$

where r_i^l denotes the computing capability of the ith IoT device and k_0 denotes the capacitance coefficient of it.

If σ_i is not executed, the ith IoT device will not need to procure computing resources and energy.

2.3   Profits of DM and ESP

When σ_i is successfully completed, the corresponding DM receives a reward associated with D_i. The DM's profit is determined by subtracting the costs of computing resources and energy from this reward. In local mode, the cost of the IoT device only includes the energy required to execute σ_i, and it does not encompass the computing resources of the edge server. In MEC mode, the DM incurs costs for both transmitting the data of σ_i via the IoT device and utilizing computing resources on the edge server for σ_i. However, if σ_i is not executed, the

DM does not need to purchase computing resources or energy, and no reward is obtained. Therefore, for σ_i, the DM's profit can be calculated as follows:

$$U_i^{DM} = \begin{cases} \alpha D_i - \rho E_i^l & o_i = -1 \\ \alpha D_i - \rho E_i^{t,l} - v_i r_i & o_i = l \\ 0 & o_i = 0 \end{cases} \tag{8}$$

Where α represents the reward factor and ρ is the constant energy price, we have $\mathbf{v} = [v_1, \cdots, v_n]$ as the pricing structure for computing resources in MEC. Here, v_i signifies the cost of computing resources used by σ_i for each IoT device. This study delves into a pricing strategy that enables the ESP to set distinct prices for computing resources for each task on individual IoT devices, thereby optimizing the profit of ESP. The following equation represents the time consumed to complete the task, while another optimization objective in this study is time optimization.

$$T_i = \begin{cases} \frac{C_i}{r_i^l} & o_i = -1 \\ \frac{D_i}{R_i^l} + \frac{C_i}{r_i} & o_i = l \\ 0 & o_i = 0 \end{cases} \tag{9}$$

The profit of the ESP results from subtracting the cost of energy consumption from the revenue obtained by selling computing resources. Consequently, the profit for σ_i in the context of ESP can be expressed as follows:

$$U_i^{ESP} = \begin{cases} 0 & o_i = 0, -1 \\ v_i r_i - v_0 E_i^c & o_i = l. \end{cases} \tag{10}$$

Therefore, the total profit of ESP is calculated as:

$$U^{ESP} = \sum_{i \in \mathcal{N}} U_i^{ESP}. \tag{11}$$

2.4   Problem Description

Figure 1 provides a visual representation of the dynamic interaction between the ESP and the DMs, which is structured as a bilevel problem. Within this formulation, the ESP is responsible for optimizing the vector $\mathbf{v}$, while, in parallel, considerations are made for the optimal computing offloading strategies for all tasks $\mathbf{o}$ and the allocation of computing resources, given the vector $\mathbf{v}$. The computing offloading strategy pertains to the selection of the execution mode for each task. This complex interaction is encapsulated within a bilevel problem, and its formulation can be expressed as follows:

$$\mathcal{P} : \max \quad U^{ESP}(\mathbf{v}, \mathbf{o}, \mathbf{r})$$
$$\min \quad T(\mathbf{v}, \mathbf{o}, \mathbf{r})$$

$$C1 : v_{i,\min} \leq v_i \leq v_{i,\max}, i \in \mathcal{N}$$

$$\{\mathbf{o}, \mathbf{r}\} = \arg\max_{i \in \{1,2,\cdots,n\}} U_i^{DM}(\mathbf{v}, \mathbf{o}, \mathbf{r})$$

$$C2 : o_i = \{-1, 0, l\}, \forall i \in \mathcal{N}, l \in \{1, 2\}$$

$$C3 : r_i \geq 0, \forall i \in \mathcal{N}$$

$$C4 : \sum_{i \in \mathcal{N}} r_i \leq r_{\max}$$

3   Description of Algorithms

Within the context of $\mathcal{P}$, a scenario emerges in which the execution of a considerable volume of tasks in the MEC mode can lead to significant interference, consequently resulting in a pronounced reduction in the uplink data rate. This situation is compounded by the inherent resource sharing nature of the edge server, where computing resources are allocated to serve multiple DMs. Consequently, this gives rise to an apparent interdependency among the DMs. It is noteworthy that each individual DM within this system is primarily driven by the desire to maximize their own profit. This self-interest inherently frames the lower-level optimization problem as a non-cooperative game. In essence, the non-cooperative game establishes a strategic environment, where the participants' behavior is guided by individual, self-interested decisions, operating within a framework that prioritizes autonomous decision-making.

3.1   Lower-Level Optimization

The lower-level optimization problem is formally characterized as a multi-user computing offloading game, denoted by $G = (\mathcal{N}, \mathbf{o}, U_i^{DM})$, with each DM serving as an individual player within this strategic framework. In this context, $\mathcal{N}$ signifies the total number of participants or players in this game, and $\mathbf{o} = \{o_1, o_2, \cdots, o_n\}$ represents the decision set related to task σ_i for each player.

Furthermore, I_i^{DM} corresponds to the profit function tailored to the ith player in this multi-user computing offloading game. The primary objective of this game is the optimization of individual player profits, as illustrated in the following expression:

$$\max_{i \in \mathcal{N}} U_i^{DM}(\mathbf{v}, \mathbf{o}). \tag{12}$$

A Nash equilibrium for the game is achieved when no player within a set of computing offloading strategies can independently alter their strategy to yield a better outcome for themselves. This pivotal concept of Nash equilibrium is formally defined as follows:

Theorem 1. *A set of computing offloading strategies* $\mathbf{o}^* = (o_1^*, o_2^*, \cdots, o_n^*)$ *is the Nash equilibrium of the game, if no player unilaterally modifies the computing offloading strategy to make themselves more profitable, i.e.*

$$U_i^{DM}(o_i^*, o_{-i}^*) \geq U_i^{DM}(o_i, o_{-i}^*), \forall o_i \in \{-1, 0, l\}, \, i \in \mathcal{N}, l \in \{1, 2\}. \quad (13)$$

Within the confines of the game, the strategy vector $\mathbf{o} = (o_i, o_{-i})$ characterizes the computing offloading strategies employed by all players. All participants endeavor to reach a state where they have each adopted the best response strategy, resulting in a mutually beneficial outcome. At this point, there is no incentive for any player to deviate from their initial strategy.

In this specific context, it becomes imperative to address and examine the interference generated by tasks executed in the MEC mode. This interference issue warrants separate discussion. If the profit gained from executing task σ_i in the MEC mode surpasses that attainable in the local mode, it is crucial to satisfy the following equation based on (9):

$$\rho \frac{P_i D_i}{R_i^l} + v_i r_i \leq \rho k_0 (r_i^l)^2 C_i. \quad (14)$$

Then, we can obtain

$$R_i^l \geq \frac{\rho P_i D_i}{\rho k_0 (r_i^l)^2 C_i - v_i r_i}. \quad (15)$$

According to (1), for given P_i and h_i, when (15) holds, we can obtain the following equation.

$$\sum_{n \neq i} I_{\{o_n = l\}} P_n h_n \leq L_i \triangleq \frac{P_i h_i}{2^{\frac{\rho P_i D_i}{W(\rho k_0 (r_i^l)^2 C_i - v_i r_i)} - 1}} - \omega. \quad (16)$$

The selection of the mode by the DM depends on specific conditions. When the interference experienced by the IoT device, stemming from interactions with other devices, is below the interference threshold, and there is a positive profit associated with the MEC mode, the DM opts for the MEC mode. Conversely, if the interference exceeds the defined threshold, and the local mode yields a positive profit, the DM selects the local mode. In cases where neither condition is met, the DM resorts to the non-execution mode. Hence, the best response of the ith player can be expressed as follows:

$$o_i^* = \begin{cases} l, \left(\sum_{n \neq i} I_{\{o_n = l\}} P_n h_n < L_i \right) \wedge (U_{\{o_i = l\}} > 0) \\ -1, \left(\sum_{n \neq i} I_{\{o_n = l\}} P_n h_n \geq L_i \right) \wedge (U_{\{o_i = -1\}} > 0) \\ 0, \text{ otherwise.} \end{cases} \quad (17)$$

In order to establish the presence of a Nash equilibrium in the multi-user computing offloading non-cooperative game that we have introduced, it is essential to demonstrate that this game qualifies as a potential game. It is the key observation that any potential game inherently possesses a Nash equilibrium and exhibits the finite improvement property.

Theorem 2. *For $\forall i \in \mathcal{N}$, $\forall o_i \in \{-1, 0, l\}$, $l \in \{1, 2\}$, if there exists a potential equation ψ in a game such that the following equation holds:*

$$U_i^{DM}(o_i, o_{-i}) > U_i^{DM}(o_i', o_{-i}). \tag{18}$$

Then, the expression $\psi(o_i, o_{-i}) > \psi(o_i', o_{-i})$ is true.

The concept of a potential function is instrumental in our analysis, and it shares the same monotonic characteristics as the original function. Specifically, if the ith player autonomously alters their strategy in a manner that leads to an increase or decrease in their profit, this change corresponds to an increase or decrease in the value of the potential function ψ.

To demonstrate that the multi-user computing offloading non-cooperative game indeed qualifies as a potential game, we introduce and define a potential function:

$$\psi(o_i, o_{-i}) = \sum_{m=1}^{N} P_m h_m I_{\{o_m \neq l\}} L_m - \frac{1}{2} \sum_{m=1}^{N} \sum_{n \neq m} P_m h_m I_{\{o_m = l\}} P_n h_n I_{\{o_n = l\}} \tag{19}$$

Based on the above potential function, it can be known that when the computing offloading strategy of the ith player is changed from the local model to the MEC model, the value of the potential function changes as follows:

$$
\begin{aligned}
\psi(l, o_{-i}) - \psi(-1, o_{-i}) = & \sum_{m \neq i}^{N} P_m h_m I_{\{o_m \neq l\}} L_m + P_i h_i L_i \\
& - \frac{1}{2} \sum_{m \neq n \neq i} \sum P_m h_m I_{\{o_m = l\}} P_n h_n I_{\{o_n = l\}} \\
& - \frac{1}{2} P_i h_i \sum_{m \neq n \neq i} P_m h_m I_{\{o_m = l\}} \\
& - \frac{1}{2} P_i h_i \sum_{m \neq n \neq i} P_m h_m I_{\{o_m = l\}} \\
& + \frac{1}{2} \sum_{m \neq n \neq i} \sum P_m h_m I_{\{o_m = l\}} P_n h_n I_{\{o_n = l\}} \\
& - \sum_{m \neq i}^{N} P_m h_m I_{\{o_m \neq l\}} L_m \\
= & P_i h_i L_i - P_i h_i \sum_{m \neq i} P_m h_m I_{\{o_m = l\}} > 0
\end{aligned}
\tag{20}
$$

Algorithm 1: Lower-level of DBGOA

Input: $\mathbf{v}, \mathbf{r}$
Output: $\mathbf{o}$

1 Calculate the profit of each mode of the DMs according to the inputs. Initialize the execution mode (i.e., o_i) for each task with the more profitable one between the local mode and the non-execution mode;
2 **repeat**
3   Obtain the best response $o_i^*, \forall i \in \mathcal{N}$, according to (17) ;
4   **if** $o_i^* \neq o_i, \forall i \in \mathcal{N}$ **then**
5   Calculate the profit improvement (i.e., Δ_i^{DM}) of the DM from o_i^* against o_i;
6   **end**
7   Initialize $\lambda = 0$;
8   **while** $\lambda = 0$ **do**
9         $k = \arg\max_i \Delta_i^{DM}$;
10   **if** $\Delta_k^{DM} > 0$ **then**
11             $o_k \leftarrow o_k^*$;
12   Broadcast o_k to all DMs;
13   **if** *the constraints are satisfied* **then**
14                 $\lambda = 1$;
15   **else**
16                 $\lambda = 0$;
17                 $\Delta_k^{DM} = 0$;
18   **end**
19   **end**
20   **end**
21 **until** *termination conditions are met*;

Based on the preceding derivation, when transitioning the computing offloading strategy of the ith player from the local mode to the MEC mode, it must adhere to the interference threshold condition $\sum_{n \neq i} I_{\{o_n = l\}} P_n h_n \leq L_i$. Consequently, this leads to $\psi(l, o_{-i}) - \psi(-1, o_{-i}) \geq 0$. Given that $U_i^{DM}(l, o_{-i}) > U_i^{DM}(-1, o_{-i})$ is already established, and in line with the definition of a potential game, we can affirm the nature of the multi-user computing offloading game as a potential game. Through the proof above, we establish the existence of a Nash equilibrium, achievable within a finite number of iterations in the lower-level problem. Subsequently, we propose a distributed computing offloading optimization algorithm leveraging the finite improvement property of the potential game to attain mutually satisfactory offloading strategies for DMs.

As outlined in Algorithm 1, our process begins with the initialization of the execution mode for each DM. This initialization is executed by selecting the most profitable mode from between the local mode and the non-execution mode, taking into account the provided price vector $\mathbf{v}$. Following this initialization, we proceed to derive the best response strategies for the DMs as determined by

(17). Then, we calculate the profit improvement obtained by the best response of each DM in comparison to their existing strategy:

$$\Delta_i^{DM} = U_{o_i^*}^{DM} - \max\left\{U_{o_i}^{DM}, 0\right\}, \forall j \in \mathcal{N} \tag{21}$$

where $U_{o_i^*}^{DM}$ and $U_{o_i}^{DM}$ represents the profit obtained by the ith DM when σ_i is executed in the best response and the existing strategy, respectively. In subsequent loops, Δ_i^{DM} is the profit improvement of the best response relative to the existing strategy.

By analyzing the profit improvements denoted by $\Delta_i^{DM}, i \in \mathcal{N}$, we identify the DM that has the highest profit enhancement. When this profit improvement surpasses zero, we proceed to update the DM's strategy to reflect its best response. However, it's imperative to ensure that the updated strategy satisfies all relevant constraints. If the update violates any constraints, it is considered invalid and is not incorporated. This process is repeated, checking each DM in turn, until either a valid update is accepted, or all DMs have been examined. Subsequently, once an update is accepted, we notify all other DMs of the change, prompting them to re-evaluate their own profits and potentially adjust their strategies.

3.2   Upper-Level Optimization

Traditional methods often perform poorly when dealing with situations where the decision variables may be discontinuous in $\mathcal{P}$. Known for its simplicity and efficiency, NAGA-II is an attractive alternative, especially when dealing with multi-objective optimization problems in mobile edge computing, which explores the decision space and seeks optimal configurations that satisfy multiple objectives in the upper-level. This is especially important when trying to strike a balance between various conflicting objectives such as cost reduction, resource allocation, or improving network performance. The overall flow of the proposed algorithm is described in Algorithm 2.

Algorithm 2: The upper-level of DBGOA

Input: D, C

Output: Optimal price $\mathbf{v}_{Best}$ with corresponding set of execution modes $\mathbf{o}$; The computing resource allocation scheme $\mathbf{r}$ of the ESP.

1   Initialize populations $\mathbf{T}_1 = \{\mathbf{v}_1, \mathbf{v}_2, \cdots, \mathbf{v}_{NP}\}$ and the computing resource allocation scheme $\mathbf{T}_2 = \{\mathbf{r}_1, \mathbf{r}_2, \cdots, \mathbf{r}_{NP}\}$ randomly, NP represents the population size of the differential evolution algorithm. Each individual $\mathbf{v}_i$, $i \in \mathcal{N}$ in $\mathbf{T}_1$ represents the prices of computing resource for all DMs. Each individual $\mathbf{r}_i$, $i \in \mathcal{N}$ in $\mathbf{T}_2$ represents The computing resource allocation scheme of the ESP;

2   **for** $i = 1 : NP$ **do**

3   Perform Algorithm 1 to obtain a optimal set $\mathbf{o}$ corresponding to $\mathbf{v}_i$ and $\mathbf{r}_i$;

4   Evaluate the performance of this optimal set $\mathbf{o}$ at the upper level of $\mathcal{P}'$

5   **end**

6   $Iter = NP$;

7   **while** $Iter < MaxIt$ **do**

8   Implement the mutation and crossover operators of DE to generate offspring population $\mathbf{T}'_1 = \{\mathbf{v}'_1, \mathbf{v}'_2, \cdots, \mathbf{v}'_{Np}\}$, $\mathbf{T}'_2 = \{\mathbf{r}_1, \mathbf{r}_2, \cdots, \mathbf{r}_{NP}\}$;

9   **for** $i = 1 : NP$ **do**

10   Perform Algorithm 1 to obtain a optimal set $\mathbf{o}'$ corresponding to $\mathbf{v}'_i$;

11   Evaluate the performance of this optimal set $\mathbf{o}'$ at the upper level of $\mathcal{P}'$;

12   **end**

13   Perform the non-dominated sort of NSGA- II to select the NP individuals which are better than other individuals, and treat them as the new parent $\{\mathbf{T}_1, \mathbf{T}_2\}$;

14      $Iter = Iter + NP$

15   **end**

Output: the best individual in $\mathbf{T}_1, \mathbf{T}_2$

The prices of computing resources for each DM are set randomly during the initialization process. The population of $\mathcal{P}$ is represented by $\{\mathbf{T}_1, \mathbf{T}_2\}$. Then, Algorithm 1 is implemented to search for the optimal computing offloading strategy for the ith individual $\{\mathbf{v}_i, \mathbf{r}_i\}$ in $\{\mathbf{T}_1, \mathbf{T}_2\}$ and evaluate its performance in the upper level. New offspring $\{\mathbf{T}'_1, \mathbf{T}'_2\}$ are then generated from the parent $\{\mathbf{T}_1, \mathbf{T}_2\}$. Subsequently, Algorithm 1 is repeated in the new offspring to identify the optimal computing offloading strategy for each individual and assess its performance at the upper level. Following this, the non-dominated sort selects the NP individuals which are better than other individuals and treats them as the new parent $\{\mathbf{T}_1, \mathbf{T}_2\}$. This iterative process continues until the stopping criterion is met, specifically when the maximum number of iterations in the upper level, denoted as $MaxIt$, is reached. Finally, the best individual in $\{\mathbf{T}_1, \mathbf{T}_2\}$ is outputted.

4   Experimental Studies

In this section, we conduct a comprehensive performance evaluation of our proposed distributed bilevel game optimization algorithm (DBGOA) through a series of numerical experiments. We establish a scenario where the edge server cover a circular area of $1\,\text{km} \times 1\,\text{km}$, and IoT devices are randomly distributed throughout this region, actively engaged in executing tasks to generate profits. The channel gain of each IoT device is denoted as $h_i = d_i^{-s}$, where d_i is the distance between the IoT device and the edge server, and s is the path loss factor, which is set to 4 [27]. The assessment of DBGOA performance involves the utilization of ten distinct instances, each varying in the number of IoT devices, represented by $n = 5, 10, \ldots, 50$. These diverse instances offer a robust examination of the algorithm's effectiveness across a range of deployment scales and complexities. To facilitate the experimental process, we have detailed the parameter settings of the MEC system under investigation, as succinctly summarized in Table 1. It is noteworthy that all experiments are conducted using the MATLAB environment, and they are executed on a personal computer equipped with an Intel Core i7-11700 CPU operating at $2.50\,\text{GHz}$ and 8 GB of RAM.

To assess the effectiveness of our proposed algorithm, we selected three existing algorithms as benchmarks and fine-tuned them:

- Bilevel Greedy Offload Algorithm (BIGOA): Determine the maximum profit mode for DMs according to (9). The upper level determines the price to maximize the profit for the ESP, and randomized DMs in the lower level who select the execution of the most profitable mode to maximize their profitability.
- Random component assignment scheme (RCAS): In the upper level, the ESP randomly allocates computing resources to DMs, and each DM executes its task.
- All Edge Computation Offload (AECO): In the upper level, the ESP selects a pricing scheme that maximizes its own profit, and DM selects the MEC mode unless the profit of the MEC mode is less than zero.

In our comparative study, we maintained uniformity with the settings for the mutation and crossover operators of DBGOA and its three competitor algorithms, setting both parameters at 0.9. Additionally, for the upper levels of DBGOA, BIGOA, RCAS, and AEBO, we utilized a consistent population size, setting it at $Np = 30$. Each algorithm underwent optimization at the upper level for a total of 30,000 iterations ($Iter$), denoted as $MaxIt = 30,000$.

To assess the performance of these algorithms, we conducted a rigorous evaluation, spanning 30 independent runs, and employed the profit as the primary performance metric. However, it is crucial to acknowledge that the accuracy of this performance evaluation hinges on the optimal computing offloading strategy, denoted as $\mathbf{o}$, being in alignment with the specific $\mathbf{v}_i$. Any misalignment in this respect could potentially yield an inaccurate evaluation of the ESP's performance. To mitigate this potential issue, we conducted 30 separate lower-level optimizations for each $\mathbf{v}_i$, ensuring the accurate recording of optimal computing offloading strategies, represented as $\mathbf{o}^*$.

Table 1. Parameter settings of the studied MEC system

Parameter	Value
$D_i, i \in \mathcal{N}$	$[1\,\text{KB}, 100\,\text{KB}]$
$C_i, i \in \mathcal{N}$	$[1, 1000]$ MCycles
r_i^l	$[0.4, 0.6]$ GCycles/s
α	5e$-$4
ω	-100 dbm
k_1	1e$-$10
$T_{i,\max}$	$2\,\text{s}$
v_0	0.1 per J
W	$20\,\text{MHz}$
P_i^t	$1\,\text{W}$
$r_{\max}$	20 GCycles
k_0	1e$-$25
$v_{i,\min}$	1 per GCycles
$v_{i,\max}$	20 per GCycles

In this paper, we evaluate the performance of the three algorithms relative to DBGOA over 30 runs using the Nikaido function values [28] as performance metrics. We conduct a performance comparison between DBGOA and three alternative algorithms, namely BIGOA, RCAS, and AECO, using Nikaido function values as the evaluation metric. The results, outlined in Table 2, present the average Nikaido function values over 30 runs. Among them, the content in [] indicates the number of cases in which the comparison algorithm performs poorly relative to the algorithm proposed in this paper in every 30 runs. It is worth noting that although BIGOA achieves the optimal solution when n = 10, it still yields poorer results compared to DBGOA in 16 out of 30 runs, and its optimal solution for other number of devices is still not feasible. In addition, none of the ten instances of BIGOA and RCAS produced a NASH equilibrium in any of the 30 runs. Thus, DBGOA shows a greater ability to obtain NASH balanced solutions compared to the other three algorithms.

Figure 2 and 3 show the change in ESP's profit during the iteration of the DBGOA and the result of the time optimization. As can be seen from the figures, with the Nash equilibrium found in the lower level, the ESP's profit in the upper level increases as the number of devices increases and always ensures that the time optimization for the IoT device uploading task remains within a relatively low range.

Theorem 3. *Let f_i^{DM} be the profit function of the ith player, then the Nikaido-Isoda function $\psi : (o_1 \times o_2 \times \cdots \times o_n) \times (o_1 \times o_2 \times \cdots \times o_n) \to R$ is*

$$\Psi(\mathbf{o}, \mathbf{o}') = \sum_{i=1}^{n} \left[f_i^{DM}(o_i' \mid \mathbf{o}) - f_i^{DM}(\mathbf{o}) \right]. \tag{22}$$

Each term of the series in the Nikaido-Isoda function represents the change of one player's profit from strategy o_i to o'_i, while the strategies of the others are held fixed. Nikaido-Isoda function represents the sum of the change in profit or loss of all players.

Table 2. The average Nikaido function values of DBGOA relative to the three compared algorithms

n	BIGOA	RCAS	AECO
10	2.33E+01[16]	−9.12E+01[30]	−4.06E+00[29]
20	−9.14E+00[30]	−3.26E+02[30]	−2.62E+01[30]
30	−5.55E+01[30]	−7.03E+02[30]	−9.09E+01[30]
40	−7.25E+01[30]	−8.25E+02[30]	−1.22E+02[30]
50	−8.46E+01[30]	−1.04E+03[30]	−1.49E+02[30]
60	−9.09E+01[30]	−1.35E+03[30]	−1.54E+02[30]
70	−7.85E+01[30]	−1.71E+03[30]	−1.94E+02[30]
80	−7.25E+01[30]	−1.97E+03[30]	−1.90E+02[30]
90	−8.53E+01[30]	−2.37E+03[30]	−2.57E+02[30]
100	−7.10E+01[30]	−2.81E+03[30]	−2.22E+02[30]

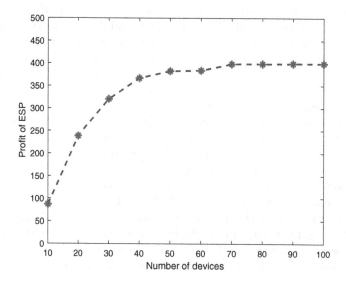

Fig. 2. Average total profit of ESP.

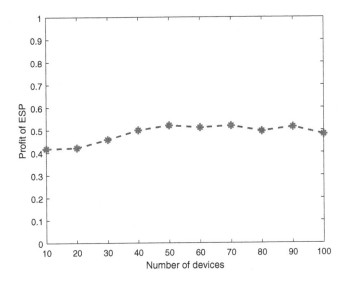

Fig. 3. The optimization of task average completion time for DM.

5   Conclusion

The study centers on the Mobile Edge Computing (MEC) system within the context of wireless interference. The primary objective is to optimize the allocation and pricing of computing resources provided by the Edge Service Provider (ESP) at the upper level, concurrently optimizing the computing offloading strategies of Device Managers (DMs) at the lower level. The upper-level ESPs are optimized for higher profit and faster task completion times on the IoT devices, and the lower-level DMs are optimized to enable higher profit to be obtained. Furthermore, we formulate the computing offloading strategies of DMs as a multi-user computing offloading game. To address this problem, we introduce a bilevel optimization algorithm named DBGOA. DBGOA leverages the NSGA- II in the upper level and a distributed algorithm based on Nash equilibrium in the lower level. We apply DBGOA to 10 instances, each with varying numbers of IoT devices, and compare its performance with two alternative optimization methods. The numerical results convincingly establish that our proposed DBGOA algorithm surpasses the other methods in effectively solving the bilevel optimization problem.

Based on the current research, for better addressing the optimization problem of task offloading and resource allocation in mobile edge computing, future work can focus on expanding the applicability of algorithms. Considerations may include the mobility of IoT devices, variations in network topology, fluctuations in real-time interference, and incorporating energy efficiency as an optimization objective.

Acknowledgements. The authors is supported by the National Natural Science Foundation of China under 61903003 and supported by Anhui Provincial Natural Science Foundation (Grant 2308085MF199 and 2008085QE227) and Scientific Research Projects in Colleges and Universities of Anhui Province (Grant 2023AH051124) and Supported by the Open Project of Anhui Province Engineering Laboratory of Intelligent Demolition Equipment (Grant APELIDE2022A007) and Supported by the Open Project of Anhui Province Key Laboratory of Special and Heavy Load Robot (Grant TZJQR001-2021).

References

1. Ouyang, H., Liu, K., Zhang, C., Li, S., Gao, L.: Large-scale mobile users deployment optimization based on a two-stage hybrid global HS-DE algorithm in multi-UAV-enabled mobile edge computing. Eng. Appl. Artif. Intell. **124**, 106608 (2023). https://doi.org/10.1016/j.engappai.2023.106608, https://www.sciencedirect.com/science/article/pii/S0952197623007923
2. Craig, A.B.: Understanding Augmented Reality: Concepts and Applications (2013)
3. Burdea, G.C., Coiffet, P.: Virtual Reality Technology. Wiley, Hoboken (2003)
4. Yurtsever, E., Lambert, J., Carballo, A., Takeda, K.: A survey of autonomous driving: common practices and emerging technologies. IEEE Access **8**, 58443–58469 (2020)
5. Yang, Z., Ding, Y., Jin, Y., Hao, K.: Immune-endocrine system inspired hierarchical coevolutionary multiobjective optimization algorithm for IoT service. IEEE Trans. Cybern. **50**(1), 164–177 (2018)
6. Li, H., Shou, G., Hu, Y., Guo, Z.: Mobile edge computing: progress and challenges. In: 2016 4th IEEE International Conference on Mobile Cloud Computing, Services, and Engineering (MobileCloud), pp. 83–84. IEEE (2016)
7. Safavat, S., Sapavath, N.N., Rawat, D.B.: Recent advances in mobile edge computing and content caching. Digit. Commun. Netw. **6**(2), 189–194 (2020)
8. Fernando, N., Loke, S.W., Rahayu, W.: Mobile cloud computing: a survey. Futur. Gener. Comput. Syst. **29**(1), 84–106 (2013)
9. Lin, J., Pan, L.: Multiobjective trajectory optimization with a cutting and padding encoding strategy for single-UAV-assisted mobile edge computing system. Swarm Evol. Comput. **75**, 101163 (2022)
10. Vaquero, L.M., Rodero-Merino, L.: Finding your way in the fog: towards a comprehensive definition of fog computing. ACM SIGCOMM Comput. Commun. Rev. **44**(5), 27–32 (2014)
11. Lin, J., Yu, W., Zhang, N., Yang, X., Zhang, H., Zhao, W.: A survey on internet of things: architecture, enabling technologies, security and privacy, and applications. IEEE Internet Things J. **4**(5), 1125–1142 (2017)
12. Vimal, S., Khari, M., Dey, N., Crespo, R.G., Robinson, Y.H.: Enhanced resource allocation in mobile edge computing using reinforcement learning based MOACO algorithm for IIoT. Comput. Commun. **151**, 355–364 (2020)
13. Tong, Z., Deng, X., Ye, F., Basodi, S., Xiao, X., Pan, Y.: Adaptive computation offloading and resource allocation strategy in a mobile edge computing environment. Inf. Sci. **537**, 116–131 (2020)
14. Zhang, Y., Xiu, S., Cai, Y., Ren, P.: Scheduling of graph neural network and Markov based UAV mobile edge computing networks. Phys. Commun. **60**, 102160 (2023). https://doi.org/10.1016/j.phycom.2023.102160, https://www.sciencedirect.com/science/article/pii/S1874490723001635

15. Tang, H., Jiao, R., Dong, T., Qin, H., Xue, F.: Edge computing energy-efficient resource scheduling based on deep reinforcement learning and imitation learning. In: Pan, L., Cui, Z., Cai, J., Li, L. (eds.) BIC-TA 2021. CCIS, vol. 1566, pp. 222–231. Springer, Singapore (2022). https://doi.org/10.1007/978-981-19-1253-5_16
16. Huang, Y., Luo, A., Zhang, M., Bai, L., Song, Y., Li, J.: Task location distribution based genetic algorithm for UAV mobile crowd sensing. In: Pan, L., Zhao, D., Li, L., Lin, J. (eds.) BIC-TA 2022. CCIS, vol. 1801, pp. 165–178. Springer, Singapore (2023). https://doi.org/10.1007/978-981-99-1549-1_14
17. Wang, Y., Sheng, M., Wang, X., Wang, L., Li, J.: Mobile-edge computing: partial computation offloading using dynamic voltage scaling. IEEE Trans. Commun. **64**(10), 4268–4282 (2016). https://doi.org/10.1109/TCOMM.2016.2599530
18. Dinh, T.Q., Tang, J., La, Q.D., Quek, T.Q.S.: Offloading in mobile edge computing: task allocation and computational frequency scaling. IEEE Trans. Commun. **65**(8), 3571–3584 (2017). https://doi.org/10.1109/TCOMM.2017.2699660
19. Wu, Y.C., Dinh, T.Q., Fu, Y., Lin, C., Quek, T.Q.S.: A hybrid DQN and optimization approach for strategy and resource allocation in MEC networks. IEEE Trans. Wireless Commun. **20**(7), 4282–4295 (2021). https://doi.org/10.1109/TWC.2021.3057882
20. Du, Y., Yang, K., Wang, K., Zhang, G., Zhao, Y., Chen, D.: Joint resources and workflow scheduling in UAV-enabled wirelessly-powered MEC for IoT systems. IEEE Trans. Veh. Technol. **68**(10), 10187–10200 (2019)
21. Chen, X., Jiao, L., Li, W., Fu, X.: Efficient multi-user computation offloading for mobile-edge cloud computing. IEEE/ACM Trans. Netw. **24**(5), 2795–2808 (2015)
22. Lyu, X., Tian, H., Sengul, C., Zhang, P.: Multiuser joint task offloading and resource optimization in proximate clouds. IEEE Trans. Veh. Technol. **66**(4), 3435–3447 (2016)
23. Bi, S., Zhang, Y.J.: Computation rate maximization for wireless powered mobile-edge computing with binary computation offloading. IEEE Trans. Wireless Commun. **17**(6), 4177–4190 (2018)
24. Zhou, J., Zhang, X.: Fairness-aware task offloading and resource allocation in cooperative mobile-edge computing. IEEE Internet Things J. **9**(5), 3812–3824 (2021)
25. Zhang, H., Liu, X., Bian, X., Cheng, Y., Xiang, S.: A resource allocation scheme for real-time energy-aware offloading in vehicular networks with MEC. Wireless Commun. Mob. Comput. **2022** (2022)
26. Huang, P.Q., Wang, Y., Wang, K.: A divide-and-conquer bilevel optimization algorithm for jointly pricing computing resources and energy in wireless powered MEC. IEEE Trans. Cybern. **52**(11), 12099–12111 (2021)
27. Chen, X., Jiao, L., Li, W., Fu, X.: Efficient multi-user computation offloading for mobile-edge cloud computing. IEEE/ACM Trans. Netw. **24**(5), 2795–2808 (2016). https://doi.org/10.1109/TNET.2015.2487344
28. He, F., Zhang, W., Zhang, G.: A differential evolution algorithm based on Nikaido-Isoda function for solving Nash equilibrium in nonlinear continuous games. PLoS ONE **11**(9), e0161634 (2016)

Expert-Guided Deep Reinforcement Learning for Flexible Job Shop Scheduling Problem

Wenqiang Zhang[1,2,3](✉) ⓘ, Huili Cong[3], Xuan Bao[3], Mitsuo Gen[4], Guohui Zhang[5], and Miaolei Deng[1,2]

[1] Key Laboratory of Grain Information Processing and Control (Henan University of Technology), Ministry of Education, Zhengzhou 450001, Henan, China
zhangwq@haut.edu.cn
[2] Henan Key Laboratory of Grain Photoelectric Detection and Control (Henan University of Technology), Zhengzhou 450001, Henan, China
[3] College of Information Science and Engineering, Henan University of Technology, Zhengzhou 450001, Henan, China
[4] Fuzzy Logic Systems Institute, Tokyo University of Science, Shinjuku City, Japan
[5] School of Management Engineering, Zhengzhou University of Aeronautics, Zhengzhou 450046, China

Abstract. Flexible job shop scheduling (FJSP) is crucial for automated production, ensuring efficiency and flexibility. In recent years, deep reinforcement learning (DRL) has achieved success in solving sequence decision-making problems. However, the efficiency of the generated scheduling plans is often constrained by the dependence of most DRL algorithms on priority dispatching rules (PDR). In order to enable the agent trained in DRL to autonomously choose operations and machines, this paper proposes an expert-guided deep reinforcement learning framework (EGDRL). Based on the representation of scheduling states using a disjunctive graph and an operation-machine topology graph, a graph neural network (GNN) is used to capture the complex relationships between operations and machines. More importantly, in the early stages of training, this paper introduces expert-guided solutions by using PDR to guide the action selection of reinforcement learning, which greatly improves the quality of decision-making. Experimental results consistently show that the proposed method outperforms traditional PDRs and other DRL algorithm. Notably, this superiority is observed across various scales, including larger instances. Additionally, the method exhibits robust performance on instances not encountered during training.

Keywords: Flexible job shop scheduling · Priority dispatching rules · Deep reinforcement learning · Graph neural network · Expert-guided

1 Introduction

In manufacturing, production scheduling significantly influences efficiency and profitability. The job shop scheduling problem (JSP) is a classic issue, optimiz-

L. Pan et al. (Eds.): BIC-TA 2023, CCIS 2062, pp. 50–60, 2024.
https://doi.org/10.1007/978-981-97-2275-4_4

ing job sequences across multiple machines. The FJSP is a well-known extension of JSP, breaking the uniqueness constraint of production resources. In FJSP, each operation can be allocated to one or more available machines, and different machines have varying processing times [2]. It's a more complex NP-hard problem than JSP and has diverse applications, such as semiconductor manufacturing, automotive assembly, and mechanical production systems [5]. In recent years, exact algorithms, heuristic methods, and meta-heuristics have been extensively investigated to address FJSP [7]. In previous literature, researchers have put forth several exact algorithms for solving the FJSP, the majority of which are comprised of integer linear programming or mixed-integer linear programming models. In practical scheduling systems, PDR [6] is widely employed as a heuristic approach. For instance, Shanker and Tzen [14] proposed four scheduling rules: first in first out (FIFO), shortest processing time (SPT), LPT, and MOPR. Meta-heuristic algorithms [12], like genetic algorithms (GA) [13] and particle swarm optimization (PSO) [11], use iterations for searching scheduling solutions, but have long optimization times unsuitable for real-time needs.

Since the 1990s, DRL has been used for scheduling problems [4,17]. Zhang et al. [18] introduced a solution for JSP using end-to-end DRL agents that autonomously learned PDR and integrated GNN into the DRL process. Regarding RL in FJSP, Luo designed a Double DQN to solve dynamic FJSP with new job insertions by minimizing total lateness [9]. Chen et al. [3] proposed a self-learning genetic algorithm based on RL to minimize makespan in FJSP. Most of these studies train with different RL algorithms to adaptively choose scheduling strategies based on the current state. This typically yields scheduling results quickly, outperforming general heuristic rules. However, this RL approach operates within the heuristic search space rather than the solution search space, making solution quality reliant on heuristic algorithm selection [16]. Fully unlocking DRL's potential for superior scheduling solutions requires further research, more complex decision frameworks, and richer representation techniques. This paper proposes a hybrid reinforcement learning method utilizing GNN, demonstrating outstanding performance in both solution quality and efficiency. Key contributions include introducing a novel end-to-end learning framework that effectively leverages disjunctive and operation-machine topology graphs for optimized scheduling solutions. Additionally, a mixed training strategy is implemented, combining early-stage guidance using PDR solutions with a gradual transition to autonomous action selection for improved exploration and exploitation balance.

The subsequent sections are structured as follows: Sect. 2 presents the necessary preliminaries for addressing FJSP. Section 3 explains our method, EGDRL. Section 4 provides experimental results and discussions. Finally, Sect. 5 reveals the conclusions and prospects.

2   Preliminaries

2.1   Problem Description

The research considers a set of n jobs, denoted as $J = \{J_1, \ldots, J_n\}$, and a set of m machines, denoted as $M = \{M_1, \ldots, M_m\}$. Each job J_i is composed of n_i sequential and specifically ordered operations $O_i = \{O_{i1}, O_{i2}, \ldots, O_{in}\}$. Operations represented by O_{ij} can be processed on a subset of machines $M_{ij} \subseteq M$ that meet specific conditions. Furthermore, the processing time of operation O_{ij} on machine M_k is denoted as p_{ijk}. The optimization scheduling problem involves assigning suitable processing equipment for each operation and determining the processing sequence on each machine and the completion time for each operation. This study aims to optimize the minimization of the makespan. An example gantt chart for the problem is shown in Fig. 1.

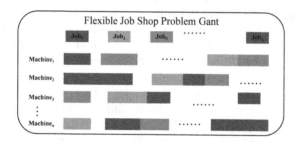

Fig. 1. The example gantt chart for FJSP.

2.2   Disjunctive Graph and Operation-Machine Topology Graph

The FJSP problem can be defined by a disjunctive graph $G = (O, C, D)$. Here, O represents the set of nodes corresponding to all operations, as shown in Fig. 2a and 2b. Additionally, C is a set of connecting arcs that depict precedence constraints between consecutive operations on the same job. D represents a set of disjunctive arcs, where each arc connects a pair of operations that can be processed by the same machine. In Fig. 2b, all disjunctive arcs have been assigned directions, indicating the sequence of processing operations on the machine.

The disjunctive graph model for FJSP often becomes highly dense due to the simultaneous representation of operation nodes, conjunction arcs, and disjunctive arcs. This can lead to a significantly large graph size, potentially increasing the computational burden. Due to these issues, we believe that a combination of disjunctive graphs and operation-machine topological graphs simultaneously is more beneficial for addressing FJSP problems.

We define the operation-machine topological graph model for FJSP as $F = (O, M, C, \varphi)$, where O and C have the same meanings as in the disjunctive graph. M represents the set of machine nodes, and φ is the edge set connecting all feasible machines to their corresponding operation nodes, where each element

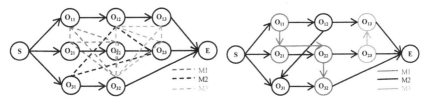

(a) Disjunctive graph for an FJSP in-   (b) Example of a feasible solution
stance

Fig. 2. Disjunctive graph representation of FJSP.

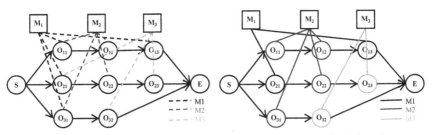

(a) O-M topological graph for an FJSP   (b) Example of a feasible solution
instance

Fig. 3. Operation-machine topological graph representation of FJSP.

can represent an operation-machine pair, as illustrated in Fig. 3. In the agent's
action selection, each scheduling action corresponds to determining an O-M edge
at each step.

3   EGDRL Algorithm

In this section, we will introduce the framework of EGDRL. First, we conceptual-
ize FJSP as a Markov Decision Process (MDP). Next, we propose a GNN-based
network model that serves as an encoder to extract features from the disjunctive graph, the operation-machine topological graph, and machine features for
scheduling tasks at decision points. Finally, we design a Proximal Policy Optimization (PPO) method based on an actor-critic framework to train the proposed
network model. It's important to note that this approach incorporates expert-
guided deep reinforcement learning strategies. The workflow of this method is
illustrated in Fig. 4.

3.1   Markov Decision Process Formulation

To implement FJSP as a reinforcement learning task and train an intelligent
agent, it is necessary to transform FJSP into a Markov Decision Process (MDP).
Specifically, modeling FJSP as an MDP involves designing states, actions, state
transitions, reward schemes, and the employed policy.

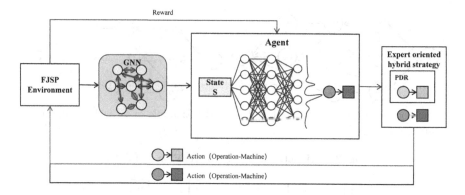

Fig. 4. The workflow of the proposed method.

State At time step t, the global state s_t is composed of two local states, s_{to} and s_{tm}.

Action At time step t, the action $a_t \in A_t$ comprises both operation and machine selection action, forming a compound decision.

Transition At time step t, based on s_t , the agent selects a operation-machine pair a_t and the environment transitions deterministically to a new state s_{t+1}.

Reward The reward is defined as the difference in the maximum time span of a partial schedule between s_t and s_{t+1}

Policy The policy $\pi(a_t|s_t)$ defines a probability distribution over the action set a_t for each state s_t.

3.2   Graph Neural Network

GNN, a deep neural network that excels at learning representations of graph-structured data. Analyzing the operation-machine topology graph reveals that machine nodes exhibit more neighbors with a single type of connection. Typically, machine nodes have multiple process nodes available for processing, connected solely by processing relationships. In real-world scheduling scenarios, the significance of late-stage processes and imminent processes to the machines differs. Thus, there is a need to strike a balance between short-term and long-term goals. To address this, the application of attention mechanisms [8] in GNN becomes a natural choice. Unlike machine nodes, operation nodes have fewer neighbors, but their connections exhibit diversity in terms of types. Consequently, we leverage the original disjunctive graph to extract features for process nodes. The overall architecture of GNN is illustrated in Fig. 5.

3.3   Expert-Guided Hybrid Strategy

In reinforcement learning, achieving effective learning requires a balance between two critical factors: exploration and exploitation. If an agent emphasizes exploration too much in decision-making, it may take a long time to acquire an effective strategy, let alone reach an optimal one. On the other hand, if the agent

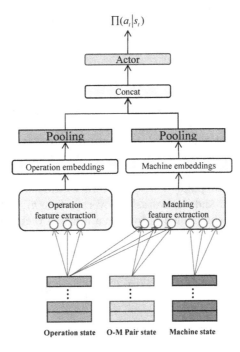

$\Pi(a_t|s_t)$

Fig. 5. The architecture of the graph neural network.

focuses too much on exploitation, it may get stuck in local optima, making it challenging to further optimize the current strategy. Therefore, in reinforcement learning, striking the right balance between these two factors is crucial for effective learning and optimization.

EGDRL employs a clever hybrid strategy where ϵ is used as a probability to choose between priority scheduling rules and the policy network. This means that in each decision, with a probability of ϵ, the solution from priority scheduling rules is adopted, and with a probability of $1-\epsilon$, the solution generated by the policy network is adopted. This probability ϵ gradually decreases after each action selection, encouraging the model to gradually transition from priority scheduling rules to solutions generated by the policy network during training, ultimately learning a better decision-making strategy. The advantage of this method lies in its ability to expedite the learning process, especially when dealing with complex environments or problems, as it leverages the knowledge and expertise of external experts.

3.4 Training Method

PPO is a promising variant of policy gradient methods that enhances the efficiency of network training. Its core idea involves using the stochastic gradient ascent method to perform small, batch-wise parameter updates, continuously optimizing the objective function, ultimately achieving the goal of updating the optimal policy. The final computation of PPO is represented as follows:

$$L_t^{CLIP}(\theta) = \hat{\mathbb{E}}_t \left[\min \left(\frac{\pi_\theta(a_t \mid s_t)}{\pi_{\theta_{\text{old}}}(a_t \mid s_t)} \hat{A}_t, \right. \right.$$
$$\left. \left. \text{clip} \left(\frac{\pi_\theta(a_t \mid s_t)}{\pi_{\theta_{\text{old}}}(a_t \mid o_t)}, 1 - c_e, 1 + c_e \right) \hat{A}_t \right) \right] \tag{1}$$

As the final part of this neural architecture, a policy network is designed below. An action represents a feasible operation-machine pair, denoted as (O_{ij}, M_k). At scheduling time step t, the policy $\pi(a_t)$ selects action based on the current state s_t. The policy network leverages the embeddings of operation o'_{ij}, machine nodesm'_{ij}, and the workshop overall state h_t, acquired from GNN. These embeddings are concatenated to form a working vector and are simultaneously fed into an MLP to compute action priorities, as described in Eq. (2). The probabilities for selecting each a_t given s_t are calculated by applying softmax to all $P(s_t, a_t)$.

$$\text{P}(a_t, s_t) = \text{MLP}_\omega \left[o'_{ij} \middle\| m'_k \middle\| h_t \right] \tag{2}$$

4   Computational Results and Analysis

The study evaluates the performance of the proposed algorithm through a series of experiments and compares it with other algorithms. Randomly generated datasets, following specified distributions, are used for training and validation to enhance the model's generalization ability. Testing is conducted on both small and large-scale test sets with the same distribution to assess the model's solving capabilities comprehensively. Studies [1,10] have shown that Most Operations Remaining (MOR), SPT, and FIFO rules often yield good solutions to scheduling problems. Therefore, this paper selects four scheduling rules, namely MOR, SPT, FIFO, and Most Work Remaining (MWKR), for comparison, in addition to a deep reinforcement learning method [15]. Since the optimal solutions for randomly generated datasets are unknown and difficult to obtain through exact algorithms, this paper uses the results obtained by the OR-Tools solver as the optimal solutions for the generated datasets. The relative error δ of the solution results is calculated as follow: $\delta = \left(\frac{C_{\max}}{C_{\max}^*} - 1 \right) \times 100\%$.

4.1   Parameter Settings

We trained EGDRL with fixed hyperparameters. Through analysis, we found the following hyperparameter values suitable for this paper. The dimensions of the embedding vectors for both machine and job nodes are set to 8. Both the policy network and the value network consist of two fully connected layers, with each layer containing 128 neurons. In the PPO algorithm, the hyperparameters for the three loss functions, c_p, c_v, and c_e, are set to 0.2, 0.01, and 0.2, respectively. The

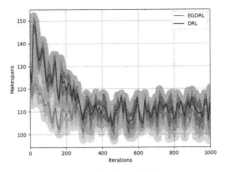

(a) The trends in 10*5 iterations

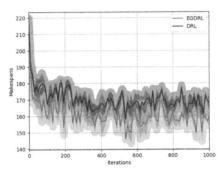

(b) The trends in 15*10 iterations

Fig. 6. The comparative trend of makespan over iterations for EGDRL and DRL.

proposed framework is optimized using the well-known Adam optimizer with a learning rate set to 0.0002. In the hybrid strategy, set ϵ to 0.6 with a decrement value of 0.001.

4.2   Comparison of Experimental Results

For each training task on different scales, regular validation on the validation set was conducted during training, and the best model was saved. The training curves demonstrate that the training process of the algorithm in this paper is relatively stable. The training curves, as depicted in Fig. 6.

For each training scale, Table 1 reports the average makespan and the gap between the proposed method and OR-Tools solutions for 100 test instances randomly generated from the same distribution as the training data. The proposed method consistently outperforms all baseline PDRs for all training scales and other DRL algorithm. Combining the method with expert strategy guidance yields superior solutions.

Table 1. Results for instances of different training sizes.

Size		OR-Tools	MOR	SPT	FIFO	MWKR	DRL	**OURS (EGDRL)**
10 * 5	$Cmax$	96.59	116.69	129.06	119.62	115.29	105.61	**104.37**
	Gap		20.89%	33.57%	23.94%	19.44%	9.38%	**8.05%**
15 * 10	$Cmax$	145.42	173.40	198.20	185.08	169.18	160.36	**160.16**
	Gap		15.29%	20.05%	14.73%	15.17%	10.13%	**10.13%**
20 * 05	$Cmax$	188.43	217.17	229.89	216.13	216.95	207.50	**204.77**
	Gap		19.29%	36.30%	27.27%	16.34%	10.31%	**8.66%**
20 * 10	$Cmax$	197.24	221.86	254.59	234.21	220.85	214.87	**215.85**
	Gap		12.53%	29.09%	18.82%	11.99%	8.97%	**9.43%**

Additionally, we tested the generalization performance of the models trained on the 10×5 task on 30×10 and 40×10 test instances. As shown in Table 2, EGDRL consistently produces high-quality solutions for large-scale problems. These results indicate that the EGDRL algorithm is an effective approach for solving FJSP, providing robust support for practical applications.

Table 2. Application results of training small instances for large-scale problems.

Size		OR-Tools	MOR	SPT	FIFO	MWKR	DRL	**OURS (EGDRL)**
30 * 10	$Cmax$	294.10	320.20	347.40	328.50	319.89	312.20	**311.20**
	Gap		8.90%	18.12%	11.74%	8.79%	6.18%	**5.81%**
40 * 10	$Cmax$	397.36	425.19	443.30	427.22	425.70	415.14	**413.91**
	Gap		7.02%	11.57%	7.53%	7.15%	4.49%	**4.16%**

5 Conclusion

This study proposes a novel end-to-end learning framework for addressing the FJSP with the goal of minimizing makespan. EGDRL combines GNN and DRL while maintaining scalability, meaning it can be trained on small-scale problems and deployed on larger-scale problems. Notably, the framework incorporates expert-guided strategies, harnessing the power of PDR to enhance early-stage learning. This hybrid methodology, combining DRL with expert knowledge, demonstrates superior performance over traditional heuristic approaches. The results from testing on a large dataset show that this method outperforms baseline PDR with reasonable efficiency and generalizes well to larger unknown instances. In future work, this method could be applied to real-time FJSP scenarios with dynamics and uncertainties, such as random job arrivals and unpredictable machine failures.

Acknowledgements. This research work is supported by the National Natural Science Foundation of China (62276091, U1904167), Science & Technology Research Project of Henan Province (232102211049, 222102210140), Key Research and Development Special Program of Henan Province (231111221200), Zhengzhou Science and Technology Collaborative Innovation Project (21ZZXTCX19), Open Fund of Key Laboratory of Grain Information Processing and Control, Ministry of Education (KFJJ2023005), Innovative Research Team (in Science and Technology) in University of Henan Province (21IRTSTHN018), and Scientific Research (C) of Japan Society of Promotion of Science (JSPS) (19K12148).

References

1. Brandimarte, P.: Routing and scheduling in a flexible job shop by tabu search. Ann. Oper. Res. **41**(3), 157–183 (1993)
2. Brucker, P., Schlie, R.: Job-shop scheduling with multipurpose machines. Computing (1990)
3. Chen, R., Yang, B., Li, S., Wang, S.: A self-learning genetic algorithm based on reinforcement learning for flexible job-shop scheduling problem. Comput. Industr. Eng. **149**, 106778 (2020)
4. Gabel, T., Riedmiller, M.: Distributed policy search reinforcement learning for job-shop scheduling tasks. Int. J. Prod. Res. **50**(1), 41–61 (2012)
5. Gao, K.Z., Suganthan, P.N., Chua, T.J., Chong, C.S., Cai, T.X., Pan, Q.K.: A two-stage artificial bee colony algorithm scheduling flexible job-shop scheduling problem with new job insertion. Expert Syst. Appl. **42**(21), 7652–7663 (2015)
6. Haupt, R.: A survey of priority rule-based scheduling. Oper.-Res.-Spektr. **11**(1), 3–16 (1989)
7. Jiang, J., Hsiao, W.: Mathematical programming for the scheduling problem with alternate process plans in FMS. Comput. Industr. Eng. **27**(1), 15–18 (1994). 16th Annual Conference on Computers and Industrial Engineering
8. Lee, J.B., Rossi, R.A., Kim, S., Ahmed, N.K., Koh, E.: Attention models in graphs: a survey. ACM Trans. Knowl. Discov. Data (TKDD) **13**(6), 1–25 (2019)
9. Luo, S.: Dynamic scheduling for flexible job shop with new job insertions by deep reinforcement learning. Appl. Soft Comput. **91**, 106208 (2020)
10. Montazeri, M., Van Wassenhove, L.: Analysis of scheduling rules for an FMS. Int. J. Prod. Res. **28**(4), 785–802 (1990)
11. Nouiri, M., Bekrar, A., Jemai, A., Niar, S., Ammari, A.C.: An effective and distributed particle swarm optimization algorithm for flexible job-shop scheduling problem. J. Intell. Manuf. **29**, 603–615 (2018)
12. Panwalkar, S.S., Iskander, W.: A survey of scheduling rules. Oper. Res. **25**(1), 45–61 (1977)
13. Pezzella, F., Morganti, G., Ciaschetti, G.: A genetic algorithm for the flexible job-shop scheduling problem. Comput. Oper. Res. **35**(10), 3202–3212 (2008). Part Special Issue: Search-based Software Engineering
14. Shanker, K., Tzen, Y.J.J.: A loading and dispatching problem in a random flexible manufacturing system. Int. J. Prod. Res. **23**(3), 579–595 (1985)
15. Song, W., Chen, X., Li, Q., Cao, Z.: Flexible job-shop scheduling via graph neural network and deep reinforcement learning. IEEE Trans. Industr. Inf. **19**(2), 1600–1610 (2022)
16. Sutton, R.S., Barto, A.G.: Reinforcement Learning: An Introduction. MIT Press, Cambridge (2018)

17. Wang, Y.F.: Adaptive job shop scheduling strategy based on weighted Q-learning algorithm. J. Intell. Manuf. **31**(2), 417–432 (2020)
18. Zhang, C., Song, W., Cao, Z., Zhang, J., Tan, P.S., Chi, X.: Learning to dispatch for job shop scheduling via deep reinforcement learning. Adv. Neural. Inf. Process. Syst. **33**, 1621–1632 (2020)

Eating Behavior Analysis of Cruise Ship Passengers Based on K-means Clustering Algorithm

Tao Zhang[1], Wei Cai[2], Min Hu[2], Guangzhao Yang[2(✉)], and Wenchu Fu[3]

[1] Science Education Innovation Park, Wuhan University of Technology, Sanya 572000, China
[2] Green and Smart River-Sea-Going Ship, Cruise Ship and Yacht Research Center, Wuhan University of Technology, Wuhan 430063, China
guangzhao@whut.edu.cn
[3] China Harzone Industry Corp., Ltd., Wuhan 430200, China

Abstract. In order to discover the real demand of cruise ship passengers and find the connection between their demand and cruise ship space design, we conduct a questionnaire survey of the passengers' characteristics and their eating behavior. Through the analysis on the passengers' eating behavior by using K-means clustering algorithm on the basis of the survey data, different types of their behavior characteristics are recognized, and the connections between the passengers' characteristics and their eating behavior are discussed. The research results are helpful to understand cruise ship passengers' eating demand and improve the design concept of cruise ship.

Keywords: K-means · Cruise Ship Passengers Eating Behavior · Cluster Analysis

1 Introduction

Modern luxury cruise ship is an urban complex integrating hotel, catering, entertainment, health, shopping, leisure and many other functions. And its design is an extremely complicated and systematic project for the passengers' large amount and variety. The designer has to consider both the space division, decoration style and facilities and equipment on the decks according to passenger's needs, and the passengers' characteristics and behavior. Studying cruise ship passengers' behavior patterns is both necessary to higher ship security and economic in ship management. Since passengers' eating behavior is a main part of their tour, by analyzing it, the manager of cruise ship may provide personalized recommendation to the passengers and strategically arrange the shops, enhancing its profit potential.

In recent years, behavior analysis has been a research hotspot, and many scholars conducted in-depth researches on different types of people's behavior patterns by different methods. Peng et al. [1] conduct load curve clustering analysis on typical daily load characteristics of different users, realizing an intelligent analysis of electricity consumption. Lu et al. [2] extracted user behavior patterns from the perspectives of temperature

L. Pan et al. (Eds.): BIC-TA 2023, CCIS 2062, pp. 61–73, 2024.
https://doi.org/10.1007/978-981-97-2275-4_5

sensitivity, electricity price sensitivity and power consumption stability, and established multi-attribute user consumption behavior graph. Benevenuto et al. [3] collected users' click data on social network platforms and extracted relevant fields to study user duration and regularities. In addition, in order to increase the proportion of users' interactive behaviors, he proposed a new modeling method to identify the type and frequency of interactive behaviors based on user click data. Zeng et al. [4] analyzed pedestrian behavior characteristics such as collective or individual avoidance with surrounding pedestrians, collision avoidance with conflicting vehicles, reaction to signal control and pedestrian crossing boundary, and simulated pedestrian crossing behavior by developing a micro-simulation model for pedestrian behavior analysis at signalized intersections. Bandyopadhyay et al. [5] analyzed the influence of vehicle characteristics, road types, traffic accidents and other factors on driving behavior, and classified driving behavior based on this. Ennahbaoui et al. [6] integrate user behavior analysis performed by mobile agents, which senses and tracks user behavior to describe potential vulnerabilities and abnormal activity on sensitive resources. Chen et al. [7] distributed the behavioral data of users browsing and placing orders on e-commerce websites, and then made use of the cosine distance between the behavioral distributed representations in different situations to accurately determine when users will browse or place orders again. Devineni et al. [8] used users' social data as a data set to analyze users' social behaviors over different periods of time, and based on this, implemented a visual framework to display users' personalized behaviors. Guimaraes et al. [9] used the network data generated in social networking to extract user characteristics, and classified and analyzed the age group of users through deep learning algorithms.

As a key step in information retrieval system, clustering algorithm has been applied in many fields and is still in the stage of continuous research. Among them, K-means clustering algorithm has been widely used in engineering practice [10–13] for its high computing speed and good clustering performance. Hu et al. [14] proposed a new k-means clustering algorithm (LK-means) based on Levy flight trajectory. In the iterative process of the LK-means algorithm, Levy flights are used to search for new positions to avoid premature convergence in the clustering. It is also used to increase the diversity of clustering, enhance the global search capability of K-means algorithm, and avoid falling into local optimal values prematurely. Ghazal et al. [15] used three different mathematical measures to evaluate the K-means clustering algorithm according to the execution time of different data sets and different cluster numbers. The results show that the implementation of the Manhattan distance measure achieves the best results in most cases. In addition, distance metrics can affect the execution time and the number of clusters created by the K-means algorithm. Madadizadeh et al. [16] used the characteristics of population density, area, distance from the original epicenter (Qom province), altitude from sea level, and Human Development index (HDI) in different provinces to study their correlation with the cumulative frequency of new coronavirus disease cases. On this basis, Spearman correlation coefficient and K-means cluster analysis (KMCA) were used to rank and cluster coronavirus disease (COVID-19) cases in Iranian provinces. Singh [17] used the K-means clustering algorithm to extract different RGB (red, green, blue) colors existing in the leaves. Du [18] et al. selected six economic, social and climate indicators that affect the production of Municipal Solid Waste: population, per capita GDP (PCGDP), environmental sanitation investment (ESI), average temperature, average precipitation and average humidity as classification indicators. All cities are divided into four clusters by K-means algorithm. Mohammadi et al. [19] used the K-means algorithm to determine the gold grade with a correlation coefficient of 91%, and derived the gold grade estimation equation based on the four parameters of arsenic content, antimony content, sampling point length and width.

However, previous studies have shown that traditional cruise ship designers tend to focus on the cruise ship itself, while ignoring the spatial relationship between passengers and the cruise ship. The above research did not consider the influence of passengers' eating behavior on the spatial scale and spatial composition of cruise ships. Though the overall layout design plan is reasonable from the perspective of engineering technology, it's deficient in satisfying passengers' real demand.

In this paper, data of cruise ship passengers are collected by a questionnaire survey firstly. Secondly, the questionnaire results are sorted and analyzed. Then the eating behavior is analyzed by K-means clustering algorithm to divide the passengers into different types. Finally, through further sensitivity analysis, the connections between passengers' characteristics and their eating behavior is discussed.

The rest of this paper is as follows. The Sect. 2 describes the questionnaire design focused on passengers' characteristics and their eating behavior. The Sect. 3 shows the establishment of the mathematical model of cluster analysis based on K-means algorithm, and the cluster analysis on the passenger eating behavior. The Sect. 4 is the analysis of the connections between passengers' characteristics and their eating behavior. The last section is the conclusion of the paper.

2 K-means Algorithm

Cluster analysis refers to the classification based on data similarity. Among them, the K-means algorithm proposed by J.B. Macqueen in 1967 is one of the simple and effective clustering algorithms widely used in science and industry [20]. The essence of the K-means algorithm is to divide n sample points into k clusters, so that the sample points within each cluster have high similarity, while the sample points between each cluster have lower similarity. The similarity calculation is based on the average value of the sample points in a cluster. According to the principle of minimizing clustering performance index, the clustering criterion function usually used is the minimum sum of squares of errors from each sample point in the cluster to the center of the cluster. Figure 1 shows the process of K-means algorithm [21], and the specific steps are shown as follows in Fig. 1.

(1) Select k sample points in sample data D, and assign the values of k sample points to the initial clustering center respectively $(\mu_1^{(1)}, \mu_2^{(2)}, \ldots, \mu_k^{(k)})$.

(2) At the j-th iteration, the Euclidean distance $d(t, i)$ between all points $P_t(t = 1, \ldots, n)$ in sample D and the center of each cluster $\mu_i^{(j)}$ are calculated successively. The Equation is shown in (1).

$$d(t, i) = \sqrt{\left(P_t - \mu_i^{(j)}\right)^2} \tag{1}$$

(3) Find out the minimum distance from P_t to $\mu_i^{(j)}$, then assign P_t to the closest cluster from $\mu_i^{(j)}$.

(4) Update the clustering center of each cluster. The Equation is shown in (2).

$$\mu_i^{(j+1)} = \frac{1}{n} \sum_{t=1}^{n_i} P_{it} \tag{2}$$

(5) Calculate the squared error E_i of all samples in data set D and compare it with the previous error E_{i-1}. The Equation is shown in (3).

$$E_i = \sum_{i=1}^{k} \sum_{t=1}^{n_i} \left| P_{it} - \mu_i^j \right|^2 \tag{3}$$

If $|E_{i1} - E_i| < \delta$, the algorithm ends, otherwise go to (2) for another iteration.

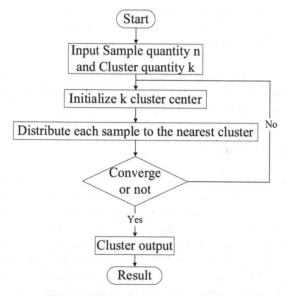

Fig. 1. Flowchart of K-means algorithm

In the K-means clustering algorithm, the quantity of cluster k plays an important role in the final clustering result, and this value needs to be determined before clustering [22]. For practical cases with a large amount of data, if the value of k is too small, the data difference in the cluster will be great. On the other hand, if the value of k is too big, the difference between clusters will be very small. The main methods to determine the value of k include elbow method and contour coefficient method. Elbow method is easy to realize, it can provide obvious result and be more efficient when processing large quantities of data, so in this paper elbow method is adopted to select the value of k. The elbow method uses the comparison between the value of k and the sum of squared errors (SSE) to determine k's value [23]. The Equation of SSE is shown in (4).

$$E_{SSE} = \sum_{i=1}^{k} \sum_{s \in C_i} \|s - x_i\|_2^2 \qquad (4)$$

In Eq. (4), E_{SSE} is the sum of squared errors (SSE), k is the quantity of clusters, C_i is the i-th cluster, s is the sample point in C_i, x_i is the cluster center in C_i.

Equation (4) sums the square sum of Euclidean distance from each data point in all clusters to the cluster center. The result is the basis for judging the density of data points in each cluster after clustering. The smaller the value of E_{SSE}, the higher the data points density in each cluster. During the iteration of the k value, the E_{SSE} value of each generation can be calculated. When k increases, the data points are divided into more clusters and the distance between clusters is smaller. At the same time, the value of E_{SSE} decreases as the value of k increases until the value of k equals the quantity of data points, and the calculation ends. In addition, the E_{SSE} value is not the smaller the better, when the k value is less than the real quantity of clusters that the data set should reasonably cluster, the E_{SSE} value will decrease significantly with the increase of the k value. When E_{SSE} tends to be flat, the corresponding k value is the best value, and the value of k can be determined by the $E_{SSE} - k$ simulation curve.

3   Questionnaire Design and Data Collection

In order to accurately obtain the information of cruise ship passengers' eating behavior in different scenarios for further analysis, this paper uses questionnaire survey. The questionnaire consists of two parts, namely, the passenger characteristic survey and the eating behavior survey.

Table 1. Passenger Characteristics.

Question description	Options
Gender	Male/Female
Birth year	Before 1960/1960–1964/1965–1969/1970–1974/1975–1979/1980–1984/1985–1989/1990–1994/1995–1999/2000–2004/after 2004
Occupation	Government employee/Public institution employee/State-owned enterprise employee/Foreign company employee/Private enterprise employee/Self-employed people/Student/Retired people/Others
Monthly income	Less than ¥5000/¥5000–10000/¥10000–20000/¥20000–30000/More than ¥30000
Cabin	Inside cabin/Oceanview cabin/ Balcony cabin/ Suite
Cruise ship companion	None/Family/Partner/Colleague/Friend
Travelling with children or not	Yes/No

3.1 The Passenger Characteristic Survey

This survey mainly focuses on the social and economic characteristics of passengers on the cruise ship, which is helpful to analyze the heterogeneity of passengers' behavior in different scenarios. This survey mainly includes gender, age, occupation, income, cabin type and other indicators, as shown in Table 1.

Table 2. Survey of Eating behavior characteristic.

Question category	Question description	Options
Breakfast eating behavior	Would you chose meal delivery?	Yes/No
	Which restaurant would you prefer?	Main restaurant/Cafeteria/Characterized restaurant
	What time do you usually have breakfast?	6:30/7:00/7:30/8:00/8:30/9:00/9:30
	How long does it take you to have breakfast?	10 min/20 min/30 min/40 min/50 min/60 min
Lunch eating behavior	Which restaurant would you prefer?	Main restaurant/Cafeteria/Characterized restaurant
	What time do you usually have lunch?	11:00/11:30/12:00/12:30/13:00/13:30
	How long does it take you to have lunch?	20 min/40 min/60 min/80 min/100 min/120 min
Dinner eating behavior	Which restaurant would you prefer?	Main restaurant/Cafeteria/Characterized restaurant
	What time do you usually have dinner?	17:00/17:30/18:00/18:30/19:00/19:30
	How long does it take you to have dinner?	20 min/40 min/60 min/80 min/100 min/120 min

3.2 Survey of Eating Behavior Characteristics

In order to verify the rationality of the distribution of survey samples and analyze the influence of eating behavior characteristics on passenger behavior, the eating behavior survey mainly includes the choice of eating places, starting time and eating duration of passengers on the cruise ship. The details are shown in Table 2.

A total of 680 questionnaires were issued in this survey, 650 questionnaires were recovered, and 602 valid questionnaires were obtained after incomplete and damaged questionnaires were excluded. Thus, the proportion of valid questionnaires was 89%. The demographic characteristics of the 602 respondents are shown in Table 3.

4 Case Calculation

4.1 Determination of the Value of k

In this experiment, the iteration number of K-means clustering algorithm is set to be 300 times, each k value is run 30 times and the corresponding SSE value is obtained, and then the average value of 40 SSE values are calculated, which is the final SSE value. As shown in Fig. 2, the latter half of the curve is relatively flat, so it is inferred that the k value is around here. Combined with Table 4, it can be seen that when k > 6, the slope of $SSE - k$ is small, indicating that the k value has reached the "elbow" value, resulting in a significant reduction in the "return" brought by the reduction of k value. Therefore, the most reasonable k value is 6, that is, six clusters should be selected for the data set obtained by the questionnaire.

4.2 Results Analysis

The K-means algorithm was used to cluster the cruise ship passenger data, and the 602 passengers were clustered into 6 categories. The clustering results are shown in Table 5. The characteristics of the six categories of passengers are as follows:

(1) C1 passengers

There are 99 C1 passengers, accounting for 16.45% of the total number. They usually choose to go to the restaurant for breakfast, and they go there at relatively similar time which is around 8:00 am, then they have their meal for 40 min or so. They have lunch at about 12:00 pm with the duration of less than 40 min. And they usually have dinner at about 6:30 pm with the duration of less than 60 min.

(2) C2 passengers

There are 74 C2 passengers, accounting for 12.29% of the total number. They choose not to go to the restaurant for breakfast, some of them ask for meal delivery, and some skip breakfast. The time they go to the restaurant is rather scattered, ranging from 12:00 pm to 1:00 pm, and they enjoy their lunch for usually an hour or even longer such as over 80 min. They usually go for dinner after 6:30 pm, with a mealtime of 100 min or so.

(3) C3 passengers

There are more C3 passengers with the number of about 133, reaching 22.09% of the total number. Their breakfast habit is similar to those of C2 passengers, that they prefer meal delivery or skipping breakfast over going to restaurant. Nevertheless, they don't spend too much time on eating, with a short duration of lunch and dinner: no more than 40 min.

(4) C4 passengers

There are 92 C4 passengers, accounting for 15.28% of the total number. Similar to C1 kind, they all choose to go to the restaurant for breakfast in person. Compared with C1 passengers, they go to the restaurant earlier, usually around 7:00 am with

Table 3. Characteristics of Survey Participants.

Type	Characteristic	Number of people	Pecentage
Gender	Male	315	52.33%
	Female	287	47.67%
Age	Under 18	30	4.98%
	18–25	228	37.87%
	26–30	149	24.75%
	31–40	140	23.26%
	41–50	34	5.65%
	51–60	18	2.99%
	Over 60	3	0.50%
Monthly Income	Less than ¥5000	276	45.85%
	¥5000–10000	219	36.38%
	¥10000–15000	68	11.30%
	¥15000–20000	19	3.15%
	¥20000–25000	9	1.50%
	¥25000–30000	1	0.16%
	More than¥30000	10	1.66%
Occupation	Government employee	51	8.47%
	Public institution employee	74	12.29%
	State-owned enterprise employee	47	7.81%
	Foreign company employee	34	5.65%
	Private enterprise employee	84	13.95%
	Self-employed people	109	18.11%
	Student	53	8.80%
	Retired people	9	1.50%
	Others	141	23.42%
Companion number	0 person	90	14.95%
	1 person	163	27.08%
	2 persons	173	28.74%
	3 persons	77	12.79%
	4 persons	21	3.48%
	5 or more persons	78	12.96%
Companion	Family (including children)	172	28.57%

(*continued*)

Table 3. (*continued*)

Type	Characteristic	Number of people	Pecentage
	Family (without children)	91	15.12%
	Partner	91	15.12%
	Friend	99	16.45%
	Colleague	65	10.79%
	Others	84	13.95%
Number of tours	0 times	197	32.72%
	1 time	208	34.55%
	2 times	100	16.62%
	3 times	44	7.31%
	4 times	15	2.49%
	5 or more times	38	6.31%
Cabin choice	Inside cabin	283	47.01%
	Oceanview cabin	151	25.08%
	Balcony cabin	96	15.95%
	Suite	72	11.96%
Tour duration	1–3 days	331	54.98%
	4–7 days	150	24.92%
	8–14 days	81	13.46%
	More than 14days	40	6.64%

the duration of about 20 min. 12:00 pm is most common for them to have lunch, and their eating duration is about 40 min. They usually have dinner at around 6:00 pm, but compared with C1, they have shorter dinner duration which is generally about 40 min.

(5) C5 passengers

There are fewer people in C5 group with the number of 35, accounting for 5.82% of the total number. They go to the restaurant for breakfast at about 8:00 am. Compared with C1 and C4 kind, they take longer time enjoying their breakfast for around an hour. Their lunch time is relatively scattered, generally from 12:00 pm to 1:00 pm with a longer duration of about 80 min. Their dinner time is usually at 6:30 pm with a longer duration of about 80 min too.

(6) C6 passengers

The number of C6 passengers is the biggest, totaling 169 people, reaching 28.07% of the total number. Similar to C2 and C3 passengers, they prefer meal delivery or skipping breakfast. They have lunch at scattered time from 11:30 am to 12:30 pm with a duration of no more than an hour. They have dinner also at scattered time from 6:00 pm to 7:00 pm with a duration of about an hour.

Table 4. Value of $SSE - k$.

k	SSE	k	SSE
1	2092.864608759984	11	1233.3777342770313
2	1733.3530860019046	12	1210.0343036566376
3	1559.2342773079906	13	1206.257796317133
4	1474.2450197266878	14	1181.807919581603
5	1445.7152324331535	15	1179.601831860945
6	1361.591878910245	16	1178.713454443838
7	1325.5984322517693	17	1142.146945304543
8	1314.4147357668026	18	1144.7651523979216
9	1277.462151986639	19	1123.6968210978175
10	1250.237668137771	20	1110.9173248632594

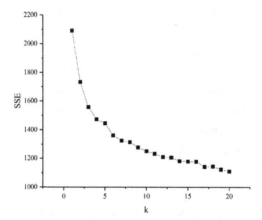

Fig. 2. Value of $SSE - k$

Table 5. K-means passenger clustering results.

Category	Number of passengers	Percentage
C1	99	16.45%
C2	74	12.29%
C3	133	22.09%
C4	92	15.28%
C5	35	5.82%
C6	169	28.07%

5 Discussion

The eating behavior of passengers on cruise ships is affected by different factors, such as their ages and consumption levels. In order to find the correlation between cruise ship passengers' eating behavior and their characteristics, on the basis of the current cluster analysis of eating behavior characteristics, this paper analyzes the connection between passengers' characteristics and eating behaviors. The results show that the three factors of age, cabin choice and tour duration have a great impact on the passengers' eating behaviors. The specific analysis is shown as follows.

(1) C2, C3 and C6 passengers have similar breakfast habits for choosing meal delivery or skipping breakfast over going to the restaurant in person. Through further analysis of the passenger characteristics, it can be seen that people over 30 years old prefer breakfast in a restaurant, while people over 60 prefer meal delivery for their limited mobility and people under 30 prefer meal delivery or skipping breakfast. The results show that younger passengers won't spend too much time on breakfast, while older ones prefer to eat breakfast in the restaurant.

(2) Inside cabin passengers prefer having their breakfast, lunch and dinner in the main eating room. And passengers of oceanview, balcony cabins and suites prefer cafeterias. Among them, suite passengers cover most of those who choose extra charge characterized restaurants. Nevertheless, the oceanview cabin guests prefer characterized restaurants when having lunch. The results show that there is a certain connection between the cabin choice and the restaurant selection. The more luxurious room they book, the more possibility they enter an extra charge characterized restaurant. Those who choose the relatively cheap inside cabins are more inclined to head to the main eating room where offers free meals.

(3) Passengers who stay for 1–3 days and 4–7 days on board tend to go to the restaurant earlier and spend shorter time in eating which is about 40 min. Passengers who stay for more than 8 days on the cruise ship are more inclined to go to the restaurant later, and spend more time (about an hour) on eating. The results showed that the shorter the tour is, the earlier the passengers have their meals. At the same time, in order to have more time to experience the entertainment activities on the cruise ship, their eating duration is relatively short. In contrast, passengers with longer tour duration are more likely to enter restaurants later and spend more time in eating.

6 Conclusion

This paper investigated the restaurant choice preference, the time to go to the restaurant and the eating duration of cruise ship passengers through questionnaires surveys. Using K-means clustering algorithm, the eating behavior of cruise ship passengers is analyzed, and the eating behavior characteristics of different categories of cruise ship passengers are further summarized. Based on the above analysis, the relationship between the characteristics of cruise ship passengers and their eating behavior is analyzed from the aspects of age, cabin choice and tour duration. These investigations and analyses can provide certain basic data for the study on the needs of cruise ship passengers, and also provide certain references for the spatial design of cruise ships.

Acknowledgement. This work was financially supported by the Hainan Special PhD Scientific Research Foundation of Sanya Yazhou Bay Science and Technology City. The authors are pleased to acknowledge these supports.

References

1. Wenhao, P., Zhe, D., Yanping, Z., Jun, L.: An analytical method for intelligent electricity use pattern with demand response. In: 2016 China International Conference on Electricity Distribution (CICED), Xi'an, China, pp. 1–4 (2016)
2. Lu, S., Jiang, H., Lin, G., Feng, X., Li, Y.: Research on creating multi-attribute power consumption behavior portraits for massive users. In: 8th International Conference on Power and Energy Systems (ICPES). IEEE (2018)
3. Benevenuto, F., Rodrigues, T., Cha, M., Almeida, V.: Characterizing user behavior in online social networks. In: Proceedings of the 9th ACM SIGCOMM Conference on Internet Measurement, pp. 49–62 (2009)
4. Zeng, W., Chen, P., Nakamura, H., Iryo-Asano, M.: Application of social force model to pedestrian behavior analysis at signalized crosswalk. Transp. Res. Part C **40**(mar.), 143–159 (2014)
5. Bandyopadhyay, S., Datta, A., Sachan, S., Pal, A.: SocialLink: unsupervised driving behavior analysis using representation learning and exploiting group-based training. ArXivLabs (2022). https://doi.org/10.48550/arXiv.2205.07870
6. Ennahbaoui, M., Idrissi, H.: A new agent-based framework combining authentication, access control and user behavior analysis for secure and flexible cloud-based healthcare environment. Concurr. Comput.: Pract. Exp. **34**(5), 1–36 (2022)
7. Hung-Hsuan, C.: Behavior2vec: generating distributed representations of users' behaviors on products for recommender systems. ACM Trans. Knowl. Discov. Data **12**(4), 43.1 (2022)
8. Devineni, P., Papalexakis, E.E., Koutra, D., Doruz, A.S., Faloutsos, M.: One size does not fit all: profiling personalized time-evolving user behaviors. In: the 2017 IEEE/ACM International Conference, pp. 331–340. ACM (2017)
9. Guimaraes, R.R., Renata, D.G., Denise, R., Demostenes, Z., Bressan, G.: Age groups classification in social network using deep learning. IEEE Access **5**, 10805–10816 (2017)
10. Lin, J., Pan, L.: Multiobjective trajectory optimization with a cutting and padding encoding strategy for single-UAV-assisted mobile edge computing system. Swarm Evol. Comput. **75**, 101163 (2022)
11. Selim, S.Z., Ismail, M.A.: K-means-type algorithms: a generalized convergence theorem and characterization of local optimality. IEEE Trans. Pattern Anal. Mach. Intell. **6**(1), 81–87 (1984)
12. Onoda, T., Sakai, M., Yamada, S.: Careful seeding method based on independent components analysis for k-means clustering. In: IEEE/WIC/ACM International Conference on Web Intelligence & Intelligent Agent Technology, pp. 51–59. ACM (2012)
13. Chitta, R., Jin, R., Havens, T.C., Jain, A.K.: Scalable kernel clustering: approximate kernel k-means. Comput. Sci. (2014)
14. Hu, H., Liu, J., Zhang, X., Fang, M.: An effective and adaptable k-means algorithm for big data cluster analysis. J. Pattern Recogn. Soc. **139**, 109404–109422 (2023)
15. Ghazal, T., Hussain, M., Said, R., Naseem, M.T.: Performances of k-means clustering algorithm with different distance metrics. Intell. Autom. Soft Comput. **30**(2), 735–742 (2021)
16. Lin, J., He, C., Cheng, R.: Adaptive dropout for high-dimensional expensive multiobjective optimization. Complex Intell. Syst. **8**(1), 271–285 (2022)

17. Madadizadeh, F., Sefidkar, R.: Ranking and clustering Iranian provinces based on covid-19 spread: k-means cluster analysis. J. Environ. Health Sustain. Dev. 6(1), 1184–1195 (2021)
18. Singh, L., Huang, H., Bordoloi, S., Garg, A., Jiang, M.: Exploring simple k-means clustering algorithm for automating segregation of colors in leaf of axonopus compressus: towards maintenance of an urban landscape. J. Intell. Fuzzy Syst.: Appl. Eng. Technol. 40(1), 1219–1243 (2021)
19. Xingyu, D., Dongjie, N., Yu, C., Xin, W., Zhujie, B.: City classification for municipal solid waste prediction in mainland China based on K-means clustering. Waste Manage. 144, 445–453 (2022)
20. Mohammadi, N.M., Hezarkhani, A., Maghsoudi, A.: Application of k-means and PCA approaches to estimation of gold grade in Khooni district (central Iran). Acta Geochim. 37(01), 104–114 (2018)
21. Modha, D.S., Spangler, W.S.: Feature weighting in k-means clustering. Mach. Learn. 52(3), 217–237 (2003)
22. Zhu, J., Wang, H.: An improved k-means clustering algorithm. Microcomput. Inf. 10(1), 193–199 (2010)
23. Wang, Y., Luo, X., Zhang, J., Zhao, Z., Zhang, J.: An improved algorithm of k-means based on evolutionary computation. Intell. Autom. Soft Comput. 26(5), 961–971 (2020)

The Bilinear-MAC Network for Visual Reasoning

Jiaxing Zeng[1,2], Wangli Zheng[3], and Yunhan Lin[1,2,4](✉)

[1] School of Computer Science and Technology, Wuhan University of Science
and Technology, Wuhan 430081, China
yhlin@wust.edu.cn
[2] Hubei Province Key Laboratory of Intelligent Information Processing
and Real-Time Industrial System, Wuhan 430081, China
[3] State Grid Electric Power Research Institute, Nanjing 211100, China
[4] Institute of Robotics and Intelligent Systems, Wuhan University of Science
and Technology, Wuhan 430081, China

Abstract. Most of the existing visual reasoning networks primarily
focus on aligning language with specific regions in images, but they
often struggle to comprehend the complex spatial logical relationships
within real-world scenes. Moreover, these networks lack interpretability.
In order to solve the above problems, this paper proposes the Bilinear-
MAC network with stronger reasoning ability. The MAC network is a
single-stream network that performs reasoning tasks. It only uses the
question representation as a control premise to extract image information. It can only understand simple images, but cannot understand complex images based on real scenes. The improved Bilinear-MAC network
in this paper is a dual-stream network that performs reasoning tasks. It
uses the problem representation as the control premise to extract image
information, and uses the image representation as the control premise
to extract problem information. As a result, it has a stronger understanding ability for real scene images with complex relationships and
rich object types. The proposed network achieves an overall accuracy
rate of 59.6% in the GQA dataset, which is 5.6% higher than that of the
MAC network, and an accuracy rate of 99.1% in the CLEVR dataset.
Experimental results demonstrate that the proposed network is capable
of a stronger understanding of complex real-world visual scenes.

Keywords: Visual reasoning · Feature fusion · Attention mechanism ·
Semantic relationship · Multimodal coding

1 Introduction

Visual reasoning is a challenging emerging research field in the field of artificial
intelligence, which is a multimodal learning task involving both computer vision

This work is supported by National Key R&D Program of China (Project No.
2022YFB4700400).

and natural language processing. Given certain visual information and a question related to the visual content, the algorithm needs to analyze the content and logical relationship in the visual information according to the information asked by the question, reason and generate the answer to the question.

Existing visual reasoning methods can be roughly divided into the following categories from the algorithm level: algorithms based on feature fusion, algorithms based on pre-trained transformer models, and algorithms based on attention mechanisms. The focus of research on visual reasoning algorithms based on feature fusion is how to efficiently fuse visual representations and problem representations. The learned features are not enough to model and capture all feature information, and there is a lack of visualization of the entire reasoning process. Affected by the success of the Transformer-based pre-training model, lxmert [1], vilbert [2], vilt [3] and other visual language cross-modal reasoning models have been proposed one after another. However, these cross-modal reasoning models can only align language and visual content, only obtain a superficial understanding of images and questions, fail to capture the reasonable perceptual process that leads to the correct answer, and lack interpretability.

The focus of the visual reasoning algorithm based on the attention mechanism is how to effectively capture the key clues in visual and language information, and remove useless redundant information, so as to increase the accuracy of visual question answering prediction. Anderson [4] proposed a top-down visual attention algorithm that only pays attention to the one-way visual attention information from the question to the visual object, ignoring the visual attention information from the visual information to the question word. Lu [5] proposed a joint attention mechanism to mine key information in vision and language by assigning weights to visual information and question words. However, it is difficult for this single-step attention mechanism to learn associations or logic with a large span. Yu [6] proposed a guided attention mechanism for visual question answering, which can accurately calculate the attention weights of each sequence element of visual information and question information. In order to improve the interpretability of the network reasoning process, Hudson [7] proposed the MAC network on the CLEVR [8] dataset. This is a novel fully differentiable neural network architecture designed to facilitate explicit and expressive reasoning. The MAC model only uses the problem representation as the control premise to perform single-flow reasoning process, and can only understand the relatively simple logical relationship of visual scenes. For real scenes with more complex images, the efficiency of the MAC model drops sharply. It cannot understand the logical relationship of complex visual scenes well, and has poor generalization performance.

Based on the above problems, this paper proposes a Bilinear-MAC network for visual reasoning question answering tasks. The Bilinear-MAC network is based on the MAC network as the basic framework. Aiming at the fact that the MAC network can only understand simple images, but cannot understand complex images based on real scenes, a dual-stream MAC reasoning module is proposed. Using a parallel dual-channel MAC inference flow, the inference operation is performed with questions and images as control information, and the final answer is predicted through an end-to-end inference network. The Bilinear-MAC network can accurately obtain the deep meaning of language and visual information for complex questions and complex visual scenes, and can capture the reasonable perception process to get the correct answer. And this paper introduces a multi-head attention mechanism in the object feature extraction module and language encoding module to obtain the correlation between image features and question features. It can help to better explain the reasoning logic process and enhance the interpretability of the reasoning process.

The main contributions of this paper are as follows:

1) The Bilinear-MAC network model for visual reasoning is proposed, use a parallel dual-channel MAC inference flow, and use questions and images as control information to perform inference operations. For complex real visual scenes, it can accurately obtain the deep meaning of language and visual information, while capturing the rational perceptual process that leads to correct answers.

2) The object feature extraction module and language encoding module are proposed, and introduce multi-head attention mechanism. Obtaining the correlation between image features and question features can help to better explain the reasoning logic process and enhance the interpretability of the reasoning process.

3) Combining the above content, experiments are carried out on the GQA and CLEVR datasets. The experimental results and visualization results prove that the proposed model can achieve more effective performance than the baseline model.

2   Bilinear-MAC Network Model

This paper proposes the Bilinear-MAC network, which uses visual features and question representations as the control premise to perform inference operations, which greatly boosts the MAC unit to understand the spatial relationship of visual features and deeper semantic logic information. This paper sets up an object feature extraction module and a language encoding module based on a multi-head attention mechanism to extract rich visual feature information and semantic information. It is then input to the MAC dual-stream reasoning module to complete iterative reasoning to generate the final memory information, and finally the memory information is sent to the output representation module to predict the final answer to the question based on visual features and question representation. The framework of the model is shown in Fig. 1. The Bilinear-MAC network consists of four parts, namely visual feature extraction module, language encoding module, MAC dual-stream reasoning module, and output representation module. The functions and detailed information of each module will be introduced in detail in the following chapters.

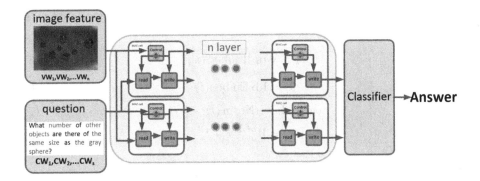

Fig. 1. Bilinear-MAC Network model frame diagram

2.1 Object Feature Extraction Module

The object feature extraction module embeds the features of the detected objects as an image embedding series $o_1, ..., o_n$, each object o_i is represented by its position feature p_i and its 2048-dimensional region of interest Roi feature f_i. Then input to the multi-head attention mechanism to generate attention-based hidden vector representation, and then output conv4 features through ResNet101, the generated tensor passes through two CNN layers with d output channels to obtain the final attention-based image representation. That is image feature information $vw_1, vw_2, ...vw_n$. The object feature information vw_i is transformed into a position-aware vector v_i by a learned linear transformation, representing the image representation associated with the i-th inference step. The process formula is expressed as follows:

$$\hat{f}_i = \text{LayerNorm}\left(W_F f_i + b_F\right) \tag{1}$$

$$\hat{p}_i = \text{LayerNorm}\left(W_P p_i + b_P\right) \tag{2}$$

$$vw_i = k^d(\hat{f}_i + \hat{p}_i) \tag{3}$$

$$v_i = W_i^{d*2d} vw_i + b_i^d \tag{4}$$

2.2 Language Encoding Module

To obtain the key information of the question, the sentence is first split by the same tokenizer into a series of word embeddings $w_1, ..., w_n$ and their corresponding indices u_i. Then project the word w_i and its index u_i (the absolute position of w_i in the sentence) to the vector through the embedding sublayer, and then generate an attention-based implicit vector representation through the multi-head attention mechanism. It is then further processed by a d-dimensional LSTM to generate attention-based context words $(cw_1, cw_2...cw_s)$. For each step i = 1, 2...n, the question q is transformed into a position-aware vector q_i by a learned linear transformation, representing the question representation associated with the i-th reasoning step. The calculation process is as follows:

$$\hat{w}_i = \text{WordEmbed}\,(w_i) \tag{5}$$

$$\hat{u}_i = \text{IdxEmbed}(i) \tag{6}$$

$$cw_i = \text{LayerNorm}\,(\hat{w}_i + \hat{u}_i) \tag{7}$$

$$q_i = W_i^{d*2d} q + b_i^d \tag{8}$$

2.3   MAC Dual-Stream Reasoning Module

The MAC dual-stream inference module is used to infer the internal relevance of visual features and semantic information, and consists of two parts, namely the MAC-Q inference stream and the MAC-V inference stream.

The work of the MAC-Q reasoning flow is based on the question as the control premise, and the visual features are extracted through the MAC reasoning unit and stored in the memory m_i. Since the problem is split into n inference steps, the input of each inference step q_i order of iterations is in the inference module consisting of n MAC inference units. Then, n times of inference operations are completed in a loop, the previous inference result is used as the input of the next inference unit, n times of inference operations are iterated, and finally the extracted visual information is stored in the memory m_p.

The work of the MAC-V inference flow is based on the image feature information as the control premise, and the semantic information of the question is extracted through the MAC inference unit, and then stored in the memory m_i as an intermediate result. The image feature information is split into n reasoning steps, and each area feature information vw_i corresponds to an image-related representation of a reasoning step, and the sequential iteration input of each reasoning step v_i is in the reasoning module composed of n MAC reasoning units. Then the intermediate reasoning result m_{i-1} of the previous step is used as the input of the next MAC unit, iterates n times of reasoning, and finally stores the extracted question information in the memory m_a.

The MAC reasoning unit is used to capture the connection between the extracted information and the control information of each step reasoning operation. Each MAC unit is composed of three operation units, namely the control unit CU, the read unit RU and the write unit WU, which jointly complete the task by performing an iterative reasoning process:

Control Unit: The control unit performs the i-th inference operation by receiving the contextual interrogative words $cw_1, cw_2...cw_s$ and the control state c_{i-1} and question location awareness q_i from the previous step. The specific steps are to first combine q_i and c_{i-1} into attention-based cq_i by linear transformation, and then project cq_i to the subspace of control information. By measuring the similarity between cq_i and each question word cw_i, and then feeding the result into a softmax layer, we finally sum the words to produce the i-th inference operation c_i. The calculation process is represented by 9–12, and the flow chart is shown in Fig. 2. The principle of the MAC-V inference flow control unit is the same as that of the MAC-Q flow, the difference is that the control information becomes image feature information $vw_1, vw_2, ...vw_n$.

$$cq_i = W^{d \times 2d} [c_{i-1}, q_i] + b^d \tag{9}$$

$$ca_{i,s} = W^{1 \times d} (cq_i \odot cw_s) + b^1 \tag{10}$$

$$cv_{i,s} = \text{softmax} (ca_{i,s}) \tag{11}$$

$$c_i = \sum_{s=1}^{S} cv_{i,s} \cdot cw_s \tag{12}$$

Reading Unit: The reading unit retrieves the required information r_i in the image according to the control information c_i given by the control unit. First, the image elements $vw_1, vw_2, ...vw_n$ are directly interacted with the memory m_{i-1} to obtain the correlation $I_{i,j}$ between the measurement elements and the previous intermediate results, and then perform inference operations. In order to enable the inference operation to align two independent visual and linguistic information, we connect $vw_1, vw_2, ...vw_n$ to $I_{i,j}$, and then pass the result through a linear transformation to obtain $ra_{i,j}$. Based on the control information, we measure its similarity $rv_{i,j}$ to each interaction vw_j, generate attention distributions, and finally obtain their weighted average r_i. The calculation process is represented by 13–17, and the process is shown in Fig. 3. The principle of the reading unit of the MAC-V inference flow is the same as that of the MAC-Q flow, the difference is that the reading unit retrieves the required information r_i from the question according to the control information c_i.

$$I_{i,j} = \left[W_m^{d \times d} m_{i-1} + b_m^d \right] \odot \left[W_k^{d \times d} vw_j + b_k^d \right] \tag{13}$$

$$I'_{i,j} = W^{d \times 2d} [I_{i,j}, vw_j] + b^d \tag{14}$$

$$ra_{i,j} = W^{d \times d} \left(c_i \odot I'_{i,j} \right) + b^d \tag{15}$$

$$rv_{i,j} = \text{softmax} (ra_{i,j}) \tag{16}$$

$$r_i = \sum_{j=1}^{S} rv_{i,j} \cdot vw_j \tag{17}$$

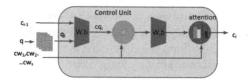

Fig. 2. Control unit

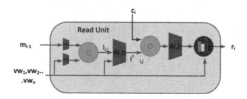

Fig. 3. Reading unit

Write Unit: The task of the write unit is to obtain the intermediate result of the i-th step reasoning operation, and then store it in the memory m_i. The specific operation is to integrate the information retrieved by the reading unit RU with the intermediate result m_{i-1} of the previous step through the guidance of the i-th reasoning operation c_i, and finally obtain the memory information m_i. The calculation process is represented by Eqs. 18–23, and the process is shown in Fig. 4.

$$m_i^{info} = W^{d \times 2d}[r_i, m_{i-1}] + b^d \tag{18}$$

$$sa_{ij} = \text{softmax}\left(W^{1 \times d}(c_i \odot c_j) + b^1\right) \tag{19}$$

$$m_i^{sa} = \sum_{j=1}^{i-1} sa_{ij} \cdot m_j \tag{20}$$

$$m_i' = W_s^{d \times d} m_i^{sa} + W_p^{d \times d} m_i^{info} + b^d \tag{21}$$

$$c_i' = W^{1 \times d} c_i + b^1 \tag{22}$$

$$m_i = \sigma(c_i') m_{i-1} + (1 - \sigma(c_i')) m_i' \tag{23}$$

2.4 Output Representation Module

The output representation module is based on the question representation q and the image feature representation $vw_1, vw_2, ...vw_n$, as well as the final memory m_p of the MAC-Q inference flow and the final memory m_a obtained by the MAC-V inference flow, through weighted summation. It is then processed by a softmax classifier and produces a distribution over the candidate answers to obtain the final answer. As shown in Fig. 5.

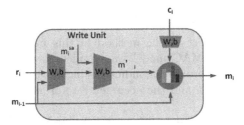

Fig. 4. Write unit

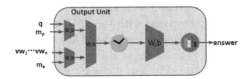

Fig. 5. Output representation module

3   Experiment

In this section, we mainly introduce the dataset used to evaluate the Bilinear-MAC model, and compare our model results with baseline models and existing model results. Multiple sets of ablation experiments are set up to demonstrate the indispensability of the Bilinear-MAC model components, and visualizations of the inference steps are shown. We use NVIDIA RTX3080Ti for training in Ubuntu18.04 system. Set the MAC unit length of MAC-Q and MAC-V inference flow to n = 12, the optimizer is Adam [9] optimizer, and the basic learning rate is 1×10^{-4}. In order to alleviate overfitting, this paper sets the weight decay size to 0.01. The batch size is set to 64, the model is trained for 25 epochs on the training-set, and the best model evaluated on the validation-set is submitted to the standard test-set to obtain the final test results.

3.1   Dataset

CLEVR contains 100k rendered images and about 1 million auto-generated questions, 853k of which are unique. It contains pictures and questions for challenging visual reasoning skills such as counting, comparing, logical reasoning, existence and other types. The GQA [10] dataset revolves around real-world reasoning, scene understanding, and synthetic question answering, and it consists of 113K images and 22 million questions. These questions, of varying types and degrees of synthesis, measure performance on a range of reasoning skills, including object and attribute recognition, transitive relationship tracking, spatial reasoning, logical reasoning, and comparison. Compared with CLEVR, the object relationship in the image scene of GQA is relatively complex, the feature types are more abundant, and the reasoning difficulty is relatively large.

3.2 Results of Empirical Comparison

In order to verify the effect of the method in this paper on the visual reasoning task, the Bilinear-MAC model was applied to the GQA and CLEVR datasets respectively. The evaluation indicators include the overall accuracy rate of answers to all questions, and the accuracy rate of answers to various single types of questions.

The experimental results show that the Bilinear-MAC method achieves an overall accuracy rate of 59.6% on the GQA test-std, which is 5.6% higher than the MAC method. Especially in the field of answering open questions, it has a significant advantage, improving the baseline model by 7.1%. In addition to comparing the accuracy of the answer, compared with the MAC method, Bilinear-MAC has achieved quite good results compared with the MAC method in terms of validity and plausibility. In terms of consistency, Bilinear-MAC has increased by 6.6% compared to the MAC method. We also compare with existing state-of-the-art methods, including GAT2R [11], MCAN [12], MCAOA [13], DMRFNET [14], RWSAN [15], BLIP-2 [16]. From Table 1, the overall accuracy of Bilinear-MAC surpasses these models, and these results verify the effectiveness of the proposed model on visual reasoning tasks.

Table 1. GQA dataset validation results

Method	Accuracy	Binary	Open	Consistency	Plausibility	Validity	Distribution
Human	89.3	91.2	87.4	98.4	97.2	98.9	–
GAT2R	54.5	–	–	83.3	85.0	96.2	–
MCAN	56.6	–	–	–	–	–	–
MCAOA	56.7	74.5	41.2	86.7	**85.5**	96.7	–
DMRFNET	57.1	74.0	41.9	87.0	84.9	**97.6**	–
RWSAN	57.4	75.6	41.5	86.5	85.1	96.7	–
BLIP-2	44.7	–	–	–	–	–	–
MAC	54.0	71.2	38.9	81.6	84.5	96.2	5.3
Bilinear-MAC	**59.6**	**77.3**	**46.0**	**88.2**	84.8	96.3	**5.0**

On the CLEVR test-set, the overall accuracy of the inference question answering task of the Bilinear-MAC method reached 99.1%, which is 0.2% higher than that of the MAC method. Bilinear-MAC can achieve the same effect as the MAC method on the problems of existence type, count type, comparison type and query type. The results are shown in Table 2.

3.3 Ablation Experiment

In order to quantitatively analyze the contribution of each module to the model, corresponding ablation experiments are carried out in this paper. All experiments use the same hyperparameters and training strategies, and use the CLEVR and

Table 2. CLEVR dataset validation results

Method	Accuracy	Count	Exist	CN	QA	CA
Human	92.6	86.7	96.6	86.5	95.0	96.0
Q-type	41.8	34.6	50.2	51.0	36.0	51.3
LSTM	46.8	41.7	61.1	69.8	36.8	51.8
CNN+LSTM	52.3	43.7	65.2	67.1	49.3	53.0
CNN+LSTM+SA+MLP	73.2	59.7	77.9	75.1	80.9	70.8
CNN+GRU+FILM	97.6	94.3	99.3	93.4	99.3	99.3
MAC	98.9	97.1	99.5	99.1	99.5	99.5
Bilinear-MAC	**99.1**	**97.5**	**99.8**	**99.2**	**99.6**	**99.6**

GQA datasets for training and testing respectively. The experimental results are shown in Table 3. In order to demonstrate the effectiveness of the object feature extraction module and the language encoding module to extract features based on the multi-head attention mechanism, this paper replaces the multi-head attention mechanism with the attention mechanism of the base network. The overall accuracy drops by 3.3% and 0.5% in GQA and CLEVR. To demonstrate the effectiveness of the dual-stream inference MAC flow, we cancel the MAC-V inference flow and only use the question as the control premise, resulting in an overall accuracy drop of 5.2% and 9.6%. We cancel the MAC-Q inference flow and only use image features as the control premise, and the overall accuracy drops by 6.3% and 11.5%, respectively. To demonstrate the contribution to the accuracy of inference results by keeping question and image information separate throughout the inference process, we cross-fused the MAC unit outputs of the Bilinear-MAC dual-stream inference module, resulting in an overall accuracy drop of 3.2% and 12.7%. This demonstrates the effectiveness of the Bilinear-MAC model to separate image information from question control information.

Table 3. Ablation experiment verification results

Method	GQA	CLEVR
Bilinear-MAC-(MHA)	56.3	98.6
Bilinear-MAC-(MAC-V)	54.4	89.5
Bilinear-MAC-(MAC-Q)	53.3	87.6
Bilinear-MAC+(cross)	56.4	86.4
Bilinear-MAC	**59.6**	**99.1**

3.4 Interpretability

In order to give us a more specific understanding of the inference process of Bilinear-MAC, we visualize the attention distribution of the model during the iterative inference calculation process, and provide an example in the inference process of the CLEVR problem in Fig. 6. By looking at the sequence of attention distributions on images and questions, the inference patterns and steps of Bilinear-MAC can be discovered. The model's linguistic and visual attention is focused on specific terms and regions in the image, and the attention maps demonstrate the model's ability to capture the underlying semantic structure of the question as well as the transitive relationship between objects.

This paper shows in Fig. 6 how the model finally answers the correct answer through iterative reasoning steps: In the first iteration, the model pays attention to the gray sphere, and updates the memory information m_1 according to the visual representation of the gray sphere. In the next step, the control unit realizes that it should now look for an object of the same size as the gray sphere and store it in the control information c_2. Then the reading unit realizes from m_1 and c_2 that it should look for an object (c_2) of the same size as the gray sphere (m_1), thus finding 5 objects such as the purple cylinder, the yellow sphere and updating m_2. A similar process is repeated in the next iteration, and the final iteration leads to a final answer of 5, thus answering the question correctly.

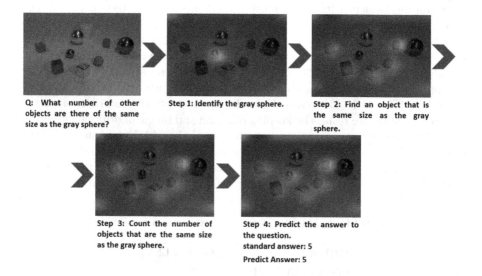

Q: What number of other objects are there of the same size as the gray sphere?

Step 1: Identify the gray sphere.

Step 2: Find an object that is the same size as the gray sphere.

Step 3: Count the number of objects that are the same size as the gray sphere.

Step 4: Predict the answer to the question.
standard answer: 5
Predict Answer: 5

Fig. 6. Interpretability of inference process

4   Conclusion

Based on the MAC network framework, this paper proposes an attention-based dual-stream inference network (Bilinear-MAC), which solves the problem that the MAC network cannot understand the spatial logic relationship of complex real scenes well. Firstly, according to the spatial relationship between image region features and semantic features, the MAC dual-stream reasoning module is designed. Using a parallel dual-channel MAC reasoning flow, the reasoning operation is performed using questions and images as control information, which realizes the complementarity of visual features and semantic features, and improves the network's ability to understand complex visual scenes and complex problems. Secondly, a multi-head attention mechanism is introduced to obtain the correlation between image features and question features, which can better explain the reasoning logic process and enhance the interpretability of the reasoning process. The proposed Bilinear-MAC network is tested on public datasets GQA and CLEVR, and the effectiveness of the network is verified by ablation experiments and visual analysis. In the GQA dataset, the overall accuracy rate is 59.6%, which is 5.6% higher than that of the MAC network, and in the CLEVR dataset, the accuracy rate is 99.1%. The Bilinear-MAC network is more understandable to complex images and questions by setting images and questions together as control information. Future work will further study more efficient cross-modal reasoning methods.

References

1. Tan, H., Bansal, M.: LXMERT: learning cross-modality encoder representations from transformers. In: Proceedings of the 2019 Conference on Empirical Methods in Natural Language Processing and the 9th International Joint Conference on Natural Language Processing (EMNLP-IJCNLP), Hong Kong, China (2019)
2. Lu, J., Batra, D., Parikh, D., et al.: ViLBERT: pretraining task-agnostic visiolinguistic representations for vision-and-language tasks. Adv. Neural Inf. Process. Syst. **32** (2019)
3. Kim, W., Son, B., Kim, I.: ViLT: vision-and-language transformer without convolution or region supervision. In: International Conference on Machine Learning, pp. 5583–5594. PMLR (2021)
4. Anderson, P., He, X., Buehler, C., et al.: Bottom-up and top-down attention for image captioning and visual question answering. In: Proceedings of the IEEE Conference on Computer Vision and Pattern Recognition, pp. 6077–6086 (2018)
5. Lu, J., Yang, J., Batra, D., et al.: Hierarchical question-image co-attention for visual question answering. Adv. Neural Inf. Process. Syst. **29** (2016)
6. Yu, Z., Yu, J., Cui, Y., et al.: Deep modular co-attention networks for visual question answering. In: Proceedings of the IEEE/CVF Conference on Computer Vision and Pattern Recognition, pp. 6281–6290 (2019)
7. Hudson, D.A., Manning, C.D.: Compositional attention networks for machine reasoning. In: International Conference on Learning Representations (2018)
8. Johnson, J., Hariharan, B., Van Der Maaten, L., et al.: CLEVR: a diagnostic dataset for compositional language and elementary visual reasoning. In: Proceedings of the IEEE Conference on Computer Vision and Pattern Recognition, pp. 2901–2910 (2017)
9. Kingma, D.P., Ba, J.: Adam: a method for stochastic optimization. In: International Conference on Learning Representations (2015)
10. Hudson, D.A., Manning, C.D.: GQA: a new dataset for real-world visual reasoning and compositional question answering. In: Proceedings of the IEEE/CVF Conference on Computer Vision and Pattern Recognition, pp. 6700–6709 (2019)

11. Miao, Y., Cheng, W., He, S., et al.: Research on visual question answering based on GAT relational reasoning. Neural Process. Lett. 1–14 (2022)
12. Yu, Z., Yu, J., Cui, Y., et al.: Deep modular co-attention networks for visual question answering. In: Proceedings of the IEEE/CVF Conference on Computer Vision and Pattern Recognition, pp. 6281–6290 (2019)
13. Rahman, T., Chou, S.H., Sigal, L., et al.: An improved attention for visual question answering. In: Proceedings of the IEEE/CVF Conference on Computer Vision and Pattern Recognition, pp. 1653–1662 (2021)
14. Zhang, W., Yu, J., Zhao, W., et al.: DMRFNet: deep multimodal reasoning and fusion for visual question answering and explanation generation. Inf. Fusion **72**, 70–79 (2021)
15. Qin, B., Hu, H., Zhuang, Y.: Deep residual weight-sharing attention network with low-rank attention for visual question answering. IEEE Trans. Multimed. **25**, 4282–4295 (2023)
16. Li, J., Li, D., Savarese, S., et al.: Blip-2: bootstrapping language-image pre-training with frozen image encoders and large language models. arXiv preprint arXiv:2301.12597 (2023)

Artificial Intelligence Analysis of State of Charge Distribution in Lithium-Ion Battery Based on Ultrasonic Scanning Data

Jie Tian[1], Jinqiao Du[1], Kai Huang[2], Xueting Liu[2], Yu Zhou[2(✉)], and Yue Shen[2(✉)]

[1] Shenzhen Power Supply, Shenzhen 518000, China
[2] School of Materials Science and Engineering, Huangzhong University of Science and Technology, Wuhan 430000, China
zhouyutech@yeah.net, shenyue1213@hust.edu.cn

Abstract. Lithium-ion batteries are the most prevelant electrochemical energy storage devices, but they often suffer from inconsistent charging/discharging speeds at different positions in the cell. This problem would lead to uneven distribution of state of charge (SOC), and may cause capacity degradation acceleration. To address this issue, this paper proposes a novel non-destructive method for characterizing the distribution of SOC within the battery: by using ultrasonic C-scan technology to collect ultrasonic transmission waveforms at different positions inside the battery, establishing the correlation between battery SOC and ultrasonic waveforms using convolutional neural networks, and further analyzing the non-uniformity of the ultrasound signals to infer the SOC differences at different locations within the battery. The research findings in this paper provide valuable insights for understanding the failure mechanisms of lithium-ion batteries and guiding battery fabrication process optimization.

Keywords: lithium-ion battery · state of charge · ultrasound · artificial intelligence · feedforward neural network

1 Background

Lithium-ion batteries have the advantages of high energy density, long cycle life, and high energy efficiency, and are widely used in electric vehicles, energy storage stations, mobile electronic products, and other fields [1, 2]. Lithium-ion batteries rely on the migration of ions in the positive and negative electrode materials to achieve charging and discharging. In order to shorten the migration path of ions, the electrodes and separators inside the battery are usually made into large-area thin film structures, which are then wound or stacked and inserted into a casing made of aluminum alloy or aluminum-plastic composite film to form a battery cell [3]. In practice, it has been found that due to differences in coating thickness, stacking density, local temperature, and distance to the electrode tab, there are often inconsistencies in charging and discharging speeds and uneven state of charge (SOC) in different positions inside the battery [4]. The charging and discharging speed in certain local areas of the battery is always higher than in other

© The Author(s), under exclusive license to Springer Nature Singapore Pte Ltd. 2024
L. Pan et al. (Eds.): BIC-TA 2023, CCIS 2062, pp. 87–93, 2024.
https://doi.org/10.1007/978-981-97-2275-4_7

areas, which may lead to accelerated aging in that area, resulting in problems such as decreased battery capacity and aging failure.

However, although the existence of uneven SOC distribution in lithium-ion battery has been confirmed by battery engineers via cell disassembly, quantitative studies on this issue have always faced difficulties [5, 6]. Lithium-ion batteries have a sealed shell made of aluminum plastic film or aluminum alloy, which hinders the transmission of many detection signals such as visible light, electron beams, terahertz waves, and infrared rays. Therefore, It difficult to accurately characterize the state of charge (SOC) at different positions within the cell. Even X-rays have limited penetration depth for typical commercial large cells, thus rapid imaging of SOC distribution was impossible previously [7–10].

To address the above problem, Huang et al. proposes that the degree of lithium insertion in the electrode materials varies at different states of charge (SOC), resulting in different levels of particle packing density [11]. This, in turn, affects the characteristics of ultrasonic wave propagation. Based on this principle, a focused ultrasonic beam can be used as a detection signal to scan the battery. By comparing the differences in ultrasonic conduction characteristics at different SOC levels, the SOC of different positions in the battery can be estimated. However, the challenge of this approach lies in the lack of a linear correspondence between the ultrasonic conduction characteristics of the battery and SOC. It is difficult to directly identify SOC based on a single transmitted ultrasonic wave feature [12, 13].

In order to achieve high-precision SOC estimation, this study focuses on a soft-packed battery composed of $LiNi_{0.8}Co_{0.1}Mn_{0.1}O_2$ (NCM811) cathode and graphite anode. The complete ultrasonic transmitted waveforms at various positions are scanned as input signals. A convolutional neural network is employed to establish the correspondence between ultrasonic transmitted waveforms and SOC. Based on the ultrasonic transmitted signals from different positions, the SOC distribution image of the battery can be inferred. This method can achieve SOC recognition with a relative error of less than 5% and a spatial resolution within the SOC distribution plane of approximately 0.5 mm.

2 Experimental

2.1 Detailed Description of the Tested Cells and the Charging Procedure

The tested cells were pouch-type cells with NCM811 as the cathode material and artificial graphite (AG) as the anode material. The size of the cells was 65 mm × 35 mm × 5 mm. The rated capacity of the cell was 4.15Ah. The cells were charged with constant current of 1.2 A for 10 min and then rested for 25 min and then scanned with the ultrasonic beam. This procedure was repeated to reach the maximum capacity of 4.15 Ah.

2.2 Ultrasonic Scanning Process

The ultrasonic scanning of the tested cell was conducted with an instrument produced by Wuxi Topsound Tech. Co. Ltd., of which the model number was UBSC-LD50. During testing, the cells were immersed in silicone oil to improve ultrasonic beam transmission. Two ultrasonic focusing transducers (frequency 2 MHz, focal length 30 mm, focusing diameter <1 mm) were mounted on the front and rear sides of the battery. The progressive scanning step length was set at 0.5 mm, so that the resolution of the SOC distribution map was 0.5 mm. The driving source used in this study is a 200 V pulse signal with a pulse width of 250 ns and an average power of the ultrasonic transducer of 25 mW. A false-color diagram is drawn according to the intensity of the ultrasonic transmission signal at different positions. The color scale is defined by the peak-to-peak value of the waveform detected by the receiving transducer. The color scale from blue to red corresponds to a peak-to-peak value of 0 ~2.5 A.U. That is, blue represents low ultrasound transmittance area, red is the high transmittance area. The whole set-up is as shown in Fig. 1a. Besides the ultrasonic intensity diagram, for each pixel of the ultrasonic diagram, the entire wave form of the transmitted ultrasound was recorded.

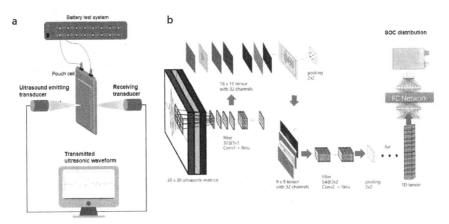

Fig. 1. (a) Scheme of the ultrasonic scanning set-up. (b) The procedure for processing ultrasonic data and predicting battery SOC distribution using a convolutional neural network.

2.3 Training of the Feedforward Neural Network

Ultrasonic data were collected from 6 cells, each containing 20 different SoCs. The scanning range of cells was both 40 * 75 mm, the raw acoustic data was collected and saved as a 30 * 60 matrix for excluding non-battery ultrasound data. Each element in the matrix is a 10000-length vector, i.e., it $[\alpha 1, \alpha 2, ..., \alpha 10000]$ also corresponds to the transmitted ultrasonic signal of 1 mm * 1 mm area on cell. The data of 4 cells were selected as the training set, and the remaining two were used as the test set to verify the model.

The regions of the ultrasound data that were most relevant (400-length vector) to SoC changes were used as samples. To match the 2D convolutional layer of the neural network model, we raised dimensions to a 2D matrix (20×20) of the 400-length vector. The pre-processed data were labeled with the corresponding SoC labels and used as the input of the neural network after being out of order.

The extraction rate of 0.3 from the training set were taken as the validation set in the model training process, then the model was applied to constantly monitor the convergence of the training set and the validation set loss function. In the process of multiple training, the learning rate is constantly adjusted to obtain the global optimal solution according to the performance of the model on the test set, and the dropout layer is appropriately added to prevent overfitting.

After several parameter adjustments, the model goes through about 1000 iterations and reaches convergence. Ultrasonic data of different SOC states on the test set were randomly selected as input to the model and evaluated the estimation accuracy.

3 Results and Discussion

3.1 The Variation of Transmitted Ultrasound Waveform with Changes in SOC

Figure 2a shows the ultrasound intensity image of a tested cell at 50% SOC. The uniform green color in the battery area indicate that the intensity of the transmitted ultrasound does not vary much. According to previous literature, the electrolyte wetting status of the battery is good.

Figure 2c shows the charge curve of the battery. Its electrochemical performance is also good. At low SOC region (5–50%), the slopes of the curve (dash line) were relatively small. Considering the voltage is also influenced by the current and temperature, it is not easy to predict accurate SOC if only according to the voltage value.

At a central position of the cell, the transmitted ultrasound waveforms at different SOCs are shown in Fig. 2d. The waveform changes with SOC, but for simple characters of the waveform, such as amplitude and time of flight, their relation with SOC is not linear, not even monotone. Therefore, it is impossible to use these simple characters as the mono indicator to predict SOC. It is necessary to use feedforward neural network to extract SOC value from complicated ultrasound waveform.

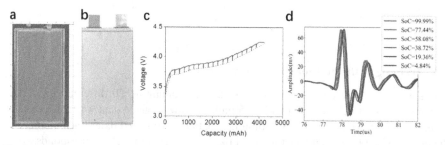

Fig. 2. (a) Ultrasonic transmission intensity image of the pouch cell. (b) Digital image of the pouch cell. (c) charge curve of the cell. (d) Transmitted ultrasound waveforms at different SOCs.

3.2 The Training Result of the Feedforward Neural Network for SOC Estimation

Figure 3a shows the structure of the FNN for SOC estimation. After several parameter adjustments, this model went through about 1000 iterations and reaches convergence.

The convergence trend of the loss function on both the training and validation dataset are shown as Fig. 3b&c. Then, we picked 114 sets of ultrasonic data that were collected at 19 different SOCs, and compared the AI estimated SOC with the real SOC. The results showed the error of the AI SOC estimation method is less than 5%, as shown in Fig. 3d.

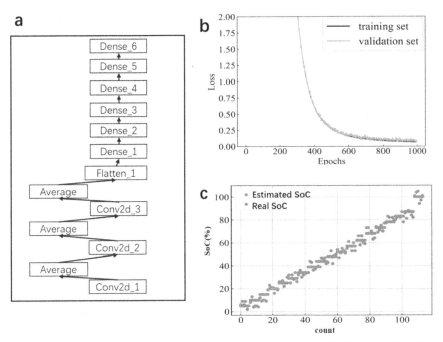

Fig. 3. (a) Structure of the FNN. (b) Loss convergence trend on the training dataset and validation dataset. (c) Comparison between the AI estimated SOC and real SOC. The real SOC values were measured via high precision charge-discharge test.

3.3   Estimated SOC Distribution of the Cell

Based on the trained FNN algorithms, the SOC at different positions of the cell can be obtained via progressive scanning, as shown in Fig. 4a. It should be noticed that the SOC maps only cover the central part of the cell, instead of the whole cell. This is because at the near edge areas, the electrode may not be perfectly flat, which may significantly affect the ultrasound transmission. The training of the FNN was conducted in the central area. Therefore, the trained FNN does not work well in the near edge areas. But still, we can get some interesting results:

1. During the charging process, the SOC increases faster in the upper parts of the cell. This is because these parts are closer to the tabs, so that the electronic resistance is smaller and the electrochemical reaction conducts more smoothly.
2. The SOC of the cell was quite uniform at the initial discharged state. Then the SOC distribution broadens after charging started. This is because the charging current is

now uniform and the nonuniformity accumulates. But as the charging goes on, in the relatively high SOC region (SOC > 40%), the SOC distribution became narrow again. This is because the slope of the voltage-capacity curve significantly raised in this region. At a certain voltage, the value of SOC became more certain. The widest distribution of SOC appears when the general SOC of the cell is 29%. At that moment, the difference between the highest and lowest local SOC was as large as 22%.

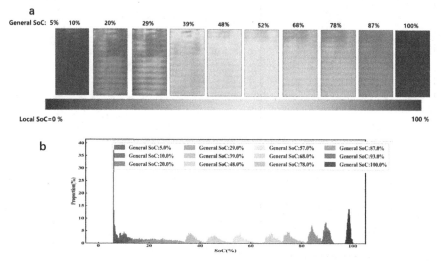

Fig. 4. (a) SOC maps of a cell at different charging status; (b) SOC distribution in the cell at different charging status.

4   Conclusions

Taking NMC811-graphite pouch cell as an example, we demonstrate an AI method to analyze SOC distribution in lithium-ion battery based on ultrasonic scanning data. A feedforward neural network is trained to recognize local SOC value according to complicated ultrasonic waveform which transmitted across the cell. Combined with progressive scanning technique with focused ultrasound beam, it is possible to obtain SOC map with 0.5 mm resolution. For the cell we tested, the SOC distribution firstly broadened and then narrowed during a charge process, which can be explained from the slope of the voltage-capacity curve. The research findings in this paper provide new insights for failure analysis of lithium-ion batteries. It gives a visualized guide to optimize the structure of advanced battery design.

References

1. Goodenough, J.B., Kim, Y.: Challenges for rechargeable li batteries. Chem. Mater. **22**(3), 587–603 (2009)
2. Tarascon, J.M., Armand, M.: Issues and challenges facing rechargeable lithium batteries. Nature **414**(6861), 359–367 (2001)
3. Cabana, J., Monconduit, L., Larcher, D., Palacín, M.R.: Beyond intercalation-based li-ion batteries: the state of the art and challenges of electrode materials reacting through conversion reactions. Adv. Mater. **22**(35), E170–E192 (2010)
4. Davies, G., et al.: State of charge and state of health estimation using electrochemical acoustic time of flight analysis. J. Electrochem. Soc. **164**(12), A2746–A2755 (2017)
5. Hu, C., Youn, B.D., Chung, J.: A multiscale framework with extended Kalman filter for lithium-ion battery SOC and capacity estimation. Appl. Energy **92**, 694–704 (2012)
6. Dai, H.F., Wei, X.Z., Sun, Z.C., Wang, J.Y., Gu, W.J.: Online cell SOC estimation of Li-ion battery packs using a dual time-scale Kalman filtering for EV applications. Appl. Energy **95**, 227–237 (2012)
7. Robert, D., et al.: Multiscale phase mapping of LiFePO4-based electrodes by transmission electron microscopy and electron forward scattering diffraction. ACS Nano **7**(12), 10887–10894 (2013)
8. Holtz, M.E., et al.: Nanoscale imaging of lithium ion distribution during in situ operation of battery electrode and electrolyte. Nano Lett. **14**(3), 1453–1459 (2014)
9. Mühlbauer, M.J., Dolotko, O., Hofmann, M., Ehrenberg, H., Senyshyn, A.: Effect of fatigue/ageing on the lithium distribution in cylinder-type Li-ion batteries. J. Power. Sources **348**, 145–149 (2017)
10. Petz, D., et al.: Lithium heterogeneities in cylinder-type Li-ion batteries – fatigue induced by cycling. J. Power. Sources **448**, 227466 (2020)
11. Huang, Z.Y., et al.: Precise state-of-charge mapping via deep learning on ultrasonic transmission signals for lithium-ion batteries. ACS Appl. Mater. Interfaces **15**, 8217–8223 (2023)
12. Deng, Z., et al.: Ultrasonic scanning to observe wetting and "unwetting" in li-ion pouch cells. Joule **4**(9), 2017–2029 (2020)
13. Liu, X., et al.: Decoupling of the anode and cathode ultrasonic responses to the state of charge of a lithium-ion battery. Phys. Chem. Chem. Phys. **25**(32), 21730–21735 (2023)

Sequence-Based Deep Reinforcement Learning for Task Offloading in Mobile Edge Computing: A Comparison Study

Xiang-Jie Xiao[1], Yong Wang[1(✉)], Kezhi Wang[2], and Pei-Qiu Huang[1(✉)]

[1] School of Automation, Central South University, Changsha 410083, China
{xjxiao,ywang,pqhuang}@csu.edu.cn
[2] Department of Computer Science, Brunel University, London, UK
kezhi.wang@brunel.ac.uk

Abstract. Mobile edge computing aims to extend cloud services to the network edge, thereby reducing the computational burden on mobile devices and enabling even simple devices to perform computationally intensive tasks within a reasonable time frame. An important issue in mobile edge computing is whether to offload interdependent and heterogeneous tasks on user equipment to edge servers. In recent years, with the emergence of neural combinatorial optimization, numerous learning-based approaches have been applied to address this problem. One significant approach is to model the task offloading problem as a sequential decision problem and implement sequence-based networks to derive offloading decisions. However, the selection of appropriate sequence-based networks remains insufficiently explored. To this end, this paper compares the performance of three mainstream sequence-based networks (i.e., recurrent neural network, long short-term memory network, and transformers), aiming to offer a concise guideline for future researchers in selecting sequence-based networks for task offloading. The experimental results suggest that the long short-term memory network outperforms other sequence-based networks.

Keywords: Mobile edge computing · Task offloading · Deep reinforcement learning · Sequence-based network

1 Introduction

Recent years have witnessed the rapid proliferation of smart mobile devices. Additionally, this trend has driven the development of various innovative applications [4,14,15], such as virtual reality, face recognition, and health monitoring, which require significant computing and storage resources. However, currently, most devices face challenges in meeting the demands of these computation-intensive applications due to unbearable latency and energy consumption. Mobile edge computing (MEC) has been proposed to address this issue. It leverages

L. Pan et al. (Eds.): BIC-TA 2023, CCIS 2062, pp. 94–106, 2024.
https://doi.org/10.1007/978-981-97-2275-4_8

cloud (remote) devices at the network edge to alleviate the computational burden on user equipment (UE).

Task offloading is a significant technique in MEC. It involves offloading computationally intensive tasks from the UE to MEC servers, thereby reducing latency and energy consumption during task execution and improving the Quality of Service (QoS). Heuristic methods have been used to obtain high-quality offloading decisions, due to the NP-hard characteristic [2,5,8,18,24]. However, these methods often consume a considerable amount of time. It is worth noting that real-time requirements are crucial in task offloading.

Due to the aforementioned issue, some researchers have shifted their focus to learning-based methods that exhibit strong real-time capabilities [12,28]. They model the task offloading problem as the sequential decision problem and then utilize sequence-based networks to make offloading decisions [19,22,23,25,27]. Representative sequence-base networks include recurrent neural networks (RNNs) [7], long short-term memory networks (LSTMs) [26], and transformers [10,13]. Despite the significant achievements of these networks in addressing sequence decision problems, a definitive answer regarding the most suitable network for task offloading has not yet been established. Therefore, it is imperative to compare these representative networks to offer guidance in selecting the appropriate networks. To this end, this paper compares RNN, LSTM, and Transformer for task offloading, intending to reduce latency and energy consumption.

The rest of this paper is organized as follows. Section 2 presents the problem description. Section 3 describes the details of the network framework and training procedure. Experimental results are presented and discussed in Sect. 4. Finally, Sect. 5 concludes the paper.

2   Problem Description

In general, various applications are composed of multiple interdependent tasks that can be represented by a directed acyclic graph (DAG). We denote the DAG as $G = (\mathcal{T}, \mathcal{E})$, where each vertex $t_i \in \mathcal{T}$ represents task t_i and directed edge $e(i, j) \in \mathcal{E}$ represents the dependency between task t_i and task t_j. The task can only be executed after its predecessors have been completed, and each task can only be executed once. Additionally, the germinal task has no predecessors, and the terminal task has no successors. The offloading decision is represented by the set $\mathcal{A} = \{a_1, \ldots, a_n\}$ where n denotes the number of tasks in an application, and $a_i = 0$ and $a_i = 1$ indicate the local and remote execution of task t_i, respectively.

2.1   System Model

In this paper, each task t_i can be executed in two ways, i.e., local execution and remote execution. Local execution refers to tasks directly processed in the local UE. Remote execution involves three stages: 1) The UE transmits task t_i to the MEC server via the uplink channel; 2) The MEC server performs the execution of task t_i; 3) The UE receives the execution result of task t_i from the MEC server via the downlink channel.

Three tuples, namely $cycle_i$, $data_i^u$, and $data_i^d$, are associated with task t_i. In this context, $cycle_i$ represents the number of clock cycles to execute task t_i; $data_i^u$ represents the amount of data that needs to be uploaded from UE to the MEC server before task execution; and $data_i^d$ represents the amount of data to be downloaded from the MEC server to the UE after task completion.

Moreover, we use T_i^l to represent the latency of task t_i for local execution, T_i^u, T_i^e, and T_i^d to respectively represent the latency of task t_i in the sending stage, the execution stage, and the receiving stage. All the defined latencies can be calculated using the following formulas:

$$T^l = \frac{cycle_i}{f^l}, \quad T^u = \frac{data_i^u}{tr}$$
$$T^r = \frac{cycle_i}{f^e}, \quad T^d = \frac{data_i^d}{tr} \tag{1}$$

where tr represents the transmission rate between the UE and the MEC server, and f^l and f^e denote the CPU clock speed of the UE and the MEC server, respectively.

We also define the finish time (FT) of local excution stage, sending stage, remote execution stage, and receiving stage as FT_i^l, FT_i^u, FT_i^e, and FT_i^d, respectively. Similarly, for a given offloading decision $\mathcal{A}$, calculating the latency requires knowledge of the available time (AT) for each channel. We use AT_i^l, AT_i^u, AT_i^e, and AT_i^d to represent AT for the local processor, the uplink channel, the MEC server, and the downlink channel when executing task t_i. Subsequently, we will elaborate on local execution and remote execution.

Local Execution: When task t_i is executed locally, the local processor on the UE is limited to handling one task at a time, necessitating a waiting period until the completion of the preceding task. Moreover, considering task dependencies, it is crucial to verify the completion of all predecessor tasks of t_i before its execution. In summary, when executing t_i locally, FT_i^l and AT_i^l can be defined as follows:

$$FT_i^l = \max_{j \in pred_i} \{AT_i^l, FT_j^l, FT_j^d\} + T_i^l$$
$$AT_i^l = \max\{AT_p^l, FT_p^l\} \tag{2}$$

where $pred_i$ is the immediate predecessors of task t_i, FT_p^l and AT_p^l represent FT and AT of the previously executed task t_p on the local processor. Note that, if task t_p is offloaded to the MEC server, FT_p^l is set to zero.

Regarding the energy consumption of local execution, it suffices to calculate the energy usage of the local processor. Therefore, the energy consumption of task t_i when executed locally is given by:

$$E_i^l = c(f^l)^\beta T_i^l \tag{3}$$

where $c > 0$ represents a constant that is dependent on the average switched capacitance and average activity factor, and $\beta \geq 2$ denotes another constant (usually close to 3) [5].

Remote Execution: When task t_i is remotely executed, it includes three stages: transmission, execution, and receiving stages. It is important to note that both the uplink and downlink channels, as well as the MEC server, can only execute one predecessor task at a time. During the transmission stage, similar to Eq. (2), it is essential to verify the availability of the uplink channel and the completion of all predecessor tasks of t_i before task execution. We calculate FT and AT of task t_i as follows:

$$FT_i^u = \max_{j \in pred_i} \{AT_i^u, FT_j^l, FT_j^d\} + T_i^u$$
$$AT_i^u = \max\{AT_p^u, FT_p^u\}. \tag{4}$$

During the execution stage, there exists a slight distinction from previous stages. As the transmission stage guarantees the completion of predecessors before task execution, it solely necessitates the availability of the MEC server and the completion of the sending stage. Therefore, FT and AT are given by

$$FT_i^e = \max\{AT_i^e, FT_i^u\} + T_i^e$$
$$AT_i^e = \max\{AT_p^e, FT_p^e\}. \tag{5}$$

Likewise, FT and AT during the receiving stage can be determined by the following equations:

$$FT_i^d = \max\{AT_i^d, FT_i^e\} + T_i^d$$
$$AT_i^d = \max\{AT_p^d, FT_p^d\}. \tag{6}$$

It is worth noting that if task t_i is executed locally, FT_i^u, FT_i^e, and FT_i^d should all be set to zero.

With respect to remote execution, it is sufficient to solely consider the energy consumption of the UE, specifically referring to the energy consumption incurred during the sending and receiving stages:

$$E_i^r = P^u T_i^u + P^d T_i^d \tag{7}$$

where P^u and P^d represent the power of sending task t_i and receiving the results, respectively.

2.2 Optimization Objective

Based on our aforementioned system model, for a given $\mathcal{A} = \{a_1, \ldots, a_i, \ldots, a_n\}$ in an application, we can separately compute the latency and energy consumption:

$$FT(\mathcal{A}) = \max_{i=1,\ldots,n} \{(1 - a_i)FT_i^l + a_i FT_i^d\} \tag{8}$$

$$E(\mathcal{A}) = \sum_{i=1}^{n}((1 - a_i)E_i^l + a_iE_i^r).\qquad(9)$$

In this paper, we aim to compare the performance of different sequence-based models in handling task offloading in MEC. Therefore, it is necessary to choose a suitable objective function that can accurately measure the performance of the offloading decision. Drawing inspiration from [23] and [3], we utilize a comparable QoS as the optimization objective. The calculation of this QoS metric involves determining the weighted sum of the normalized differences between the latency and energy consumption:

$$\max_{\mathcal{A}} J(\mathcal{A}) = 0.5\frac{FT^l - FT(\mathcal{A})}{FT^l} + 0.5\frac{E^l - E(\mathcal{A})}{E^l}.\qquad(10)$$

3   Method

3.1   Task Ranking

Since we employ a sequence-based approach to address the problem, it is necessary to sequence the tasks, considering their dependencies, before inputting them into the network. We utilize the *upward rank* algorithm [20] to calculate the ranking value of each task. This value determines the priority order for task offloading decisions. The ranking value is given as follows:

$$rank(t_i) = \begin{cases} T_i^r, & \text{if task } t_i \text{ is the terminal task} \\ T_i^r + \max_{j \in \mathcal{S}(i)} rank(j), & \text{otherwise} \end{cases}\qquad(11)$$

where $\mathcal{S}(i)$ is the immediate successors of task t_i and $T_i^r = T_i^u + T_i^e + T_i^d$.

3.2   General Network Framework

We use the encoder-decoder structure as the basis of the network, which is a widely used network structure in the field of combinatorial optimization [9,10, 13]. We specifically design a general sequence-based network framework for task offloading. To ensure accurate performance comparison of different sequence-based networks, we maintain a consistent overall structure for the framework, only modifying the structure of the Seq-based Net component while keeping the remaining parts of the network consistent in terms of structure and parameters.

Encoder: As shown in Fig. 1, we begin by inputting the tasks into the network in sequence. Each task $t_i \in \mathcal{T}$ has initial features comprising three parts: 1) the task's origin index i; 2) T_i^l, T_i^u, T_i^e, and T_i^d; 3) the indices of the predecessors and successors to indicate task dependencies (padded with −1 if there

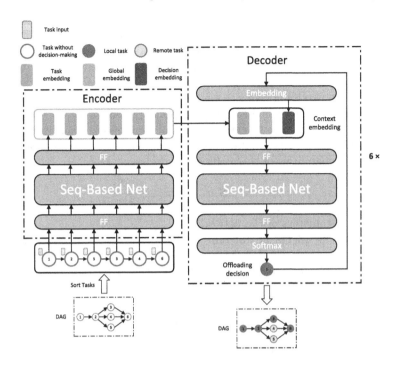

Fig. 1. An example of an encoder-decoder structure to make the offloading decision for a DAG consisting of tasks 1 to 6

are not enough indices, with a maximum of 6 for each). Subsequently, the 17-dimensional input features will undergo computation to generate a *task embedding* h_i with high-dimensional features using two trainable feed-forward layers and a sequence-based network. Additionally, the encoder will compute an aggregated *global embedding*: $\bar{h} = \frac{1}{n} \sum_{i=1}^{n} h_i$. The task embeddings and the global embedding will serve as inputs to the decoder.

Decoder: Decoding is performed sequentially based on the ranking order. The decoder generates the offloading decision for the current task t_i by considering the embeddings obtained from the encoder and the offloading decision from the preceding task, represented as $a_i \in \{0, 1\}$. Inspired by Kool et al. [9], we design a *context embedding* $\hat{h}_i$ to represent the context of the decoder when decoding task t_i in the task offloading scenario:

$$
\hat{h}_i = \begin{cases} \left[\bar{h}, h_i, \dot{h}(a_{i-1}) \right], i > 1 \\ \left[\bar{h}, h_i, \dot{h}(0) \right], \quad i = 1 \end{cases} \tag{12}
$$

where the decision embedding $\dot{h}(a_{i-1})$ is obtained by passing the offloading decision a_{i-1} of the previous task t_{i-1} through a learnable embedding layer. The symbol $[\cdot, \cdot, \cdot]$ denotes the process of fusing different embeddings through sum aggregation, achieved by adjusting the number of hidden nodes in both the feedforward and embedding layers to ensure consistent dimensions of the embeddings. Using these context embeddings in application instance $\mathcal{S}$, the decoder computes the logits in an autoregressive manner for task t_i using two feedforward layers and a sequence-based network. Finally, the Softmax function is used to output the probabilities of these offloading decisions:

$$p_i = p_\theta(a_i = j | \mathcal{S}, a_{1:a-1}) = \frac{e^{logit_j}}{\sum_k e^{logit_k}} \qquad (13)$$

where θ is the parameters of the network.

To compare the effects of different sequence-based networks, we kept the structure of all feed-forward layers and embedding layers in the network unchanged and only modified the sequence-based network part. The following provides detailed explanations of the three different sequence-based networks.

Recurrent Neural Networks: RNN is a classic and effective machine learning model for processing sequential data. It comprises one or more feedback loops of artificial neurons [6]. In this study, we employ a bidirectional RNN (Bi-RNN) [17] in the sequential-based network of the encoder to effectively capture the interdependencies among tasks, given that the input to the encoder is a sorted sequence of tasks. For the decoder, since it generates decisions in an autoregressive manner, we utilize a double RNN layer as the sequential-based network of the decoder.

Long Short-Term Memory Network: LSTM is one of the most popular network architectures derived from the RNN branch. In this approach, the structure of hidden units is substituted with memory units, which are controlled by gates, thereby replacing the conventional "sigmoid" or "tanh" structures. The gates play a crucial role in regulating the information flow to the hidden neurons and maintaining the extracted features from previous time steps [11]. Similar to RNN, in this work, we employ a Bidirectional Long Short-Term Memory (Bi-LSTM) layer for the encoder and a double LSTM layer for the decoder in their respective sequential-based networks.

Transformer: Transformer is a transformative model that relies solely on self-attention to compute input and output representations without requiring sequential alignment methods like RNN or convolution. In our paper, given the predetermined input-output sequence order, we do not employ positional encoding. Furthermore, to enhance the capture of relationships between tasks during decoding, we exclude masking from the computation of multi-head attention (MHA) in the decoder. Like the first two algorithms, we use 8-head MHA and stack two layers for the encoder and decoder.

3.3   Training Procedure

The study employs the widely used Proximal Policy Optimization (PPO) algorithm in the field of reinforcement learning for training purposes. To begin, the problem is modeled as a Markov Decision Process (MDP), with the corresponding MDP defined as follows:

Algorithm 1. Training Procedure

1: Initialize the policy and value network with trainable parameters θ and ϕ;
2: **for** *iteration* $= 1, 2, \cdots$ **do**
3:    **while** Instances is not empty **do**
4:      // Sample
5:      Randomly sample instances with batch size;
6:      Extract embeddings using Encoder;
7:      **while** s_i is not terminal **do**
8:        Sample $a_i \sim p_\theta(\cdot|s_i)$;
9:        Receive reward r_i and next state s_{i+1};
10:       Obtain state value $\hat{v}_i$ with v_ϕ;
11:       $s_i \leftarrow s_{i+1}$;
12:      **end while**
13:      Compute the generalised advantage estimates $\hat{A}_i$ for each step;
14:    **end while**
15:    // Update
16:    Compute the PPO loss $\mathcal{L}$, and optimize the parameters θ and ϕ with minibatch size;
17: **end for**

State: The state of step i, denoted as s_i, consists of the applied instance S (which includes the initial features of each task and the task ranking) and the sampled offloading decisions $[a_1, \cdots, a_{i-1}]$ that have been made so far.

Action: According to the definition in Sect. 2, every task can be executed either locally or remotely. The action space, denoted by $\{0, 1\}$, signifies the choice between local $(a_i = 0)$ and remote $(a_i = 1)$ execution of the task.

Reward: Reward is defined as the difference between the FT when s_i and s_{i-1} are locally executed and the FT when they are executed based on the offloading decision. It is defined as:

$$r_i = 0.5\frac{\overline{FT}^l - \Delta FT(s_{i-1:i})}{FT^l} + 0.5\frac{\overline{E}^l - \Delta E(s_{i-1:i})}{E^l} \qquad (14)$$

where $\overline{FT}^l = \frac{FT^l}{n}$, $\overline{E}^l = \frac{E^l}{n}$, $\Delta FT(s_{i-1:i}) = FT(\mathcal{A}_{1:i}) - FT(\mathcal{A}_{1:i-1})$, and $\Delta E(s_{i-1:i}) = E(\mathcal{A}_{1:i}) - E(\mathcal{A}_{1:i-1})$.

Policy: According to the definition of (13), we can compute the policy $p_\theta(\mathcal{A}|\mathcal{S})$ of the obtained offloading decision $\mathcal{A}$ given the applied input instance $\mathcal{S}$ under the network with parameters θ:

$$p_\theta(\mathcal{A}|\mathcal{S}) = \prod_{i=1}^{n} p_\theta(a_i|\mathcal{S}, a_1, \ldots, a_{i-1}). \tag{15}$$

The PPO algorithm is a reinforcement technique that implements the Actor-Critic structure. In our paper, an encoder-decoder structure is utilized as the policy network π_θ to make offloading decisions. Furthermore, we designed a multi-layer perceptron (MLP) with four feedforward layers to serve as the value network v_ϕ, responsible for predicting the value $v(s_i)$ of the state s_i. The training procedures of the network are presented in Algorithm 1.

4   Experimental Studies

This section will initially introduce the parameters settings within the framework of MEC. Subsequently, we will present the findings from our experimental studies.

4.1   Experimental Settings

In our experimental setup, we evaluated RNN, LSTM, and Transformer as the primary architectures for sequence-based networks. We ensured consistency by employing the same environment, network structures, and training settings across all problems with varying task-specific input sizes. For each application size, we generated 5000 instances and conducted training for 200 epochs. The models were trained on a single RTX3080 GPU, with training times ranging from 10 to 90 s per epoch, depending on the problem size and model complexity. The parameters settings of MEC system in our experiments are provided below.

Table 1. Parameters settings of MEC system

Notation	Value	Notation	Value
$cycle$	10^7 to 10^8 cycles/sec	f^l	1 GHz
f^e	10 GHz	c	3
$data^d$, $data^d$	5 to 50 KB	β	$1.25 * 10^{-26}$
tr	6 Mbps	fat	0.3 to 0.8
P^u	1.258 W	$density$	0.3 to 0.8
P^d	1.818 W	ccr	0.3 to 0.5

Environmental Parameter Settings: We address the scenario where various applications with diverse DAG topologies are executed either on UE or MEC servers [1], depending on the offloading decisions in the MEC system. The relevant environmental parameters are shown in Table 1, where fat, $density$, and ccr represent the parameters associated with DAG generation.

Network and Training Parameter Settings: The main objective of this paper is to compare the effectiveness of various widely-used sequential networks in the context of the task offloading problem. Consequently, we maintain consistent network structures, as depicted in Fig. 1, except the Seq-based Net section. Furthermore, we adopt the settings described in PPO [16] and utilize an identical set of hyperparameters to train all networks. All hyperparameters settings of network and training are shown in Table 2.

Table 2. Hyperparameters settings of Network and Training

Network		Training	
Hyperparameter	Value	Hyperparameter	Value
Hidden Dim. (RNN)	128	Learning Rate	10^4
Hidden Dim. (LSTM)	128	Optimization Method	Adam
Hidden Dim. (Transformer)	128	Value Coefficient	0.5
Embedding Dim.	256	Entropy Coefficient	0.01
Hidden 1 Dim. (Critic)	128	Clipping Constant	0.2
Hidden 2 Dim. (Critic)	256	Discount Factor	0.99
Hidden 3 Dim. (Critic)	128	Adv. Discount Factor	0.95

4.2 Results and Analysis

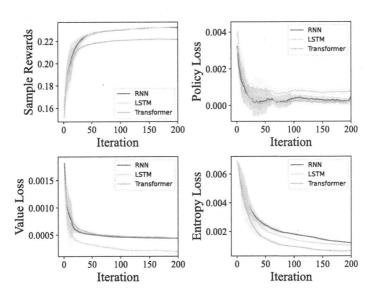

Fig. 2. Convergence of different sequence-based networks with 20 tasks

We first investigate the convergence characteristics and inference time of our proposed algorithm framework. In our experiments, we create five groups with task numbers ranging from 10 to 50. As an illustration, we present the numerical curves of rewards, policy loss, value loss, and entropy loss in Fig. 2 for three types of sequence-based networks with 20 tasks. We can observe that our task offloading network framework, regardless of the type of sequence-based network employed, achieves convergence within 150 epochs through PPO training. Moreover, Table 3 demonstrates the inference time of the three trained sequence-based networks under various task size configurations. It is evident that, regardless of the task size, the recurrent networks (RNN and LSTM) exhibit noticeably lower inference times compared to the Transformer-based network. Furthermore, as the task number increases, the inference time increases linearly.

Table 3. The inference time of a batch (100 instances) for different networks under varying task numbers.

Task Num.	RNN	LSTM	Transformer
10	0.006	0.006	0.0232
20	0.012	0.013	0.049
30	0.0172	0.0175	0.0725
40	0.022	0.023	0.0995
50	0.0279	0.0284	0.1155

We assess the performance of different models by examining the QoS of task offloading decisions. Table 4 illustrates the performance of the three sequence-based networks in task offloading scenarios with different task quantities. It is evident that, the offloading decisions derived from the two networks based on recurrent networks outperform the Transformer-based network. Additionally, LSTM shows a slightly superior performance compared to RNN.

Table 4. QoS of different networks under different task numbers

Task Num.	RNN	LSTM	Transformer
10	0.18277	**0.18428**	0.16623
20	0.19502	**0.19655**	0.17813
30	**0.18976**	0.18857	0.17537
40	0.25833	**0.25915**	0.24741
50	**0.22819**	0.22689	0.21722

5 Conclusion

In this paper, we assessed the performance of three widely used sequence-based networks for task offloading in MEC. Specifically, we proposed an encoder-decoder structure tailored for task offloading scenarios, facilitating the comparison of various sequence-based networks by integrating a plug-and-play module. Additionally, we employed the PPO algorithm to train these sequence-based networks. The experimental studies demonstrated that LSTM surpassed other networks in terms of both inference time and the quality of task offloading decisions. In future work, we will further explore the effectiveness of LSTM-based deep reinforcement learning methods for task offloading.

References

1. Arabnejad, H., Barbosa, J.G.: List scheduling algorithm for heterogeneous systems by an optimistic cost table. IEEE Trans. Parallel Distrib. Syst. **25**(3), 682–694 (2013)
2. Asim, M., Wang, Y., Wang, K., Huang, P.Q.: A review on computational intelligence techniques in cloud and edge computing. IEEE Trans. Emerg. Top. Comput. Intell. **4**(6), 742–763 (2020)
3. Cao, H., Cai, J.: Distributed multiuser computation offloading for cloudlet-based mobile cloud computing: a game-theoretic machine learning approach. IEEE Trans. Veh. Technol. **67**(1), 752–764 (2017)
4. Ding, Y., et al.: Online edge learning offloading and resource management for UAV-assisted MEC secure communications. IEEE J. Sel. Top. Signal Process. **17**(1), 54–65 (2022)
5. Dinh, T.Q., Tang, J., La, Q.D., Quek, T.Q.: Offloading in mobile edge computing: task allocation and computational frequency scaling. IEEE Trans. Commun. **65**(8), 3571–3584 (2017)
6. Haykin, S.: Neural Networks: A Comprehensive Foundation. Prentice Hall PTR, Hoboken (1998)
7. Hu, R., Xu, J., Chen, B., Gong, M., Zhang, H., Huang, H.: TAP-Net: transport-and-pack using reinforcement learning. ACM Trans. Graph. (TOG) **39**(6), 1–15 (2020)
8. Huang, P.Q., Wang, Y., Wang, K., Liu, Z.Z.: A bilevel optimization approach for joint offloading decision and resource allocation in cooperative mobile edge computing. IEEE Trans. Cybern. **50**(10), 4228–4241 (2019)
9. Kool, W., Van Hoof, H., Welling, M.: Attention, learn to solve routing problems! arXiv preprint arXiv:1803.08475 (2018)
10. Kwon, Y.D., Choo, J., Kim, B., Yoon, I., Gwon, Y., Min, S.: POMO: policy optimization with multiple optima for reinforcement learning. Adv. Neural Inf. Process. Syst. **33**, 21188–21198 (2020)
11. Le, Q.V., Jaitly, N., Hinton, G.E.: A simple way to initialize recurrent networks of rectified linear units. arXiv preprint arXiv:1504.00941 (2015)
12. Lee, D.H., Liu, J.L.: End-to-end deep learning of lane detection and path prediction for real-time autonomous driving. SIViP **17**(1), 199–205 (2023)
13. Lin, X., Yang, Z., Zhang, Q.: Pareto set learning for neural multi-objective combinatorial optimization. arXiv preprint arXiv:2203.15386 (2022)

14. Lu, W., et al.: Secure NOMA-based UAV-MEC network towards a flying eaves-dropper. IEEE Trans. Commun. **70**(5), 3364–3376 (2022)

15. Mazza, D., Pages-Bernaus, A., Tarchi, D., Juan, A.A., Corazza, G.E.: Supporting mobile cloud computing in smart cities via randomized algorithms. IEEE Syst. J. **12**(2), 1598–1609 (2016)

16. Schulman, J., Wolski, F., Dhariwal, P., Radford, A., Klimov, O.: Proximal policy optimization algorithms. arXiv preprint arXiv:1707.06347 (2017)

17. Schuster, M.: Bi-directional recurrent neural networks for speech recognition. In: Proceeding of IEEE Canadian Conference on Electrical and Computer Engineering, pp. 7–12 (1996)

18. Song, M., Lee, Y., Kim, K.: Reward-oriented task offloading under limited edge server power for multiaccess edge computing. IEEE Internet Things J. **8**(17), 13425–13438 (2021)

19. Tang, M., Wong, V.W.: Deep reinforcement learning for task offloading in mobile edge computing systems. IEEE Trans. Mob. Comput. **21**(6), 1985–1997 (2020)

20. Topcuoglu, H., Hariri, S., Wu, M.Y.: Performance-effective and low-complexity task scheduling for heterogeneous computing. IEEE Trans. Parallel Distrib. Syst. **13**(3), 260–274 (2002)

21. Vaswani, A., et al.: Attention is all you need. Adv. Neural Inf. Process. Syst. **30** (2017)

22. Wang, J., Hu, J., Min, G., Zhan, W., Ni, Q., Georgalas, N.: Computation offloading in multi-access edge computing using a deep sequential model based on reinforcement learning. IEEE Commun. Mag. **57**(5), 64–69 (2019)

23. Wang, J., Hu, J., Min, G., Zhan, W., Zomaya, A.Y., Georgalas, N.: Dependent task offloading for edge computing based on deep reinforcement learning. IEEE Trans. Comput. **71**(10), 2449–2461 (2021)

24. Wang, Y., Ru, Z.Y., Wang, K., Huang, P.Q.: Joint deployment and task scheduling optimization for large-scale mobile users in multi-UAV-enabled mobile edge computing. IEEE Trans. Cybern. **50**(9), 3984–3997 (2019)

25. Zhan, Y., Guo, S., Li, P., Zhang, J.: A deep reinforcement learning based offloading game in edge computing. IEEE Trans. Comput. **69**(6), 883–893 (2020)

26. Zhang, H., Li, W., Gao, S., Wang, X., Ye, B.: ReLeS: a neural adaptive multipath scheduler based on deep reinforcement learning. In: IEEE INFOCOM 2019-IEEE Conference on Computer Communications, pp. 1648–1656. IEEE (2019)

27. Zhu, S., Gui, L., Zhao, D., Cheng, N., Zhang, Q., Lang, X.: Learning-based computation offloading approaches in uavs-assisted edge computing. IEEE Trans. Veh. Technol. **70**(1), 928–944 (2021)

28. Zhu, X., Su, W., Lu, L., Li, B., Wang, X., Dai, J.: Deformable DETR: deformable transformers for end-to-end object detection. arXiv preprint arXiv:2010.04159 (2020)

A Sample Reuse Strategy for Dynamic Influence Maximization Problem

Shaofeng Zhang[1], Shengcai Liu[2], and Ke Tang[1(✉)]

[1] Southern University of Science and Technology, Shenzhen 518055, China
12232415@mail.sustech.edu.cn, tangk3@sustech.edu.cn
[2] Centre for Frontier AI Research (CFAR), Agency for Science, Technology
and Research (A*STAR), Singapore 138632, Singapore
liu_shengcai@cfar.a-star.edu.sg

Abstract. Dynamic influence maximization problem (DIMP) aims to maintain a group of influential users within an evolving social network to maximize the influence scope at any given moment. A primary category of DIMP algorithms focuses on updating reverse reachable (RR) sets designed for static social network scenarios to accelerate the estimation of influence spread. The generation time of RR sets plays a crucial role in algorithm efficiency. However, their update approaches require sequential updates for each edge change, leading to considerable computational costs. In this paper, we propose a strategy for batch updating the changes in network edge weights to maintain RR sets efficiently. We retain those with a high probability by calculating the probability that previous RR sets can be regenerated at the current moment. This method can effectively avoid the computational cost of updating and sampling these RR sets. Besides, we propose a resampling strategy that generates high-probability RR sets to make the final distribution of RR sets approximate to the sampling probability distribution under the current social network. The experimental results indicate that our strategy is both scalable and efficient. On the one hand, compared to the previous update strategies, the running time of our approach is insensitive to the number of changes in network weight; on the other hand, for various RR set-based algorithms, our strategy can reduce the running time while maintaining the solution quality that is essentially consistent with the static algorithms.

Keywords: Influence Maximization Problem · Dynamic Social Network · Sample Reuse

1 Introduction

In recent years, with the rapid development of online social platforms, there is a vast potential for application analysis of social networks [7]. The influence maximization problem (IMP) is a representative problem within social network analysis. It aims at discovering a part of influential users, through which the spread of products or opinions could be maximized in the whole social network [8]. The IMP has a wide range of applications, such as viral marketing [14],

L. Pan et al. (Eds.): BIC-TA 2023, CCIS 2062, pp. 107–120, 2024.
https://doi.org/10.1007/978-981-97-2275-4_9

election campaigns [12], and fake news blocking [4]. However, user relationships constantly evolve in real-world scenarios, indicating that social networks are not static. The dynamic nature of social networks can affect the spread of influence [15,20]. Every time the social network changes, the algorithms tailored for IMP must be restarted from scratch, resulting in substantial computational overhead. Therefore, the dynamic influence maximization problem (DIMP) has been formulated to solve the IMP efficiently within dynamic social networks.

The primary existing approaches [15,17,25] involve modifying RR set-based algorithms, which are efficient algorithms designed for IMP, adapting them to dynamic social networks. In RR set-based algorithms, the efficiency heavily depends on the generation of RR sets. Therefore, they accelerate the evaluation of influence spread by updating previously generated RR sets, which can reduce the time costs associated with resampling. However, their methods require sequential updates for each edge change in the network. For example, when the number of updated edge weights is a, their methods would require a repetitive runs to process each change. In real-world scenarios, the dynamics of social networks are often captured through snapshots over a period, which can involve a substantial number of updates between two snapshots. In such instances, the overhead of updating RR sets with their methods would exceed the cost of regenerating them from scratch, making the updated algorithms less efficient than the static ones.

In this paper, we propose a new update strategy for RR set-based algorithms tailored to network edge weight change scenarios, which can process multiple dynamic changes simultaneously. Our strategy can efficiently reduce the time costs of redundant sampling by preserving partial historically generated RR sets, which are highly likely to be regenerated at the next moment. This method does not require updating the RR sets for each updated weight, achieving the purpose of batch-processing dynamic changes. Besides, we propose a resampling strategy that generates new high-probability RR sets with previously rejected low-probability RR sets. This approach makes the final distribution of RR sets approximate the probability distribution derived from sampling in the current social network. The experiments indicate that our strategy is scalable. When the number of updated edge weights increases by 9 times, the running time of our algorithms only grows by 0.3 times, whereas the running time of the DynIM algorithm [15] increases by more than 60 times. Moreover, our strategy is efficient; it can help RR set-based algorithms, IMM [22] and SUBSIM [6], achieve up to a 19% and 12% reduction in running time, respectively.

2   Preliminaries and Related Work

2.1   Problem Definition

Dynamic influence maximization problem aims to maintain a group of influential users in dynamic environments, specifically within dynamic social networks, ensuring that their influence spread remains maximized at every moment. In this section, we will provide a formal definition of the DIMP.

Social Network: The social network $G = (V, E, P)$, where $V = \{v_1, \ldots, v_n\}$ represents the set of nodes, $E \subseteq V \times V$ represents the edges between nodes, and $P = \{p_{(u,v)} | (u, v) \in E\}$ represents the weights of the edges, reflecting the probability of influence propagation between nodes. Here, $p_{(u,v)} \in (0, 1]$ indicates the influence strength of node u on node v.

Seed Set: The seed set refers to the set of seed nodes $S \subseteq V$ that are chosen to be activated in the initial state.

Budget: The budget k constrains the size of the seed set, meaning that at most k seed nodes can be selected.

Diffusion Model: The diffusion model M captures the random process of information dissemination by the seed set in the social graph [11]. The diffusion model adopted in this paper is the independent cascade (IC) model [8]. In the IC model, each node has two states: activated and inactivated. The influence diffusion process unfolds in the following discrete steps:

- In the initial step $a = 0$, the nodes in the seed set S are activated, while other nodes remain in the inactive state.
- At step $a \in [1, n]$, when node u is activated for the first time, it is considered contagious and has a single chance to activate each of its inactive out-neighboring nodes v. The probability that node v gets activated is $p_{(u,v)}$. Then, node u remains activated but becomes non-contagious.
- The influence diffusion process terminates when no new nodes are activated; that is, no more contagious nodes exist.

Influence Function: Given a social network $G = (V, E, P)$, a seed set $S \subseteq V$, and a diffusion model M, the influence function is defined as $\sigma_{G,M}(S)$. It represents the expected number of nodes influenced (activated) by the seed set S when the influence diffusion process terminates [11].

Dynamic Social Network: A dynamic social network is defined as a sequence of network snapshots evolving $\mathcal{G} = (G^0, \ldots, G^T)$, where $G^t = (V^t, E^t, P^t)$ is a snapshot of the network at time t [20]. This study focuses on the updates of the network edge weight. Thus, the dynamic changes of the social network G^t are defined as $\Delta G^t = (V, E, \Delta P^t)$.

Dynamic Influence Maximization Problem: Given a budget k, the influence diffusion model M, the social network snapshot at time t denoted as G^t, and the dynamic changes ΔG^t, the dynamic influence maximization problem aims to find the seed set S^{t+1} in the social network snapshot $G^{t+1} = G^t \cup \Delta G^t$ at time $t + 1$, such that $S^{t+1} \subseteq V$ and $|S^{t+1}| \leq k$ to achieve influence maximization.

$$S^{t+1} = \arg\max_{S \in V, |S| < k} \sigma_{G^{t+1}, M}(S) \tag{1}$$

2.2 Related Work

The influence maximization problem was first proposed by Kempe [8], drawing inspiration from viral marketing strategies. They seek to spread information throughout the social network using a word-of-mouth approach. There are two main difficulties in the influence maximization problem: one is that the influence of nodes can only be evaluated by Monte Carlo simulation rather than computed analytically, and the other is that the search space grows rapidly with the network size.

Influence evaluation has been proved as #P-hard problem [2]. Over recent years, many algorithms have been proposed to estimate the influence of nodes. The simulation-based algorithms aim to reduce the overall evaluation cost by reducing unnecessary Monte Carlo sampling [10] while still upholding theoretical guarantees [5,27]. Nonetheless, these methods still require substantial computational time. The metrics-based algorithms utilize specially designed heuristic measures to approximate the influence of nodes [2,3], but such methods tend to have lower accuracy. The sketch-based algorithms, particularly the RR set-based method [1,6,22], are now widely adopted because of their efficiency and theoretical guarantees. It employs a series of randomly generated RR sets $\mathbb{R}$ to approximate the influence of multiple nodes concurrently. Let R represent the RR set. The specific generation process in the IC model is shown in Algorithm 1. At this point, the influence of the seed set S is $\frac{n}{|\mathbb{R}|} \cdot |\Lambda_{\mathbb{R}}(S)|$, where n is the number of nodes and $|\Lambda_{\mathbb{R}}(S)|$ is the number of covered RR sets by S [22]. Then, we could use the greedy-based method to select influential seed seeds, which ensures an approximation ratio of $(1 - \frac{1}{e} - \epsilon)$ [8]. During the ith iteration, with the current seed set denoted as S_{i-1}, the greedy algorithm seeks a node v to maximize the marginal coverage $|\Lambda_{\mathbb{R}}(v|S_{i-1})|$, as illustrated in Eq. 2.

$$\Lambda_{\mathbb{R}}(v|S_{i-1}) = \Lambda_{\mathbb{R}}(S_i \cup \{v\}) - \Lambda_{\mathbb{R}}(S_{i-1}) \tag{2}$$

The $\Lambda_{\mathbb{R}}(v|S_{i-1})$ denotes the random RR sets within the set $\mathbb{R}$ that are covered by node v but remain uncovered by nodes in S_{i-1}.

Algorithm 1. RR-Set-Generation [22]

Input: social network G.
Output: RR set R.
1: RR set $R \leftarrow \emptyset$;
2: Randomly select a node $r \in V$ uniformly as the root node;
3: Add node r to queue Q and RR set R;
4: Set node r as activated and other nodes as inactivated;
5: **while** Q has node **do**
6:     $v \leftarrow Q.pop()$;
7:     **for** each node u where $(u, v) \in E$ **do**
8:         **if** u is inactivated **and** $rand() \leq p(u, v)$ **then**
9:             Add node u to queue Q and RR set R;
10:            Mark node u as activated;
11:        **end if**
12:    **end for**
13: **end while**
14: **return** R;

However, in real-world scenarios, social networks are constantly evolving. There are some attempts to address the IMP in dynamic social networks. Song pioneered a clear definition of the dynamic influence maximization problem (DIMP) [20], conceptualizing the dynamic social network as a sequence of static social networks. Then, DIMP can be seen as a series of static influence maximization problems. The historical evaluation information can help the algorithm efficiently update its estimation of node influence. For the metrics-based algorithms, updates can be applied to previously calculated metrics according to the dynamic changes in the network graph [16,23]. To reduce the update costs, local update strategies aim to limit the impact of network changes heuristically, thus preventing the need to update the influence of all nodes in the social network [23,24]. For the sketch-based algorithms, especially the RR set-based algorithms, they tried to update the previously simulated RR sets, avoiding the computational cost associated with resampling [15,17,25]. However, their algorithms need to resample for each changed edge and update the affected RR sets, which can only process dynamic updates one by one to make the sampling probability of the RR sets match the new social network. Therefore, their update strategy is effective primarily for scenarios with few dynamic changes. As the number of dynamic updates increases, the cumulative time required for updates can surpass the time needed to regenerate RR sets from scratch, leading to poorer scalability of these algorithms. Besides, historical data can also assist in constructing seed sets more quickly. One approach directly updates the existing seed set. Specifically, it attempts to select new nodes from the social network to replace nodes in the existing seed set, aiming to maximize the influence gain [20,24]. This method can avoid the time cost of selecting seed sets from scratch. Additionally, when changes in the social network exhibit a specific pattern, we can leverage historical data to predict the network's change at the next moment. We can pre-select seed sets based on the predicted network dynamics, thereby accelerating the solution selection [19,26].

3   Method

3.1   Sample Reuse with Importance Mixing

In this section, we introduce a sample reuse algorithm that reuses samples based on their probability changes, aiming to avoid the computational cost caused by redundant sampling. RR set-based algorithms require generating a series of random RR sets, transforming the influence maximization problem into a maximum set coverage problem. In dynamic social networks, when the changes are relatively small, the random RR set generated at the previous time still has a high probability of being generated at the current moment. This paper takes the reverse influence sampling under the IC model as an example (as shown in Algorithm 1), treats the RR set obtained from sampling as a sample, and combines it with the Importance Mixing algorithm [18,21] to reuse samples based on their probability changes.

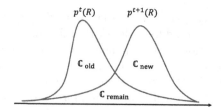

Fig. 1. Reverse reachable (RR) set R probability distribution in time t and $t+1$

At the new moment $t+1$, changes occur in the edges of the social network, that is, $G^{t+1} = G^t \cup \Delta G^t$. We could use the Importance Mixing algorithm to update the series of RR sets $\mathbb{R}_t$ generated at the previous moment, ensuring that the updated RR sets satisfy the probability distribution of sampling under the new social network G^{t+1}. Figure 1 depicts the evolution of the RR set's probability distribution. To make the previous RR sets (comprising both the $\mathbb{C}_{old}$ and $\mathbb{C}_{remain}$ parts) consistent with $p^{t+1}(R)$, the Importance Mixing algorithm first rejects the $\mathbb{C}_{old}$ part and subsequently introduces the $\mathbb{C}_{new}$ part. In detail, the Importance Mixing approach reuses previous sampling RR sets in the following step:

- **Remain Step:** For each previous RR set R_i, which was generated in G^t, the probability of reusing it is $\min\{1, \frac{p^{t+1}(R_i)}{p^t(R_i)}\}$.
- **Sample Step:** Generate a new RR set R_i' in the new social network G^{t+1}. The probability of accepting it is $\max\{0, 1 - \frac{p^t(R_i')}{p^{t+1}(R_i')}\}$.

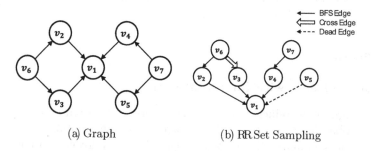

(a) Graph   (b) RR Set Sampling

Fig. 2. An example of RR set sampling.

The RR set-based algorithm for IMP only stored the activated nodes in the RR sets. The probability of the RR set is associated with the incoming edges of all activated nodes $\{(u,v)|v \in R\}$. According to the definition in [25] and the procedure of Algorithm 1, we categorize these edges as BFS edge, Cross edge, and Dead edge. For a specific edge (u,v), the BFS edge signifies that node u was activated by node v; the Cross edge indicates that when node v was activated, node u had already been activated; the Dead edge means node v tried to activate node u, but was unsuccessful, as shown in Fig. 2. The probability of generating an RR set is given by Eq. (3).

$$p(R) = \prod_{(u,v)\in E_{BFS}} p_{(u,v)} \cdot \prod_{(u,v)\in E_{Dead}} \left(1 - p_{(u,v)}\right) \tag{3}$$

However, due to the large number of Cross edges and Dead edges, storing and individually calculating probability for each RR set would lead to huge time and space costs. We develop an approximate method to efficiently compute the update probabilities for RR sets, requiring storing only the BFS edges for each RR set. When the social network changes, for each node v, we calculate the probability of all its incoming edges being Dead edges, as shown in Eq. (4), where λ is a small value introduced to avoid division by zero. The dead probability $p_{Dead}(v)$ needs to be calculated only once for each node $v \in V$ and can be applied to all RR set updates.

$$p_{Dead}(v) = \prod_{(u,v)\in E} \frac{1 - p_{(u,v)}^{t+1} + \lambda}{1 - p_{(u,v)}^{t} + \lambda} \tag{4}$$

Then, to compute the probability change for an RR set, we traverse each activated node in the RR set and update the probabilities associated with the BFS edges. The resulting updated probability is shown in Eq. 5.

$$\frac{p^{t+1}(R)}{p^{t}(R)} \approx \prod_{(u,v)\in E_{BFS}} \left(\frac{p_{(u,v)}^{t+1} + \lambda}{p_{(u,v)}^{t} + \lambda} \cdot \frac{1 - p_{(u,v)}^{t} + \lambda}{1 - p_{(u,v)}^{t+1} + \lambda} \right) \cdot \prod_{u\in R} p_{Dead}(u) \tag{5}$$

Through the reverse sampling process, as shown in Algorithm 1, apart from the seed node, every activated node u has a unique parent node v. Therefore, each activated node u corresponds to a unique BFS edge (u, v). And the space occupied by storing BFS edges is consistent with that of storing activated nodes.

3.2   New Resampling Strategy

In the **Sample Step** of the Importance Mixing algorithm, RR sets are generated but may be rejected with a certain probability, wasting time on sample generation. The purpose of the **Sample Step** is to generate RR sets with increased probability, as shown in the Fig. 1 $\mathbb{C}_{new}$ part. However, compared to resampling from scratch, updating previously rejected samples is more likely to generate RR sets with increased probabilities. Specifically, when the sample R is rejected and in the C_{old} part, it indicates that there is at least one edge $e \in \{(u, v) | v \in R\}$ with a probability decrease.

- If (u, v) is a BFS edge, $\frac{p_{(u,v)}^{t+1}}{p_{(u,v)}^{t}} < 1$. Thus, when we remove this edge, it becomes a Dead edge and makes the RR set probability changes to $\frac{1-p_{(u,v)}^{t+1}}{1-p_{(u,v)}^{t}} > 1$.

- If (u, v) is a Dead edge, $\frac{1-p_{(u,v)}^{t+1}}{1-p_{(u,v)}^{t}} < 1$, adding the edge results in a RR set probability change of $\frac{p_{(u,v)}^{t+1}}{p_{(u,v)}^{t}} > 1$.

Algorithm 2. Resample-RR-Set

Input: old RR set R, social network G^{t+1}.
Output: new RR set R'.
1: new RR set $R' \leftarrow \emptyset$;
2: Select the root node $v \in V$ in old RR set R;
3: Add node r to queue Q and RR set R';
4: Set node r as activated and other nodes as inactivated;
5: **while** Q has node **do**
6:    $v \leftarrow Q.pop()$;
7:    **for** each node u where $(u, v) \in E$ **do**
8:       **if** u is activated **then**
9:          **Continues**;
10:      **end if**
11:      **if** $p_{(u,v)}^{t} \neq p_{(u,v)}^{t+1}$ **then**
12:         **if** $rand() \leq p_{(u,v)}^{t+1}$ **then**
13:            Add node u to queue Q and RR set R';
14:            Mark node u as activated;
15:         **end if**
16:      **else**
17:         **if** $(u, v \in R)$ **or** $(v \notin R$ **and** $rand() \leq p_{(u,v)}^{t+1})$ **then**
18:            Add node u to queue Q and RR set R';
19:            Mark node u as activated;
20:         **end if**
21:      **end if**
22:   **end for**
23: **end while**
24: **return** R';

Therefore, when the sample R is rejected in **Remain Step** of Importance Mixing algorithm, in **Sample Step**, we can resample the edges with changed probabilities p^{t+1}, transitioning sample R from the $\mathbb{C}_{old}$ part to the $\mathbb{C}_{new}$ part. So, we propose a resampling strategy to generate new high-probability RR sets with previously rejected low-probability RR sets, as shown in Algorithm 2.

Besides, we need to adjust the importance sampling part, as shown in Algorithm 3. If the sample R is accepted in **Remain Step**, it suggests that the RR set R resides in the $\mathbb{C}_{remain}$, allowing us to move directly to the next iteration (Algorithm3 Lines 3–5). However, if the sample R is rejected in **Remain Step**, indicating that its position is in the $\mathbb{C}_{old}$ part, we then proceed to **Sample Step**. In this step, we resample with previous rejected low-probability RR set R, as shown in Algorithm 2, and accept it with the change probability to check whether the generated new RR set is in the $\mathbb{C}_{new}$ part (Algorithm 3 Lines 6–11). Algorithm 3 Lines 16–21 ensure that the size of $\mathbb{R}_{t+1}$ is N_R.

Algorithm 3. RR-Sets-Generation-Importance-Mixing

Input: social network G^{t+1}, set of old random RR sets $\mathbb{R}_t$, number of new RR sets N_R .
Output: set of new random RR sets $\mathbb{R}_{t+1}$.
1: $\mathbb{R}_{t+1} \leftarrow \emptyset$;
2: **for** $i = 1$ **to** $|\mathbb{R}_t|$ **do**
3:      $R_i \leftarrow \mathbb{R}_t[i]$;
4:      **if** $\min(1, \frac{p^{t+1}(R_i)}{p^t(R_i)}) \geq rand()$ **then**
5:          $\mathbb{R}_{t+1} \leftarrow \mathbb{R}_{t+1}.\text{append}(R_i)$;
6:      **else**
7:          $R_i' \leftarrow$ **Resample-RR-Set**(R_i, G^{t+1})
8:          **if** $\max(0, 1 - \frac{p^t(R_i')}{p^{t+1}(R_i')}) \geq rand()$ **then**
9:              $\mathbb{R}_{t+1} \leftarrow \mathbb{R}_{t+1}.\text{append}(R_i')$;
10:          **end if**
11:      **end if**
12:      **if** $|\mathbb{R}_{t+1}| \geq N_R$ **then**
13:          **break**;
14:      **end if**
15: **end for**
16: **while** $|\mathbb{R}_{t+1}| > N_R$ **do**
17:      remove a randomly chosen RR set R in $\mathbb{R}_{new}$;
18: **end while**
19: **while** $|\mathbb{R}_{t+1}| < N_R$ **do**
20:      $R \leftarrow$ Generate a new RR set in G^{t+1};
21:      $\mathbb{R}_{t+1} \leftarrow \mathbb{R}_{t+1}.\text{append}(R)$;
22: **end while**
23: **return** $\mathbb{R}_{t+1}$;

---

4 Experiments

4.1 Experimental Setting

In this section, we will describe the details of the experimental settings. The experiments were executed on a Linux operating system, driven by an Intel(R) Xeon(R) Silver 4310 CPU @ 2.10 GHz and equipped with 250 GB of memory. Every algorithm we tested was crafted in C++ and compiled using g++, utilizing compiler version 11.3.0.

Social Network Dataset: This paper utilizes two real-world network graph datasets for experimental evaluation, as summarized in Table 1. The HepTh and HepPh datasets [9] are both citation graphs derived from the electronic literature platform arXiv. Each node represents a paper, and edges signify citation relationships.

Table 1. Datasets overview.

Dataset	Node Number	Edge Number	Type
HepTh	27,770	352,807	Directed
HepPh	34,546	421,578	Directed

Probability Setting: This paper uses the independent cascade diffusion model for influence spread. Since the weights in the social network are unknown, we adopt the widely-used weighted cascade (WC) model [8]. Specifically, for each edge u, v in the network graph, the weight of the edge is $p_{(u,v)} = \frac{1}{d_v^{in}}$, where d_v^{in} represents the in-degree of the node v.

Algorithms: In this paper, we selected three algorithms, IMM, SUBSIM, and DynIM, as well as two algorithms proposed in this study, D-IMM and D-SUBSIM, for comparative experiments.

- **IMM [22]:** The classic RR set-based algorithm for solving the IMP.
- **SUBSIM [6]:** The state-of-the-art algorithm that uses RR sets to solve the IMP.
- **DynIM [15]:** The classic algorithm is designed to solve the DIMP.
- **D-IMM and D-SUBSIM:** Our proposed algorithms integrate the RR set-based algorithms, IMM and SUBSIM, respectively, with the sample reuse method introduced in this paper.

Parameter Setting: For the IMM, SUBSIM, D-IMM, and D-SUBSIM algorithms, the parameters are configured with values $l = 1$ and $\epsilon = 0.1$; for the DynIM algorithm, the parameter is set to $\epsilon = 0.5$. In the experimental setup, the budget for the seed set k is fixed at 50. We use the Monte Carlo simulation method to estimate the influence spread of the seed sets obtained by different algorithms, with the number of simulations $r = 10,000$. Due to the randomness inherent in the algorithms, each experiment is repeated 10 times, and the running time and influence are averaged [13]. Note that we excluded the time required to load and update the social network when evaluating running time since every algorithm needs to compute in the updated graph.

4.2   Edge Weight Update Analysis

This experiment examines the algorithms' efficiency and scalability concerning the number of dynamic updates. We focus on scenarios with edge weight changes to isolate the effect of network size variations on the algorithms' running time. The experiment involves only two snapshots of the social network; that is, the social network undergoes one change involving multiple dynamic updates. Specifically, following the settings in [15], each dynamic update represents randomly selecting an edge (u, v) and change its probability to $p^t \times 2$ or $p^t/2$. For each social network, we varied the number of dynamic updates to evaluate the algorithm's performance under various dynamic scenarios in terms of running time (as shown in Fig. 3) and the solution quality (as shown in Fig. 4).

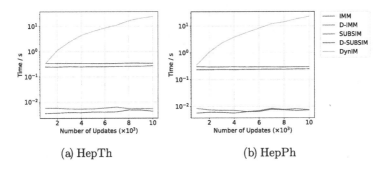

(a) HepTh   (b) HepPh

Fig. 3. The running time of the algorithms under different numbers of edge weight updates.

For the datasets HepTh and HepPh, we varied the number of updated social network edge weights within the range of $[10^3, 10^4]$ and conducted comparative experiments across all five algorithms. As shown in Fig. 3a and Fig. 3b, the average running time of the DynIM algorithm for processing 10^4 updated edges is 71.5 times and 62.3 times, respectively, of the time taken for 10^3 updated edges. In contrast, our two algorithms' running time (D-IMM and D-SUBSIM) for 10^4 updated edges are both merely 1.3 times that of 10^3 updated edges for two datasets. Therefore, compared to the DynIM algorithm, our method's running time, D-IMM and D-SUBSIM, is less sensitive to the number of updates. Moreover, for the datasets HepTh and HepPh, the time taken by the DynIM algorithm to update its RR sets is significantly higher than that of the other four algorithms. While the D-IMM algorithm achieves a 26.2% and 19.3% reduction in average running time compared to the IMM algorithm, that is $1 - \frac{t_{D-IMM}}{t_{IMM}}$, and similarly, the D-SUBSIM algorithm exhibits a 26.6% to 12.5% decrease in average running time when contrasted with the SUBSIM algorithm. Therefore, our method can effectively accelerate the IMM and SUBSIM algorithms for network edge weight change scenarios.

Figure 4 indicates that the solution quality of our algorithm is comparable to that of the static algorithms. For the two datasets, the average influence spread of the seed sets solved by the D-IMM algorithm shows a difference of less than $\frac{|\sigma(S_{IMM}) - \sigma(S_{D-IMM})|}{\sigma(S_{IMM})} \approx 0.5\%$ compared to the IMM algorithm. Besides, the average influence spread of the D-SUMSIM algorithm differs by no more than

$\frac{|\sigma(S_{SUMSIM}) - \sigma(S_{D-SUMSIM})|}{\sigma(S_{SUMSIM})} \approx 0.3\%$ from that of the SUMSIM algorithm, which is still within an acceptable range. Therefore, our dynamic algorithms, D-IMM and D-SUBSIM, do not significantly decline in solution quality compared to the static algorithms.

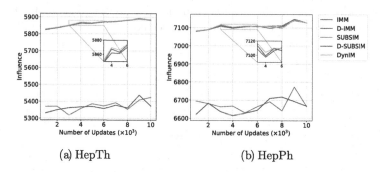

(a) HepTh   (b) HepPh

Fig. 4. The influence spread of the algorithms under different numbers of edge weight updates.

5   Conclusion

This paper proposes a new update strategy for RR set-based algorithms that can efficiently handle dynamic network batch updates. Specifically, when changes occur in the social network, it reuses historical RR sets, which still have a high probability of being regenerated at the current time, to avoid the computational cost caused by redundant sampling. Additionally, we design an efficient resampling strategy to generate new high-probability RR sets with previously rejected low-probability RR sets, which makes the final distribution of RR sets approximate to the probability distribution derived from sampling in the new social network. The experimental results demonstrate that our algorithm exhibits better scalability than the previous update strategy. Besides, our strategy can effectively reduce the running time of RR set-based algorithms. While our algorithm is efficient, the resampling strategy does not provide a theoretical guarantee that the RR sets' final probability distribution is consistent with the one derived from sampling in the new social network. In the future, we will attempt to improve the algorithm and propose a sampling strategy with stronger theoretical assurances.

References

1. Borgs, C., Brautbar, M., Chayes, J., Lucier, B.: Maximizing social influence in nearly optimal time. In: Proceedings of the 25th Annual ACM-SIAM Symposium on Discrete Algorithms, pp. 946–957. SODA'14 (2014)
2. Chen, W., Wang, C., Wang, Y.: Scalable influence maximization for prevalent viral marketing in large-scale social networks. In: Proceedings of the 16th ACM SIGKDD International Conference on Knowledge Discovery and Data Mining, pp. 1029–1038. KDD'10 (2010)

3. Chen, W., Wang, Y., Yang, S.: Efficient influence maximization in social networks. In: Proceedings of the 15th ACM SIGKDD International Conference on Knowledge Discovery and Data Mining, pp. 199–208. KDD'09 (2009)

4. Chen, W., Liu, S., Ong, Y.S., Tang, K.: Neural influence estimator: towards real-time solutions to influence blocking maximization. arXiv e-prints arXiv:2308.14012 (2023). https://doi.org/10.48550/arXiv.2308.14012

5. Goyal, A., Lu, W., Lakshmanan, L.V.: CELF++: optimizing the greedy algorithm for influence maximization in social networks. In: Proceedings of the 20th International Conference Companion on World Wide Web, pp. 47–48. WWW'11 (2011)

6. Guo, Q., Wang, S., Wei, Z., Chen, M.: Influence maximization revisited: efficient reverse reachable set generation with bound tightened. In: Proceedings of the 2020 ACM SIGMOD International Conference on Management of Data, pp. 2167–2181. SIGMOD'20 (2020)

7. Kapoor, K.K., Tamilmani, K., Rana, N.P., Patil, P., Dwivedi, Y.K., Nerur, S.: Advances in social media research: past, present and future. Inf. Syst. Front. **20**(3), 531–558 (2018)

8. Kempe, D., Kleinberg, J., Tardos, E.: Maximizing the spread of influence through a social network. In: Proceedings of the 9th ACM SIGKDD International Conference on Knowledge Discovery and Data Mining, pp. 137–146. KDD'03 (2003)

9. Leskovec, J., Krevl, A.: SNAP datasets: Stanford large network dataset collection (2014). http://snap.stanford.edu/data

10. Li, X., Liu, S., Wang, J., Chen, X., Ong, Y., Tang, K.: Data-driven chance-constrained multiple-choice knapsack problem: model, algorithms, and applications (2023). CoRR **abs/2306.14690**

11. Li, Y., Fan, J., Wang, Y., Tan, K.L.: Influence maximization on social graphs: a survey. IEEE Trans. Knowl. Data Eng. **30**(10), 1852–1872 (2018). https://doi.org/10.1109/TKDE.2018.2807843

12. Litou, I., Kalogeraki, V.: Influence maximization in evolving multi-campaign environments. In: Proceedings of the 2018 IEEE International Conference on Big Data (Big Data), pp. 448–457 (2018)

13. Liu, S., Tang, K., Lei, Y., Yao, X.: On performance estimation in automatic algorithm configuration. In: Proceedings of the 34th AAAI Conference on Artificial Intelligence, AAAI'2020, pp. 2384–2391 (2020)

14. Lu, W., Bonchi, F., Goyal, A., Lakshmanan, L.V.: The bang for the buck: fair competitive viral marketing from the host perspective. In: Proceedings of the 19th ACM SIGKDD International Conference on Knowledge Discovery and Data Mining, pp. 928–936. KDD'13 (2013)

15. Ohsaka, N., Akiba, T., Yoshida, Y., Kawarabayashi, K.I.: Dynamic influence analysis in evolving networks. Proc. VLDB Endow. **9**(12), 1077–1088 (2016)

16. Ohsaka, N., Maehara, T., Kawarabayashi, K.i.: Efficient pagerank tracking in evolving networks. In: Proceedings of the 21th ACM SIGKDD International Conference on Knowledge Discovery and Data Mining, pp. 875–884. KDD'15 (2015)

17. Peng, B.: Dynamic influence maximization. In: Ranzato, M., Beygelzimer, A., Dauphin, Y., Liang, P., Vaughan, J.W. (eds.) Advances in Neural Information Processing Systems, vol. 34, pp. 10718–10731 (2021)

18. Pourchot, A., Perrin, N., Sigaud, O.: Importance mixing: improving sample reuse in evolutionary policy search methods. arXiv e-prints arXiv:1808.05832 (2018). https://doi.org/10.48550/arXiv.1808.05832

19. Singh, A.K., Kailasam, L.: Link prediction-based influence maximization in online social networks. Neurocomputing **453**, 151–163 (2021)

20. Song, G., Li, Y., Chen, X., He, X., Tang, J.: Influential node tracking on dynamic social network: an interchange greedy approach. IEEE Trans. Knowl. Data Eng. **29**(2), 359–372 (2017)
21. Sun, Y., Wierstra, D., Schaul, T., Schmidhuber, J.: Efficient natural evolution strategies. In: Proceedings of the 11th Annual Conference on Genetic and Evolutionary Computation, pp. 539–546. GECCO'09 (2009)
22. Tang, Y., Shi, Y., Xiao, X.: Influence maximization in near-linear time: a martingale approach. In: Proceedings of the 2015 ACM SIGMOD International Conference on Management of Data, pp. 1539–1554. SIGMOD'15 (2015)
23. Wang, S., Cuomo, S., Mei, G., Cheng, W., Xu, N.: Efficient method for identifying influential vertices in dynamic networks using the strategy of local detection and updating. Futur. Gener. Comput. Syst. **91**, 10–24 (2019)
24. Yalavarthi, V.K., Khan, A.: Steering top-k influencers in dynamic graphs via local updates. In: Proceedings of the 2018 IEEE International Conference on Big Data (Big Data), pp. 576–583 (2018)
25. Yang, Y., Wang, Z., Pei, J., Chen, E.: Tracking influential individuals in dynamic networks. IEEE Trans. Knowl. Data Eng. **29**(11), 2615–2628 (2017)
26. Zhang, L., Li, K.: Influence maximization based on snapshot prediction in dynamic online social networks. Mathematics **10**(8) (2022)
27. Zhou, C., Zhang, P., Guo, J., Zhu, X., Guo, L.: UBLF: an upper bound based approach to discover influential nodes in social networks. In: Proceedings of the 2013 IEEE 13th International Conference on Data Mining, pp. 907–916 (2013)

Prediction of Rice Processing Loss Rate Based on GA-BP Neural Network

Hua Yang[1]([⊠]) [iD], Jian Li[1], Neng Liu[1], Kecheng Yi[1], Jing Wang[1], Rou Fu[1],
Jun Zhang[1], Yunzhu Xiang[1], Pengcheng Yang[1], Tianyu Hang[1], Tiancheng Zhang[2],
and Siyi Wang[2]

[1] School of Mathematics and Computer Science, Wuhan Polytechnic University,
Wuhan 430040, China
huayang@whpu.edu.cn
[2] Wuhan BisiCloud Technology Co., Ltd., Wuhan 430015, China

Abstract. Food is closely related to national economy and people's livelihood. Rice is the largest grain crop in China, it is crucial to predict the loss rate of rice during processing to reduce food waste and ensure food security. This study first obtained the loss rate of rice processing through the recovery survey form of enterprises. Then, prediction was carried out using two common models: the BP neural network and multiple linear regression. Finally, the genetic algorithm was applied to optimize the BP neural network for further prediction and com-pared with the original models. The experimental results showed that the GA-BP model had higher prediction accuracy and smaller error compared to the first two models. It is valuable in reducing processing losses and maintaining food security.

Keywords: GA-BP neural network · Prediction model · Rice processing loss

1 Introduction

China is a populous country, and how to meet the food needs of 1.4 billion people and better safe-guard food security has always been a major concern in China. According to the "China Agricultural Industry Development Report 2023" released by the Chinese Academy of Agricultural Sciences, if the loss rates in the stages of grain harvesting, storage, processing, and consumption can be reduced by 1 to 3% points respectively, the grain loss can exceed hundreds of billions of yuan, equivalent to the grain production of about 200 million mu of farmland. Food security needs to be considered from both increasing production and reducing consumption. However, in addressing the issue of ensuring food security in China, more attention is paid to the management of grain from the stages of sowing, growth, harvesting, transportation, and sales, and less attention is given to the loss in the processing of grain. Taking rice, the main grain crop in China, as an example, there are significant quantity losses and quality losses in the processing procedures such as milling and polishing. According to statistics, the losses caused by

---

H. Yang, J. Li—Equal contributions.

L. Pan et al. (Eds.): BIC-TA 2023, CCIS 2062, pp. 121–132, 2024.
https://doi.org/10.1007/978-981-97-2275-4_10

excessive processing of grain range from 3% to 10% and can reach up to about 15% at the highest. Therefore, it is crucial to predict the processing stages of grain for reducing processing losses and ensuring food security.

Currently, the focus on food security is mainly reflected in the prediction of production or crop growth conditions, and the modeling used for prediction includes linear models, multiple regression models, neural network models, etc. [1]. Deep learning, as a popular direction in artificial intelligence [2], and learning is an essential part of almost every neural network, which enables neurons to adapt to different features through external stimuli [3]. Zhu Qiannan et al. [4] combined the recurrent neural network (RNN) and convolutional neural network (CNN) in deep learning to capture the nonlinear spatial features of meteorological factors and the dynamic temporal features of historical wind power (WP) and used the particle swarm optimization algorithm to optimize the interval prediction (PLs) of wind power (WP). The results show that the model can provide high-quality PLs for decision-making and effectively reduce operating costs. Jiang Feng et al. [5] quantified and predicted the potential un-certainty of power load interval by first using singular spectrum analysis (SSA) and k-means clustering to decompose the original data, and then using the improved multi-objective path finding algorithm (IMOPATH) to optimize the Elman neural network (ELMAN). The results show that compared with other models, this method has higher coverage probability, narrower width, and lower bias degree in interval prediction of power load. Liu Tianqi et al. [6] combined harmony search algorithm with mem-brane computation (HS-MC) and used it to optimize the Elman neural network. The results show that the HS-MC algorithm has good optimization capability and practicality. Dang Lin et al. [7] introduced the RepVGG neural network structure diseases and pests and the unsatisfactory detection effects. The improved algorithm model combined with membrane computation improved the detection accuracy by about 2.7% points and the detection speed by about 2.8% points. You Wenqian [8] used a combination model based on the IOWA operator to predict China's grain production, and the results showed that the combination model had a much higher prediction accuracy than a single model. Wang Hui, Fu Hongyu et al. [9] considered that the influence of factors such as kenaf plant height, stem thickness, and number of branches on yield may not be linear, and compared and analyzed the BP neural network and regression model, revealing that the BP neural network was significantly superior to the regression model in terms of accuracy and stability. Zhang Jiuquan et al. [10] constructed models using neural networks, stepwise regression, and interpolation to predict the dates of various growth stages of soybean, and the results showed that the neural network had the best pre-diction results. Fan Chuang, Zhao Zihao [11] believed that the development rate of crops is nonlinearly related to temperature, and they used the BP neural network to predict the growth stages of early rice, with an R-squared value greater than 0.96. Zhou He et al. [12] used the BP neural network to predict the growth trends of rice during different growth stages, and the experimental results helped to further understand the growth of rice. Xing Yuqi [13] used stepwise regression to screen the key factors of wheat scab and then constructed a BP neural network to predict wheat scab, achieving an accuracy rate of 91.67%. Lei Jianyun, Ye Sha et al. [14] addressed the issue of high false-negative rates and poor detection performance in grape leaf disease. They proposed an improved identification algorithm YPLOv4-PSA-CA based on the

YOLOv4 model. This algorithm achieves multiscale feature extraction, resulting in a 4.04% increase in mean average precision (mAP). Wang Wentao, Liu Ming et al. [15] introduced a discontinuous mask layer and employed the LabelSmoothing loss function in dense convolutional neural networks to prevent overfitting, thereby enhancing the classification and detection performance of diseased leaves. Mao Tengyue, Zhu Junjie et al. [16] focused on recognizing young tea buds. They replaced the feature extraction network of CenterNet with ResNet-50, leading to improved performance compared to similar recognition methods. Yang Hua et al. [17] proposed a pyramid network with high-density advanced combination features (DHLC-FPN) and integrated it into the Detection Transformer (DETR) algorithm, achieving a high average accuracy in identifying rice leaf diseases. Wu Yongfu et al. [18] approached the issue of grain storage safety and used the SSA-BiGRU-MLP combined neural network model to predict the temperature of grain piles, and the results showed that this combined model outperformed other models and had better prediction accuracy. Most of the above studies focus on predicting grain harvest, growth conditions, storage methods, etc., but there is little research on predicting the safety of grain processing stages. At the same time, there are more studies on predicting with a large number of samples and a few indicators, while research on small samples and multiple indicators is scarce.

In order to address the issues in the aforementioned studies, taking into account the characteristics of the traditional BP neural network, such as easily falling into local extremes and slow convergence speed [19], genetic algorithms are good at solving non-linear and multi-dimensional optimization problems [20]. By incorporating genetic algorithms, the accuracy of predictions can be improved, and better prediction results can be achieved [21]. In this study, a genetic algorithm optimized BP neural network was used to predict the loss rate of rice processing stages based on only 112 samples and 12 processing steps. The results were compared with those of BP neural network and multiple linear regression in terms of R-squared, MSE, RMSE, and MAE. It was found that the BP-GA neural network model was the optimal model for predicting rice processing losses, and this model maintained high prediction accuracy even in small sample and multi-indicator scenarios. This highlights the applicability and stability of the model in addressing real-world problems, providing technical support for decision-making in rice processing stages, improving production efficiency, and promoting grain loss reduction throughout the entire food supply chain. It also contributes to maintaining food security and making contributions to conserving grain and reducing losses.

2 Prediction Model Establishment

2.1 Data Preprocessing

In this study, a total of 152 enterprises' rice processing data were collected through a survey form. In the data form, there were issues such as inconsistent units for rice quality and incomplete data entries for certain processing stages. To tackle these challenges, we eliminated incomplete data, standardized units using Excel, and computed the rice loss rates for each enterprise at every processing stage. As a result, 112 sets of valid data were obtained. A subset of the data is presented in Table 1.

Table 1. .

No	Processing Stage 1 Loss Rate (%)	Processing Stage 2 Loss Rate (%)	Processing Stage 3 Loss Rate (%)	Processing Stage 4 Loss Rate (%)	Processing Stage 5 Loss Rate (%)	...	Overall processing Loss rate (%)
1	0.400	0.402	0.202	0.000	0.120	...	8.400
2	1.000	0.030	0.000	0.230	0.000	...	0.010
3	2.000	0.050	0.000	0.050	0.170	...	1.000
4	1.510	0.170	0.000	0.030	0.130	...	0.170
5	1.500	0.510	0.000	0.150	0.000	...	4.510
6	0.999	0.000	0.000	0.130	0.000	...	14.500
7	1.000	1.010	0.000	0.099	0.000	...	13.400

2.2 BP Neural Network

The BP neural network is one of the widely used neural network models, which has high modeling ability in complex nonlinear systems and is widely used in prediction [22]. By training the original data, the trained network can acquire learning and prediction capabilities [23]. The specific process is outlined as follows:

Step 1): Network initialization. First, the number of nodes in the input, hidden, and output layers is determined, with the number of nodes in the hidden layer typically determined through experience or iterative experimentation. Next, the weights between the input and hidden layers, as well as between the hidden and output layers, are determined, denoted as w_{ij} and w_{jk}, respectively. The threshold of the hidden layer is initialized as a_j and the threshold of the output layer is initialized as b_k. Finally, the learning rate and activation function are given.

Step 2): Calculate the output of the hidden layer. Based on the input variables x_i, the weights connecting the input to the hidden layer w_{ij} and the threshold of the hidden layer a_j, calculate the output of the hidden layer H_i. See that in Eq. (1).

$$H_i = f\left(\sum_{i=1}^{n} x_i \omega_{ij} - a_j\right)(j = 1, 2, \ldots, l) \tag{1}$$

$$f(x) = \frac{1}{1 + e^{-x}} \tag{2}$$

where I represents the number of nodes in the hidden layer, and Eq. (2) represents the activation function. In this case, the sigmoid function is selected as the activation function.

Step 3): Calculate the output of the output layer. Based on the output H_i of the hidden layer, the connection weights w_{jk} between the hidden layer and the output layer, and the threshold b_k of the hidden layer, calculate the output O_k of the output layer.

$$O_k = \sum_{j=1}^{l} H_j \omega_{jk} - b_k (k = 1, 2, \cdots, m) \tag{3}$$

where m represents the number of nodes in the output layer.

Step 4) Error calculation. The prediction error e_k is determined by the predicted value O_k and the actual value Y_k. See that in Eq. (4).

$$e_k = Y_k - O_k (k = 1, 2, \cdots, m) \tag{4}$$

Step 5) Updating weights. The connection weights w_{ij} and w_{ik} are recalculated based on the error e_k.

$$\omega_{ij} = \omega_{ij} + \eta H_i (1 - H_j) x(i) \sum_{k=1}^{m} e_k \omega_{jk} \quad (j = 1, 2, \cdots, n; \; j = 1, 2, \cdots, l) \tag{5}$$

$$\omega_{jk} = \omega_{jk} + \eta H_i e_k \quad (j = 1, 2, \cdots, l; \; k = 1, 2, \cdots, m) \tag{6}$$

where η represents the given learning rate.

Step 6) Threshold updating. Update the threshold values a_j and b_k based on the error e_k.

Step 7) Determine if the algorithm is finished. Based on the error e_k, determine if the model has reached the desired error range or the maximum number of training iterations. If it has, complete the calculation; otherwise, proceed to Step 2.

The structure diagram of the BP neural network during the calculation process is shown in Fig. 1.

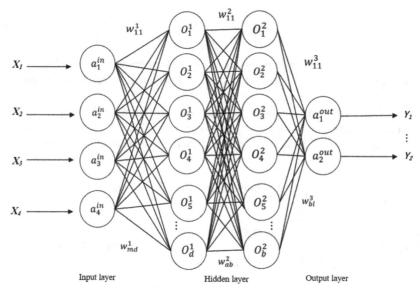

Fig. 1. Structure diagram of BP neural network.

2.3 Genetic Algorithm

Genetic algorithm simulates natural selection and mutation phenomena, and performs selection, cross-over, and mutation operations on the population to continuously evolve and obtain the best population. The basic algorithm flow is as follows [24]:

Step 1) Generate initial population. Randomly generate several data, each representing an individual, to form the initial population.

Step 2) Perform chromosome encoding. Connect the weights and thresholds of the BP neural network in a certain order to form a chromosome for the genetic algorithm.

Step 3) Select a suitable fitness function. By selecting an appropriate fitness function, calculate the fitness of each individual in the population.

Step 4) Perform genetic operations. Select suitable individuals in the population for crossover and mutation operations to determine the fitness of new individuals.

Step 5) Determine if the termination condition is met. If the termination condition is not met, repeat Step 3.

The flowchart of optimizing the BP neural network using genetic algorithm is shown in Fig. 2.

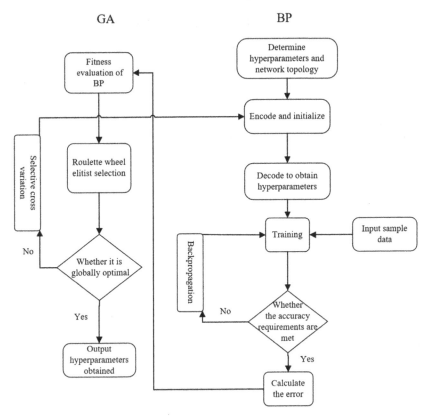

Fig. 2. Optimization flowchart

3   Results and Analysis

3.1   Prediction Results of BP Neural Network

To determine the approximate range for the number of hidden layer nodes, the 12 processing stage loss rates of rice processing are used as input layer nodes, based on the following Eq. (7).

$$l < \sqrt{(m+n)} + \alpha \tag{7}$$

where m is the number of output layer nodes, n is the number of input layer nodes, and a is a parameter between 0 and 10.

The optimal number of hidden layer nodes is determined to be 8 using a trial-and-error method. The output layer nodes represent the final loss rate of rice after undergoing the 12 processing stages and are set to 1. Other key structural parameters are shown in Table 2.

Once the convergence error meets the criteria, the training is halted. Figure 3 displays the prediction results, and Fig. 4 presents the error analysis of the BP neural network. From Fig. 3 and Fig. 4, it can be observed that the fitting degree of the BP neural network's

Table 2. Main structural parameters of the BP prediction model

Maximum number of iterations	Error threshold	Learning rate
1000	1E-6	0.01

prediction results is moderate, with significant fluctuations in prediction errors. The average error is 9.6%, the maximum error is 28.7%, and the minimum error is 0.7%.

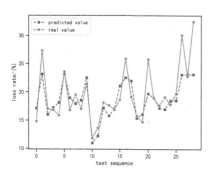

Fig. 3. BP Neural Network Prediction   **Fig. 4.** Error of BP Neural Network

3.2 Multiple Linear Regression Prediction Results

Using the loss rates of each processing stage as independent variable inputs and the total loss rate as the dependent variable output, Fig. 5 and Fig. 6 depict the outcomes of multiple linear regression along with associated errors.

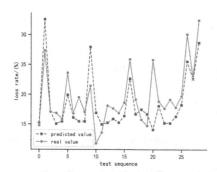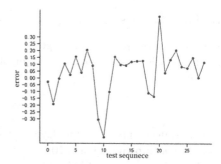

Fig. 5. Multiple Linear Regression Prediction   **Fig. 6.** Error of Multiple Linear Regression

The R-squared value obtained from the multiple linear regression prediction is only 0.25, which is lower than the accuracy of the BP neural network. Figure 5, visually demonstrates the significant fluctuation in prediction errors. Additionally, in Fig. 6, the maximum error is 45.2%, the minimum error is 0.3%, and the average error is 13.5%.

3.3 GA-BP Prediction Results

The maximum number of iterations, error threshold, and learning rate for GA-BP are the same as for BP. The other structural parameters of the GA-BP prediction model are shown in Table 3.

Table 3. Main Structural Parameters of the GA-BP Prediction Model

maximum iterations	Error threshold	Learning Rate	Genetic Iterations	Population Size	Crossover Probability/%	Mutation Probability/%
1000	1E-6	0.01	10	20	90	8

In this study, a genetic algorithm was employed to optimize the original network. Figure 7, Fig. 8 and Fig. 9 depict the iteration count, output results, and the errors between each sample and the actual values.

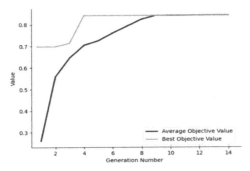

Fig. 7. Value-Generation

In Fig. 7, it can be observed that the model converges completely after around the 9th generation, as the error between the output results and the actual values becomes smaller. This indicates that the model has good convergence properties, and the model structure is reasonable at this point.

As depicted in Fig. 8 and Fig. 9, the GA-BP neural network model, optimized by a genetic algorithm, exhibits a significant enhancement in predicting processing loss rates. The average error is 6%, with a maximum relative error of 16.5% and a minimum relative error of 0.7%. Furthermore, the GA-BP model shows a higher level of agreement between the output results and the actual values compared to the single BP neural network model. The predicted values closely follow the trend of the actual values. Additionally, it can be observed that there are more instances with errors close to zero and smaller errors in the GA-BP model, indicating a more stable prediction performance. Therefore, the optimization of the BP neural network using genetic algorithm is considered successful.

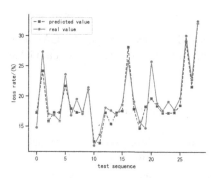

Fig. 8. GA-BP Neural Network Prediction   **Fig. 9.** Error of GA-BP Neural Network

3.4   Model Comparison

The prediction of rice processing loss rate was conducted using three different models: single BP neural network, multiple linear regression, and GA-BP combined model. The results are shown in Fig. 10 and Table 4.

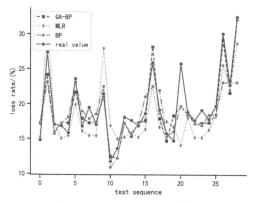

Fig. 10. Comparison of BP, Multiple Linear Regression, and GA-BP Models for Prediction

Table 4. Error Parameter Table

Model Type	MSE	RMSE	MAE	R^2
MLR	12.88	3.59	2.72	0.25
BP	8.76	2.96	2.04	0.61
GA-BP	3.61	1.90	1.44	0.86

Clearly, the GA-BP model exhibits better fitting performance compared to the multiple linear regression and single BP models. The regression fitting is good, and additionally, the R-squared value of the GA-BP model increases from 0.25 to 0.86. At the same time, the model's MSE decreases from 12.88 to 3.61, RMSE decreases from 3.59 to 1.9, and MAE decreases from 2.72 to 1.9. This indicates that the GA-BP neural network model optimized by genetic algorithm has a superior predictive capability for rice processing losses compared to the multiple linear regression and BP models.

4   Conclusions

This paper introduces a genetic algorithm into the BP neural network for predicting the loss rate in the rice processing stage. Firstly, based on the analysis of issues in traditional prediction models, predictions were made using both the BP neural network and a multiple linear regression model. The results indicate that the BP neural network performs better, suggesting that the loss rates at various stages of rice processing are not linear. Secondly, considering the characteristics of the loss rate, this paper proposes a scheme to improve the BP neural network using a genetic algorithm, addressing the problem of the BP neural network easily getting trapped in local optima. Finally, through simulation comparisons and comprehensive evaluation of various prediction models using metrics such as MAE, MSE, RMSE, and R2, it is concluded that the model improved by the genetic algorithm for the BP neural network exhibits the highest and most effective prediction accuracy.

Acknowledgements. This work is supported by the Ministry of Education's Industry-University Cooperatian and Education Project (Nos. 220900786024216), Educational Research Project in Hubei Province (No.2018368), Graduate Workstation School-Enterprise Cooperation Project in Hubei Province (No. whpu-2021-kj-762), Horizontal school-enterprise cooperation project (No. whpu-2022-kj-1586), Wuhan Polytechnic University School Established Research Project (No. XM2021015).

References

1. Li, H.Q., Zhao, Y.Q., Yan, Q., Lin, F.: Vehicle Wheel matching and vehicle reverse dynamics based on genetic neural network. J. Huazhong Univ. Sci. Technol. (Nat. Sci. Edition) **47**(05), 27–32 (2019)
2. Liu, M.C.: Road Scene Recognition of Substation Inspection Robot Based on Deep Learning. Southwest Jiaotong University (2019)
3. Yu, Q., Song, S., Ma, C., Pan, L., Tan, K.C.: Synaptic learning with augmented spikes. IEEE Trans. Neural Netw. Learn. Syst. **33**(3), 1134–1146 (2022)
4. Zhu, Q.N., Jiang, F., Li, C.S.: Time-varying interval prediction and decision-making for short-term wind power using convolutional gated recurrent unit and multi-objective elephant clan optimization. Energy **271**, 127006 (2023)
5. Jiang, F., Zhu, Q., Yang, J., Chen, G., Tian, T.: Clustering-based interval prediction of electric load using multi-objective pathfinder algorithm and Elman neural network. Appl. Soft Comput. **129**, 109602 (2022)
6. Yang, H., Liu, T.Q., et al.: Harmony search algorithm based on membrane computing framework and its application. J. Hubei Univ. (Nat. Sci. Edition) **45**(2), 171–180 (2023). https://doi.org/10.3969/j.issn.1000-2375.2022.00.089

7. Yang, H., Lin, D., Zhang, G., Zhang, H., Wang, J., Zhang, S.: Research on detection of rice pests and diseases based on improved YOLOV5 algorithm. Appl. Sci. **13**(18), 10188 (2023). https://doi.org/10.3390/app131810188

8. You, W.Q., Zhuang, K.J.: Combination prediction of grain production in china based on the IOWA operator. J. Chongqing Technol. Bus. Univ. (Nat. Sci. Edition) **37**(05), 80–87 (2020)

9. Wang, H., Fu, H.Y., Wang, J.L., et al.: Comparative study on heterogeneous estimation models of ramie yield based on multiple regression and BP neural network. China Flax **42**(5), 227–238 (2020)

10. Zhang, J.Q., Zhang, L.X., Zhang, M.H., et al.: Prediction of soybean growth and development stages using neural network and statistical models. Acta Agron. Sin. **35**(2), 341–347 (2009)

11. Fan, C., Zhao, Z.H., Zhang, X.S., et al.: Forecasting model of first-season rice development stage based on BP neural network. J. Zhejiang Agric. Sci. **35**(2), 434–444 (2023)

12. Lin, J., He, C., Cheng, R.: Adaptive dropout for high-dimensional expensive multiobjective optimization. Complex Intell. Syst. **8**(1), 271–285 (2022)

13. Xing, Y.Q., Zhou, J., Huang, W.L., et al.: Prediction model of wheat fusarium head blight based on BP neural network in rice-wheat rotation areas. Acta Agriculturae Boreali-occidentalis Sinica **32**(11), 1842–1848 (2023)

14. Lei, J.Y., Ye, S., Xia, M., et al.: Grape leaf disease detection based on improved YOLOv4. J. South-Cent. Univ. Natl. (Nat. Sci. Edition) **41**(06), 712–719 (2022)

15. Wang, W., Liu, M., Zhao, Z., et al.: Improved convolutional neural network method for classification of apple leaves. J. South-Cent. Univ. Natl. (Nat. Sci. Edition) **41**(1), 71–78 (2022)

16. Mao, T.Y., Zhu, J.J., Tie, J., et al.: Research on tea bud recognition method based on anchor-free detection network. J. South-Cent. Univ. Natl. (Nat. Sci. Edition) **42**(04), 489–496 (2023)

17. Yang, H., Deng, X., Shen, H., Lei, Q., Zhang, S., Liu, N.: Disease detection and identification of rice leaf based on improved detection transformer. Agriculture **13**(7), 1361 (2023). https://doi.org/10.3390/AGRICULTURE13071361

18. Yang, H., Wu, Y.F., at al.: Application of decomposition-based deep learning model in temperature prediction of grain piles. J. South-Central Univ. Natl. (Nat. Sci. Edition) **42**(05), 696–701 (2023). https://doi.org/10.20056/j.cnki.ZNMDZK.20230516

19. Tian, K., Sun, Y.T., Gao, H., et al.: Quantitative research on defects of BP neural network based on Bayesian algorithm. China Meas. Test. **40**(3), 5 (2014)

20. Ma, X.M., Wang, X.: Improvement of BP neural network based on genetic algorithm. J. Yunnan Univ. (Nat. Sci. Edition) **35**(S2), 34–38 (2013)

21. Liang, Y.: Research on Fermentation of Lactic Acid Bacteria and Yeast Fermented Rice Noodle Slurry and Preparation of Rice Bread. South China University of Technology (2019)

22. Lan, X.P., Chen, J.Y., Jiang, Y.J., et al.: Construction of grain storage quality prediction model based on BP neural network algorithm. J. China Cereals Oils Assoc. **35**(11), 147–151 (2020)

23. Chen, Y., Xu, J., Yu, H., Zhen, Z., Li, D.: Three-dimensional short-term prediction model of dissolved oxygen content based on pso-bpann algorithm coupled with kriging interpolation. Math. Probl. Eng. (2016)

24. Guo, L.J., Qiao, Z.Z.: Research on grain temperature prediction based on genetic algorithm optimized BP neural network. Grain Oil **36**(1), 34–37 (2023)

Research on Target Value Assessment Method Based on Attention Mechanism

Guangyu Luo[1] , Dongming Zhao[1], Xuan Guo[1(✉)], and Hao Zhou[2]

[1] School of Information Engineering, Wuhan University of Technology,
Wuhan 430070, People's Republic of China
{luoguangyu,guoxuan}@whut.edu.cn
[2] School of Transportation and Logistics Engineering, Wuhan University
of Technology, Wuhan 430070, People's Republic of China

Abstract. Under the condition of informatized joint combat system, the traditional target value analysis method based on human judgment is no longer able to quickly and accurately judge the target value from the massive battlefield information. In this paper, combining with the attention mechanism model, we propose a dual-driven target value assessment framework based on target and data, and construct data-driven and purpose-driven attention mechanism models respectively. By selecting multiple features and calculating the saliency of cue targets, the results of the two models are fused to screen out the noticed cue targets and further expanded to form a high-value target set. The experimental results show that the proposed target value assessment model based on attention mechanism is effective, and the matching rate between the actual strike targets and the high-value targets obtained by the model reaches 92%, which is in line with the actual situation of combat and can reflect the cognitive process of the commander to some extent.

Keywords: Battlefield situational awareness · Target value assessment · Attention mechanism

1 Problem Description

Under informatization conditions, assessing the value of joint combat targets and forming target strike lists is an important part of commanders' command and decision-making. The joint combat battlefield situation is complex and ever-changing, and it is no longer possible to quickly and accurately judge the value of targets from massive battlefield information by relying mainly on human resources. The speed of cognition of target value will lead to corresponding decision-making and action afterward, making the command decision-making cycle much longer than that of the opponent. Therefore, there is a need to explore computer intelligence-assisted methods to help commanders conduct target value assessment.

Fan et al [1] proposed a dynamic multi-temporal fusion target threat assessment method and designed an interval-valued intuitionistic fuzzy entropy based

© The Author(s), under exclusive license to Springer Nature Singapore Pte Ltd. 2024
L. Pan et al. (Eds.): BIC-TA 2023, CCIS 2062, pp. 133–147, 2024.
https://doi.org/10.1007/978-981-97-2275-4_11

on the cosine function to determine the target attribute weights, which effectively improves the reliability and accuracy of target threat assessment in missile defense. Zhao et al [2] proposed a new method based on the combination of interval-valued intuitionistic fuzzy sets (IVIFS), game theory, and evidential reasoning methods, and the algorithm can effectively reduce the uncertainty of target threat assessment results. Zhang et al [3] used a theoretical modeling approach, which is based on the target motion, background radar map, radar cross-section (RCS) and high-resolution ranging profile (HRRP), fused different types of damage features using bayesian network, and verified the performance of the proposed damage features in numerical experiments. Yang et al [4] established an air target threat assessment model based on PCNN neural network for the problem of air target threat assessment and classification, which can effectively realize air target. The model can effectively realize the classification of air target threat assessment, which provides ideas and methods for future battlefield combat command research and promotes the development and implementation of battlefield target threat assessment system. Chen et al [5] study target threat assessment in multi-party unmanned cluster confrontation scenarios, and improve the Adaboost algorithm by building a library of various meta-learning algorithms to reduce the error rate of the algorithm. Luo et al [6] proposed a simple and reliable threat assessment method for low-altitude slow-small (LSS) targets, which utilized the hierarchical analysis method (AHP) and information entropy to determine the subjective and objective threat factor weights of LSS targets, and used an optimization model to combine them in order to obtain more reliable evaluation weights. Li et al [7] proposed a bayesian network-based underwater target threat assessment method, which calculates the threat probability through bayesian inference, and the model has relatively good performance and generates assessment results with low time cost.

The traditional approach has the following limitations. Firstly, it lacks the problem of recognizing target value from the perspective of commanders conducting situational awareness. Previous studies have emphasized the assessment from an objective perspective, while ignoring the significant role played by the commander's subjective perception of the target value from the combat mission to the combat outcome, and therefore have not focused on the impact of subjective cognitive factors in the theory and modeling. Commander's cognitive judgment of the target value is carried out under the condition of incomplete information on the battlefield, or even true and false co-existing information, which is a highly artistic thinking activity combining macroscopic and microcosmic, experience and rationality, and a lot of the inner operation and mechanism of the role are still unclear, and need to explore the new research ideas and technical methods. Secondly, there is a lack of complex war system research features that reflect the whole, dynamic and confrontation. War system is a typical complex system, which cannot be studied simply by traditional reductionist scientific thinking. Therefore, when conducting target value assessment, it is necessary to stand in the system confrontation perspective of complex war systems. However, the traditional mostly small-sample analysis methods based on reductionist ideas also belong to the static assessment methods of simple decomposition and linear superposition, which are difficult to meet the demand for the assessment of the value of joint war targets under the conditions of informatization.

The attention mechanism of commanders is a worthwhile direction to explore in the assessment of target value. People can search for specific objects from complex environments and have the ability to selectively process information about the objects. This is the human attentional mechanism, which can help a person to grasp the focus quickly and effectively in a complex environment. The research of attention mechanism originated from psychology, and the main research is aimed at exploring the biological mechanism behind human visual attention and auditory attention, and building computer models based on it for dealing with image recognition, target localization and other problems [8]. In combat, commanders will focus their attention on certain key targets according to the combat mission, ignoring relatively minor targets. As the combat process advances, the commander will adjust and optimize his focus according to the results of combat operations. The whole process of the commander's assessment of the value of the combat target is the process of the occurrence and role of the attention mechanism, the process of the objective seen in the subjective process, and also the high-level abstract cognitive thinking. Introducing the attention mechanism in psychology into battlefield situational awareness for the problem of target value assessment is the innovation point of this paper.

Therefore, from the perspective of commanders' situational awareness in the context of joint operations, this paper studies the attention mechanism of commanders in battlefield situational awareness and improves the method of target value assessment based on the attention mechanism. This paper will focus on the following. Firstly, analyze how to realize the dynamic assessment of combat targets based on the attention mechanism. Then, the target value assessment model based on the attention mechanism is designed and validated.

2   A Target Value Assessment Model Based on a Two-Way Attention Mechanism

2.1   Modeling Framework

Commanders will consciously pay attention to the focuses in the known targets when conducting battlefield situational awareness, which is to pay attention to the focuses that have been mastered or what is required by the current combat mission, and it is a kind of top-down mode of action. At the same time, the commander will also be suddenly appeared new target attention, if the sudden appearance of the new target to help complete the current combat mission or may be a great threat to their own side, the commander will also continue to pay attention to, this is a bottom-up role. It can be seen that the battlefield situational awareness attention mechanism has two kinds of role processes, top-down purpose-driven and bottom-up data-driven.

Purpose-driven refers to an attentional mechanism role process in which the commander selectively devotes more cognitive resources to certain targets for in-depth analysis in combat, given a certain degree of relevant information about the combat targets [9]. Data-driven refers to a kind of attentional mechanism role process in which the commander in combat, without mastering any a priori information, is passively informed of the existence of combat targets through his own organization reconnaissance or received enemy intelligence briefings and other ways, so that he can master combat target-related information to a certain extent [10]. The process of analyzing combat targets based on the attention mechanism is shown in Fig. 1. The process can be formally described as Eq. 1.

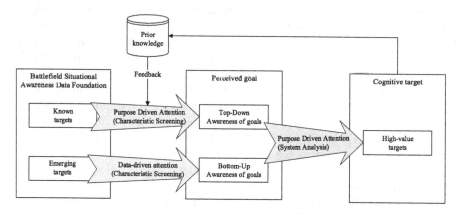

Fig. 1. The process of analyzing combat targets based on the attention mechanism

$$
\begin{cases}
E = \{E_{\text{old}}, E_{\text{new}}\} \\
E_{\text{TD}} = f_{\text{TD1}}(E_{\text{old}}, K_{\text{prior}}) \\
E_{\text{BU}} = f_{\text{BU}}(E_{\text{new}}) \\
E_{\text{sp}} = \{E_{\text{TD}} E_{\text{BU}}\}, E_{\text{sp}} \in E \\
E_{\text{HVT}} = f_{\text{TD2}}(E_{\text{sp}}), E_{\text{HVT}} \in E_{\text{sp}}
\end{cases}
\tag{1}
$$

where, E is the set of all visible combat targets of the own side, including the known target E_{old} and the just-acquired new target E_{new}; E_{old}, under the condition of a priori knowledge K_{prior}, undergoes the action of the purpose-driven attentional mechanism f_{TD1} in the awareness stage and gets the top-down awareness target E_{TD}; E_{new} undergoes the action of the data-driven attentional mechanism f_{TD1} in the awareness stage and gets the bottom-up awareness goal E_{BU}, and the set of goals E_{sp} obtained in the awareness phase is a subset of all the visible goals E on one's side; E_{sp}, after the action of the top-down attention mechanism f_{TD2} in the cognition phase, obtains the set of high-value goals f_{TD2}, which is a subset of the awareness goal E_{sp}.

2.2   Model Parameters

The mental model (MM) is the commander's comprehension and understanding of the mission. The inputs to the mental model are the results of the current mission and previous situational awareness, and the outputs are the parameters related to the operational mission and the a priori knowledge of the operational objectives. The formalization is described as Eq. 2.

$$\text{Parameter}_{\text{task}}, K_{\text{prior}} = F_{\text{MM}}(\text{BSC}, \text{Task}) \tag{2}$$

where, BSC is the result of battlefield situational awareness; Task is the current combat task; K_{prior} is the a priori knowledge of the combat target; and Parameter$_{\text{task}}$ is the parameters related to the combat task, such as the task object, the task bearer subject, the task type, and the task-related territory. The influence of the combat task on the commander's battlefield situational awareness and subsequent decision-making and action is accomplished through the mental model, because the commander will first form his own understanding of the combat task, and then he can judge the target value accordingly. Therefore, the parameters related to the combat mission and the a priori knowledge of the combat target output from the mental model affect the results of battlefield situational awareness, i.e., they are affected when screening the perceived target and assessing the target's value.

Operational targets mainly include military personnel, weapons and equipment, military facilities, transportation facilities and so on. These targets are heterogeneous, with different attributes and capabilities, and the targets are characterized by multi-dimensional information, which is closely related to the target hierarchy, location space, time and other factors. In addition, due to the existence of the fog of war, it is difficult for the commander to fully grasp all the real information of the enemy's combat targets, and what the party can grasp is based on the information mapped by intelligence perception. All of these need to be considered when describing combat targets. In view of the above description of battlefield posture and battlefield entities, the description of combat targets is as follows.

At moment t, the number of all enemy combat targets perceived by the commander is m, defined as Eq. 3.

$$E = \{e_i\}, \quad i = 1, 2, \cdots, m \tag{3}$$

The state of the target at moment t based on the perceived reality is represented by a vector, and the vector elements are target attributes, namely as Eq. 4.

$$e_i = [\text{ID}, \text{service}, \text{type}, \text{echelon}, \text{cmdID}, \text{sptID}, \text{lon}, \text{lat}, \Delta\text{cap}] \tag{4}$$

where ID means target number, service means military branch to which the target belongs, type means target type, echelon target level, cmdID means number of the target's parent unit, sptID means targeted security force numbers, lon means longitude of the target, lat means latitude of the target, Δcap means value of change in the target's combat power compared to the previous moment of observation.

Combat targets can be categorized based on the battlefield situational awareness period Δt, and the current operational moment t. What is already available at the moment $t - \Delta t$ is the known target E_{old}, and what is known at $[t - \Delta t, t)$ is the emerging target E_{new}. At the moment t, the commander's current operational task is $Task$.

2.3   Data-Driven Attention Mechanism-Based Modeling

Feature Selection. In the feature extraction process of the visual attention model, simpler and mutually independent features are generally selected. In the battlefield situational awareness stage, the commander will be attracted by unexpected situations that suddenly appear. These unforeseen situations include enemy combat targets that may pose a threat to their side, or targets that are conducive to accomplishing the combat mission. The commander should not be distracted by other irrelevant targets on the battlefield. According to actual experience, during the situational awareness phase, the commander's attention is often attracted because of some simple characteristics of the enemy target, such as being relatively close to the current combat mission territory, or being so close to the command post of his own side that it poses a threat, and so on. Considering the actual situation of joint operations, the following salient features are selected for modeling the data-driven attention mechanism in the battlefield situational awareness phase, including aggregation of targets C, distance from your side L, distance from the geographic area of the mission D, targeted Combat Cap, relevance to the mandate R.

Calculation of Feature Significance. The aggregation degree of the ith target C_i. The more the value of this feature deviates from the average, the greater the significance. This is because targets with particularly high or low aggregation, i.e., clustered and isolated targets, are more likely to attract attention. The significance of the aggregation degree of the i target is as Eq. 5.

$$\begin{cases} C_i' = \left| C_i - \bar{C} \right|, \\ S_{C_i} = 1 - e^{-\frac{C_i'}{\bar{C}_i'}} \end{cases} \tag{5}$$

The distance L_i between the i target and your side is the distance between the target location and the command post location A_1 where the main body of your battlefield situational awareness is located. The smaller the value of this feature, the greater the significance. Enemy targets that are closer to the self-side may pose a threat and therefore attract attention. The significance degree of the ith target's distance from the self-side is as Eq. 6.

$$S_{L_i} = 1 - \frac{L_i - L_{min}}{L_{max} - L_{min}} \tag{6}$$

The distance D_i of the i target from the mission territory is the distance between the location of that target and the center of gravity point A_0 of the mission territory. The smaller the value of this feature, the greater the significance. If it is an offensive mission, the closer the target is to the offensive territory, the more attention it attracts; if it is a defensive mission, the closer the enemy target is to our defensive territory, the more attention it attracts. The significance of the i target's distance from the mission territory is as Eq. 7.

$$S_{D_i} = 1 - \frac{D_i - D_{\min}}{D_{\max} - D_{\min}} \tag{7}$$

The combat power Cap_i of the i target. the domain of this eigenvalue is $[0, 100]$. The larger the eigenvalue is, the greater the significance is. According to the general rule, when the combat power is lower than the broad value, ∂ the target temporarily loses its combat power and does not need to be concerned. Therefore, the significance degree of combat power of the i target is as Eq. 8.

$$S_{Cap_i} = \begin{cases} \frac{Cap_i - 30}{100 - 30}, & Cap_i 30 \\ 0, & \text{others} \end{cases} \tag{8}$$

The i target has a relevance to the mission, R_i. The commander will focus on targets that have a high relevance to the mission, including both direct and indirect relevance. The direct relevance is the task object within the task E_obj. The indirect relevance is the target that may generate a threat E_threat. for other targets, the commander should not pay attention to them to avoid distraction. Therefore, the significance of the i target's relevance to the task is as Eq. 9.

$$S_{R_i} = \begin{cases} 1, E_i \in E_{\text{obj}} \\ 0.5, E_i \in E_{\text{threat}} \\ 0, others \end{cases} \tag{9}$$

Generate a Significant Figure. The significance degrees of the five features obtained by the above method of significance calculation are S_Ci, S_{Li}, S_Di, $S_{Cap}i$, S_{Ri}, $S \in [0, 1]$. Given that the commander always considers mission accomplishment as the primary consideration, the feature fusion is performed by giving greater weight to the saliency of geographical distance from the mission. For a target that is irrelevant to the mission or has been neutralized, the commander should not pay attention to it, regardless of the significance of the target's other features. Since the significance distributions of the individual mutually salient features are different, preprocessing is required. The significance of the target aggregation feature is processed as Eq. 10.

$$S'_{C_i} = S_{C_i} - \bar{S}_C + 1 \tag{10}$$

Then, $S'_{C_i} \in [0, 2], \bar{S}'_C = 1$. The significance degree $S_{L}i$ of the distance between the target and its own side, and the significance degree $S_{D}i$ of the distance between the target and the geographical area of the task are treated in the same way, S'_{L_i} and S'_{D_i} are obtained. Based on the above considerations, the formula for fusing the individual features into a composite saliency of the target to form a saliency map Map_1 is as Eq. 11.

$$S_{i1} = \left(S'_{C_i} + S'_{L_i} + 2S'_{D_i}\right) \times S_{R_i} \times S_{\text{Cap}_i} \tag{11}$$

2.4 Purpose-Driven Attention Mechanism-Based Modeling

Feature Selection. In response to the known targets in his possession, the commander already has a general impression in his mind about the importance of these targets, which comes from the last battlefield situational awareness cycle and is a feedback of the situational awareness results. Therefore, the commander will continue to pay attention to the targets that have been judged to be more important, but will also continue to keep track of the targets that were previously considered less important, rather than ignoring them outright. During the modeling process of the Purpose Driven Attention Mechanism, it will be similar to the Data Driven, where target salience is calculated based on some simpler features. High-value targets that have been determined to be of high value will be emphasized in terms of salience.

In the situational awareness phase, the saliency features selected in the data-driven and purpose-driven attention mechanism modeling are roughly the same because commanders are concerned about known and emerging targets for similar reasons. Considering the actual situation of joint warfare, the following salient features are selected for modeling the purpose-driven attention mechanism in the battlefield situational awareness phase, which increase the intelligence timeliness and a priori importance over the features of the data-driven attention mechanism.

Calculation of Feature Significance. For the sake of narrative convenience, differences in the way feature saliency is calculated with respect to data-driven attention mechanisms are explained, and the same ones are directly referenced above. Intelligence timeliness of the i target I_i is as Eq. 12.

$$I_i = t - t_{\text{det}} \tag{12}$$

The smaller the value of this feature, the higher the saliency. In operational reality, the commander will pay higher attention to the time-sensitive targets with higher effectiveness. The significance degree of the timeliness of the i target intelligence is as Eq. 13.

$$S_{I_i} = 1 - \tanh\left(I_i\right) \tag{13}$$

For the sake of narrative convenience, differences in the way feature saliency is calculated with respect to data-driven attention mechanisms are explained, and the same ones are directly referenced above. The a priori importance of the i target V_i. The larger the value of this feature, the greater the significance. The commander will directly focus on the target that has been predicted to be more important. The quantitative result of the target a priori knowledge on the importance of known targets is an attention weight for each known target, with a value range of $[0, 1]$, fed by the cognitive result of the last round of battlefield situational awareness. If the battle has just started, it is obtained from the result of the prediction of the enemy target in the pre-battle planning. The significance of the a priori importance of the i target is as Eq. 14.

$$S_{V_i} = w, \quad w \in K_{\text{prior}} \tag{14}$$

Generate a Significant Figure. The significance $S_c i$, $S_L i$, $S_D i$, $S_I i$, $S_V i$, $S_{C} api$, and $S_R i$ were obtained for the seven features by the above method of significance calculation. The significance of the five mutually significant features is done for the treatment. The treatment is consistent with the above, to obtain S'_{C_i}, S'_{L_i}, S'_{D_i}, S'_{I_i} and, S'_{V_i}, respectively.

Given the commander's experience in thinking during combat, feature fusion is performed to give greater weight to the saliency of features such as geographical distance from the mission, a priori importance of the target, and timeliness of intelligence. Based on the above considerations, each feature is fused into a comprehensive saliency of the target, forming the formula for the saliency map Map_2 as Eq. 15.

$$S_{i2} = \left(S'_{C_i} + S'_{L_i} + 2S'_{D_i} + 2S'_{I_i} + 2S'_{V_i}\right) \times S_{R_i} \times S_{Cap_i} \tag{15}$$

2.5   High-Value Target Screening Algorithm Based on Purpose-Data Dual Drive

The cue target saliency map Map_1 obtained by the data-driven attention mechanism contains emerging targets, assuming there are m The cue target saliency map Map_2 obtained by the purpose-driven attention mechanism contains known targets, assuming there are n. The cue target saliency map Map_2 obtained by the purpose-driven attention mechanism contains known targets, assuming there are n. The combined saliency of the targets in these two saliency maps is processed. First, they are normalized and then their mean is made uniform to 1.

$$S'_{i1} = \frac{S_{i1} - S_{1\min}}{S_{1\max} - S_{1\min}}, S'_{i2} = \frac{S_{i2} - S_{2\min}}{S_{2\max} - S_{2\min}}, \quad S'_{i1}, S'_{i2} \in [0, 1] \tag{16}$$

$$S^*_{i1} = S'_{i1} - \frac{\sum\limits_{1}^{m} S'_{i1}}{m} + 1, S^*_{i2} = S'_{i2} - \frac{\sum\limits_{1}^{n} S'_{i2}}{n} + 1, \quad S^*_{i1}, S^*_{i2} \in [0, 2] \tag{17}$$

Then for all cue targets, the significance of the combined significance map is S^*_{i1} and S^*_{i2}, and the global average significance is as Eq. 18.

$$\bar{S} = \frac{\sum\limits_{1}^{m} S^*_{i1} + \sum\limits_{1}^{n} S^*_{i2}}{m + n} \tag{18}$$

All cue targets are sorted by combined saliency S_{i1}^* and S_{i2}^*, and targets with saliency values greater than $\bar{S}$ are noted as cues E_{cue}^*. The saliency is only used to select clues, and then on the basis of the existing set of clues and targets, the clues and targets related to the clues are found to form the set of perceptual targets, which will enter the next stage of battlefield situational awareness and value assessment. Specifically, the superior unit cmdIDs of these cue targets and the combat targets with supportive relationships with these cue targets with sptIDs are added to the awareness target set. The algorithm is shown as follows.

Algorithm 1. High value target screening

Input:
　Based on a data-driven collection of cue targets $E_{\text{sp\_down}}^*$
　Based on purpose-driven cue target sets $E_{\text{sp\_up}}^*$
Output:
　Key target E_{sp}
　$\bar{S} = Aver\left(E_{\text{sp\_down}}^*.S_i, E_{\text{spLup}}^*.S_i\right)$
　$E^* = Append\left(E_{\text{sp\_down}}^*, E_{\text{sp\_up}}^*\right)$
　$E_{\text{cue}}^* = E^*\left[E^* > \bar{S}\right]$
　for e_i in E_{cue}^*　**do**
　　// Identify the set of awareness targets based on clues
　end for
　if $e_i! = E_n$ **then**
　　$E_{\text{cmd}} = FindCommand\,(e_i)$
　　$E_{\text{spt}} = FindSupport\,(e_i)$
　　$E_{\text{sp}} = Append\,(E_{\text{sp}}, E_{\text{cmd}}, E_{\text{spt}})$
　　$e_i = E_{\text{cmd}}$
　end if

3　Experimental Validation

3.1　Data Preparation

Based on the maneuver data of the military chess rehearsal system in a desired context, a complete combat time interval $[T, T + nit\Delta t)$ is selected to calculate the integrated high-value target set obtained by the attention mechanism model after iteration in this interval. Python 3.5 was used as the development tool, and Spyder was used as the development platform for data extraction, preprocessing, and model hollowing. Tableau is used as the visualization platform for the display of the situational map. In this combat phase, the air combat commander is tasked with striking an area and paralyzing the blue side's air defense and anti-missile combat system. Using the mental model, the red side's mission analysis results were obtained, as shown in Table 1. Based on the characteristics of the deduction data and the requirements of the model, the data is preprocessed, mainly including data selection, data cleaning, and data transformation.

Table 1. Red's task analysis

Parameter type	Parameter value
Type of mission m	1
Mission area Area	X region, center of gravity $A_0 = (X.15, X.72)$
Main Location A_1	Red Air Operations Command Post $A_0 = (X.24, X.25)$
Unit for implementation of the mandate E_z	Air force flying corps
Co-implementation of mandated units E_x	Missile units assisting in strikes
target of military operations E_obj	Blue army air defense and anti-missile combat system
Potential threat targets E_threat	Blue aircraft, blue anti-aircraft positions
thread target E_cue	Blue anti-aircraft firing positions, and blue air mission formations
Time interval T	Phase start time T_0, phase end time T_1
Situational awareness cycle Δt	$\Delta t = 30min$

Data Selection. Utilizing data mining tools to select from the database all the data of the blue side's targets held by the rainbow side during the combat phase, including basic data such as battlefield environment, entity attributes, combat formations, etc., as well as dynamic data on the battlefield such as target location, target combat power, target status, etc., with a total of 100,000 pieces of data.

Data Cleansing. Aiming at the possible loss in the data, as well as redundancy, the data are supplemented and smoothed by utilizing a priori knowledge and before-and-after temporal relationships, and the redundant data are deleted. At the same time, the data are divided for each situational awareness cycle, taking the 1st situational awareness cycle as an example, a total of 1000 data are obtained after integration.

Data Transformation. The individual fields of the different categories of targets are normalized using the Eq. 19 with values in the range $[0, 1]$.

$$x_i = \frac{x_i - \min}{\max - \min + \delta} \tag{19}$$

where min and max are the minimum and maximum values of the samples occurring in that dimension, respectively; δ is a very small number that prevents the denominator from being 0. The samples are normalized for each dimension in turn.

3.2 Experimental Procedures

For the commander at the moment of T_1, what has been mastered before the moment of T_0 is the known target, and what is learned at $[T_0, T_1)$ is the newly emerged target. Using the obtained data, the reconnaissance of the red side on the blue side's combat targets at the T_1 moment can be obtained. The set of all the visible targets of the blue side, as shown in Fig. 2. The rectangular box is the target display area and the black dots are the visible blue targets.

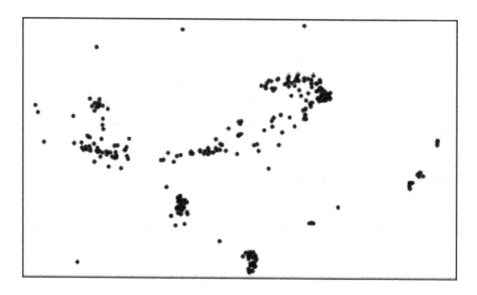

Fig. 2. All Blue visible targets at the moment of T_1 (Color figure online)

At this moment, the red side has a total of 135 blue side targets. The cue targets at that moment were analyzed using the attention model-based perceptual target analysis method, and the salient map of cue targets was obtained, as shown in Fig. 3.

It can be seen that 43 cue targets were clearly distinguished from the others due to their high salience, resulting in a total of 38 high-value targets and a greater reduction in cognitive load.

3.3 Results Analysis

Based on the strike orders given by the commander during the exercise, a list of targets actually struck was obtained, with a total of 83 targets of various types. The 38 high-value targets screened out by the model were compared with them, and 35 of them were ordered to be struck, with a match rate of 92%. The comparison between the actual strike targets and the high-value targets obtained by the model is shown in Fig. 4. The experimental results illustrate that the proposed framework of attention mechanism and the target value assessment model based on attention mechanism are effective, in line with the actual situation of combat, and able to reflect the cognitive process of commanders to a certain extent.

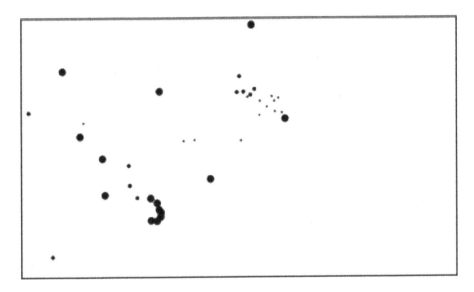

Fig. 3. Significant map of blue cue targets at the moment of T_1 (Color figure online)

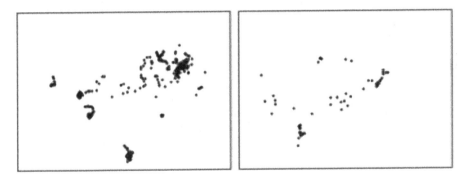

Fig. 4. Comparison of actual targets struck with high-value targets obtained from modeling

4 Conclusion

Joint warfare under informationization conditions is a confrontation between systems. Operational command decision-making is the core content of system confrontation, and target value assessment is the basis of commanders' command decision-making. In the past, most of the target value assessment adopts the isolated, static and linear method, and there is the problem of ignoring the influence of subjective cognition of commanders. Therefore, this paper introduces the attention mechanism to improve the method of target value assessment, and the main contributions are as follows.

Firstly, this paper draws on the research results of the attention mechanism in psychology and situational awareness theory, and uses the attention mechanism for the analytical framework when the target value assessment problem. The analytical framework can reflect the characteristics of complex systems as a whole, dynamics, and confrontation, and combine subjective and objective factors when commanders perceive the battlefield situation, laying a theoretical foundation for the subsequent construction of the model. Secondly, this paper models the bottom-up data-driven attention mechanism and top-down purpose-driven attention mechanism respectively, selects multiple features and calculates the saliency of cue targets, fuses the results of the two models, screens out the noticed cue targets, and further expands them to form the collection of high-value targets. Finally, this paper demonstrates through experimental validation that the proposed theoretical framework and assessment model are effective and can reflect to a certain extent the cognitive process and results of commanders' assessment of target value in command operations.

Machine learning mainly based on deep learning has achieved more than human level in many fields, and deep neural network can be used to study the problem of situational awareness that requires a priori knowledge, and the next step is to explore the use of deep learning and other methods for the construction of target value assessment model on the basis of reserving a large amount of cognitive data.

References

1. Chen, H., Cao, X., Hao, Y., Fan, J., Chen, D., Chen, D.: Multi-target threat assessment in unmanned cluster confrontation scenarios. In: 2020 3rd International Conference on Unmanned Systems (ICUS), pp. 785–790. IEEE (2020)
2. Fan, C., Fu, Q., Song, Y., Lu, Y., Li, W., Zhu, X.: A new model of interval-valued intuitionistic fuzzy weighted operators and their application in dynamic fusion target threat assessment. Entropy **24**(12), 1825 (2022)
3. Guo, M., et al.: Attention mechanisms in computer vision: a survey. Comput. Visual Media **8**(3), 331–368 (2022)
4. Jia, Y., Mingwei, L.: Air target threat assessment based on PCNN neural network. In: Fu, W., Gu, M., Niu, Y. (eds.) Proceedings of 2022 International Conference on Autonomous Unmanned Systems (ICAUS 2022), ICAUS 2022. Lecture Notes in Electrical Engineering, vol. 1010, pp. 3092–3102. Springer, Cham (2022). https://doi.org/10.1007/978-981-99-0479-2_285
5. Li, D., Liu, M., Zhang, S.: Underwater target threat assessment method based on Bayesian network. In: 2021 40th Chinese Control Conference (CCC), pp. 3363–3367. IEEE (2021)
6. Liu, L., Song, X., Zhou, Z.: Aircraft engine remaining useful life estimation via a double attention-based data-driven architecture. Reliab. Eng. Syst. Safety **221**, 108330 (2022)
7. Luo, R., Huang, S., Zhao, Y., Song, Y.: Threat assessment method of low altitude slow small (LSS) targets based on information entropy and AHP. Entropy **23**(10), 1292 (2021)

8. Rosca, E., Tate, W., Bals, L., Huang, F., Ciulli, F.: Coordinating multi-level collective action: how intermediaries and digital governance can help supply chains tackle grand challenges. Int. J. Oper. Prod. Manage. **42**(12), 1937–1968 (2022)

9. Zhang, Z., Huang, Q., Yu, J.: A method for damage assessment of orbital targets based on radar detection information. In: Liang, Q., Wang, W., Liu, X., Na, Z., Zhang, B. (eds.) CSPS 2021. LNEE, vol. 878, pp. 488–497. Springer, Singapore (2022). https://doi.org/10.1007/978-981-19-0390-8_60

10. Zhao, R., Yang, F., Ji, L., Bai, Y.: Dynamic air target threat assessment based on interval-valued intuitionistic fuzzy sets, game theory, and evidential reasoning methodology. Math. Probl. Eng. **2021**, 1–13 (2021)

An Improved Trajectory Planning Method for Unmanned Aerial Vehicles in Complex Environments

Chen Zhang[1], Moduo Yu[2], Wentao Huang[2], Yi Hu[3], Yang Chen[3], and Qinqin Fan[1(✉)]

[1] Logistics Research Center, Shanghai Maritime University, Shanghai 201306, China
forever123fan@163.com
[2] Key Laboratory of Control of Power Transmission and Conversion of the Ministry of Education, Shanghai Jiao Tong University, Shanghai 200240, China
[3] State Grid Shanghai Songjiang Electric Power Supply Company, Shanghai 200240, China

Abstract. Trajectory planning plays a crucial role in the execution of Unmanned aerial vehicle (UAV) missions. However, planning an optimal collision-free trajectory is a challenging task, especially in complex environments. To address the above issue, an enhanced Elliptical tangent graph algorithm (ETG-CPI) based on comprehensive performance indicator is proposed in the present study. In the proposed algorithm, the comprehensive performance indicator, which contains the obstacle avoidance frequency, the yaw angle and the distance from the start point to the candidate waypoint, is used to select promising waypoints. Moreover, the entropy weight method is used to integrate these performance indicators. The experimental results demonstrate that the proposed algorithm outperforms four competitive path planning methods in 26 different environments. Additionally, the results indicate that the proposed comprehensive path evaluation method can help the proposed algorithm find a high-quality path in complex environments.

Keywords: Unmanned aerial vehicle · Path planning · Autonomous flight · Elliptical tangent graph algorithm

1 Introduction

In recent years, unmanned aerial vehicles (UAVs) have attracted much attention due to their simple structure, small size, and the ability to remotely complete tasks. Initially, UAVs are primarily used in the military field for surveillance and reconnaissance missions [1]. However, with the continuous advancement of UAV technology, UAVs have gradually emerged as a prominent area of focus in various fields such as: agriculture [2], logistics [3], civil inspection [4], autonomous exploration [5] and others. The path planning plays a key role in achieving autonomous flight of UAVs, which can guide them to perform tasks in complex environments. Until now, the path planning methods can be roughly classified into three categories: (1) graph-based methods, (2) sampling-based methods, and (3) nature-based metaheuristic algorithms.

Graph-based algorithms are widely used for solve path planning problems. Some common graph-based algorithms include the A* algorithm and the tangent graph algorithm. To address the problems of low search efficiency and excessive redundant nodes in A* algorithm, Zhang et al. [6] proposed a bidirectional search strategy and an improved evaluation function. In order to address the challenge of detecting and handing explosives in unknown environments for robots, Li et al. [7] considered the aforementioned problem as the symmetric traveling salesman problem and solved it using a novel multi-objective positional path planning method. Firstly, a bidirectional dynamic weighted-A star (BD-A*) is used to search the collision-free path for detecting explosives. Subsequently, a learn memory-swap sequence particle swarm optimisation (LMSSPSO) algorithm is utilized to navigate the shortest distance for handing all explosives. Experimental results have proved that the proposed algorithm ensures that the robot efficiently disposes of all explosives. Liu et al. [8] proposed an elliptical tangent graph algorithm to generate collision-free paths in static and dynamic environments. Experimental results show that the proposed algorithm can find high-quality paths in various complex environments.

The sampling-based algorithms are applied to plan feasible paths through generating and searching nodes in space. To alleviate the limitations of the original algorithm, Cong et al. [9] proposed an improved hybrid sampling method to plan paths in complex environments. In the proposed algorithm, a target bias sampling strategy and a random sampling strategy are utilized to improve the sampling efficiency and adapt to environments with concave obstacles. Furthermore, Ye et al. [10] proposed an improved OP-PRM algorithm, which uses the obstacle potential field to effectively prevent sampling points entering the narrow region. Zhang et al. [11] proposed a novel potential function-based sampling heuristic optimal path planning considering the congestion (CCPF-RRT*) to enhance the sampling efficiency, in which the congestion intensity and path length are considered in the sampling process. Moreover, a movement cost function has been introduced to guide the extension of new nodes and update parent nodes for improving the efficiency of path planning. The experimental results demonstrate that the proposed algorithm can effectively enable robots to operate in congested environments.

Besides the above path planning methods, some metaheuristic algorithms have been applied to solve path planning problems. For example, Chan et al. [12] proposed a modified particle swarm optimization for UAV energy-efficiency path planning model. Firstly, the complex urban environment is simplified into a network model based on Voronoi diagrams. Secondly, the Dijkstra's shortest path algorithm is used to obtain the initial feasible path. Finally, path planning is performed using the modified PSO algorithm that considers energy cost. Moreover, Miao et al. [13] proposed an improved multimodal multi-objective differential evolution algorithm hybrid with a simulated annealing algorithm (IMMODE-SA) to solve the multi-robot task allocation problem. Experimental results show that the algorithm can provide feasible and flexible solutions in complex or uncertain environments. Ma et al. [14] proposed an improved NSGA-II based on multi-task optimization (INSGA-II-MTO). In the proposed algorithm, a new population initialization method is used to improve population diversity. Moreover, a multi-task optimization method is introduced that can share knowledge among different tasks. Experiments have proved that this method is an effective and competitive approach to solve the complex multi-target maritime search and rescue problem in bad weather. In Ref.

[15], a new multi-strategy self-adaptive differential sine–cosine algorithm (sdSCA) is proposed for the multi-robot path planning problem. The experimental results show that their method can find better results when compared with other competitors. To address the issue of UAV-assisted Internet of Things (IoT) for data collection, Huang et al. [16] proposed a differential evolution algorithm with a variable population size (DEVIPS). The proposed algorithm is used to optimize the number and locations of stop points for the UAV's deployment in a UAV-assisted IoT data collection system simultaneously. Firstly, the algorithm has been utilized an adaptive population size adjustment strategy to optimize the number of UAV stop points. Secondly, the DE algorithm for optimizing locations of stop points in a two-dimensional space. The experimental results show that the proposed algorithm can get better results when compared with other competitors. Moreover, Lin et al. [17] have proposed a multi-objective trajectory optimization algorithm with a cutting and padding encoding strategy (MTO-CPE) for the single-UAV-assisted mobile edge computing problem, which is used to optimize the population whose individuals have different lengths. Experimental results have proved that the effectiveness of MTO-CPE through comparing with other state-of-the-art algorithms.

To further improve the path planning capability of elliptical tangent algorithm, an improved elliptical tangent algorithm (ETG-CPI) based on comprehensive path indicators is proposed. In the ETG-CPI algorithm, three evaluation indicators are used to assess the quality of candidate waypoints. In the ETG-CPI algorithm, three evaluation indicators are used to assess the quality of candidate waypoints. Moreover, the entropy weight method is employed to integrate all performance indicators into a comprehensive performance indicator for selecting promising waypoints. The experimental results indicate that the proposed algorithm performs better than other compared algorithms in complex environments.

The rest of the paper is structured as follows. Section 2 introduces path planning problems. Section 3 describes the overall Implementation of the proposed algorithm. Section 4 demonstrates the effectiveness of the proposed algorithm compared to the state-of-the-art algorithms. The conclusion are given in Sect. 5.

2 Description of UAV Path Planning Problem

UAVs are widely used in urban environments for cargo transport missions. To ensure safety and efficiency of tasks, UAVs must plan optimal trajectories to avoid colliding with urban buildings. Moreover, this study converts the three-dimensional scene into a two-dimensional scene for path planning. To simplify the problem, the obstacles are uniformly considered as elliptical or circular.

(1) The start point of the UAV's mission is S, the end point of the UAV's mission is E.
(2) Assume that there are N obstacles in the two-dimensional environment. The set of barriers can be expressed as: $Ba = \{Ba_1, Ba_2,..., Ba_i, ..., Ba_N\}$, The center coordinates of the i-th barrier Ba_i are (x_i, y_i), where $i = 1, 2,..., N$. As shown in Fig. 1, a and b are the two semi-axes of the ellipse, r_s is the safe distance that should be maintained between a UAV and a barrier. The barrier Ba_i can be expressed as:

$$\frac{[(x - x_i) \cos \theta + (y - y_i) \sin \theta]^2}{(a + r_s)} + \frac{[(y - y_i) \cos \theta - (x - x_i) \sin \theta]^2}{(b + r_s)} = 1 \quad (1)$$

It should be noted that, if $a = b$, the barrier is a circle; otherwise, the barrier is a elliptical.

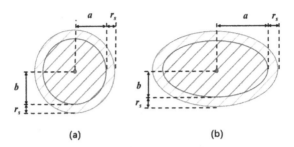

Fig. 1. Elliptical and circular obstacle models

(3) The collision-free path contains $M + 2$ path points (including start point S and end point E), and the set of collision-free path points can be expressed as: $P = \{P_0, P_1, P_2, ..., P_j, ..., P_M\}$, The center coordinate of the j-th path point P_j is (x_j, y_j), where $j = 0, 1, 2, ..., M + 1$. Therefore, the path point P_j should satisfy the following formula:

$$\frac{[(x_j - x_i)\cos \theta + (y_j - y_i)\sin \theta]^2}{(a + r_s)} + \frac{[(y_j - y_i)\cos \theta - (x_j - x_i)\sin \theta]^2}{(b + r_s)} \geq 1 \quad (2)$$

Additionally, the following constraints are also satisfied:

(1) Minimum turn radius [18]: Due to the inertial effect, change of heading requires both time and turning radius. Therefore, it is important to consider the minimum turning radius for safety.
(2) Yaw angle constraint [19]: The constraints on the yaw angle α_h prevent UAVs from deviating, allowing it to navigate closely along the designated guidance line. Therefore, the α_h should satisfy the following constraint.

$$\alpha_{min} \leq \alpha_h \leq \alpha_{max} \quad h = 0, 1, 2, ..., M \quad (3)$$

where, α_{min} and α_{max} are denoted as the minimum yaw angle and the maximum yaw angle, respectively.

3 The Proposed Algorithm

3.1 Elliptical Tangent Graph Algorithm

The elliptical tangent algorithm [8] is a competitive graph-based algorithm. Compared with traditional graph-based path planning algorithms, it can avoid the construction and search for the large-scale graph. The main steps of elliptical tangent graph algorithm are shown in Fig. 2. The algorithm performs real-time path planning for UAVs from the start point to the end point. If an obstacle is encountered in the process of guidance, two tangent lines are derived from the start point O_t and the end point D_t. Meanwhile, the intersection of two tangent lines are the candidate trajectory points P_t and P_t'. It should be noted that the security of trajectory points to be guaranteed through Eq. (2). Next, comprehensive evaluation indicators were used to assess and select the more superior trajectory point.

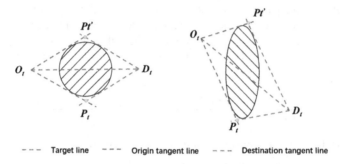

‑ ‑ ‑ Target line   ‑ ‑ ‑ Origin tangent line   ‑ ‑ ‑ Destination tangent line

Fig. 2. Elliptical tangent graph

3.2 Entropy Weight Method

The entropy weight method [20] is a multi-indicator comprehensive evaluation method based on information entropy, which determines the weight by calculating the entropy value of each indicator. The smaller entropy value indicates that the indicator has a greater impact among all performance indicators, and its weight should be greater.

It is supposed that q evaluation indicators are used to assess g paths. The information entropy values of all q indicators are calculated as follows:

$$W_{uv} = \frac{f_{uv}}{\sum\limits_{u=1}^{g} f_{uv}} \tag{4}$$

$$e_v = -\frac{1}{\ln g} \sum\limits_{u=1}^{g} W_{uv} \ln W_{uv} \tag{5}$$

where f_{uv} denotes the value of the v-th performance indicator on the u-th path. The weights of q performance indicators can be calculated as follows:

$$\omega_v = \frac{1 - e_v}{q - \sum e_v} \tag{6}$$

where e_v represents the information entropy value of the v-th indicator.

3.3 Overall Implementation of the ETG-CPI Algorithm

Based on the above description, the comprehensive evaluation indicators used in the elliptical tangent algorithm is employed to evaluate and select candidate sub-paths. The symbols required in the algorithm are detailed in Table 1. Moreover, the main steps of the ETG-CPI algorithm are described as follows:

Table 1. Description of algorithm related symbols

Symbol	Definition
S, E	The start point and the end point
T	The maximum iterations
t	The t-th iteration, $t = 1, 2, \dots, T$
O_t, D_t	The start point and the end point of the t-th iteration sub-path
$P_t, P_t{}'$	The candidate trajectory points of the t-th iteration
PT	The trajectory points set
CT	The sub-path trajectory points set
Ba_t	The barrier to be avoided in the t-th iteration
P	The waypoint in the t-th Iteration

Step 1: Initialize $t = 1$, $PT(end) = S$, $CT(end) = E$. Moreover, the set Ba_t is initialized to an empty set.

Step 2: Take the last points in the set PT and CT as O_t and D_t, respectively. Determine obstacles on the guide-path from O_t to D_t, and record the barrier to be avoided into Ba_t. If Ba_t is non-empty, execute Step3. Otherwise, add Dt to PT and delete DT from CT. Moreover, the algorithm goes to Step 5.

Step 3: Obtain the obstacle closest to O_t, and derive four tangents to this barrier from both O_t and D_t. Meanwhile, the intersection points of the tangents are the candidate trajectory points P_t and $P_t{}'$. Then, use the comprehensive performance indicators to select the promising candidate waypoint. Three performance indicators can be described as follows:

(1) The obstacle avoidance frequency ($q_1(P)$): Because fewer obstacles may reduce the probability of taking a detour, the path quality is assessed by the number of obstacles on the $O_t \rightarrow P \rightarrow D_t$ in the current study.

(2) The distance from the start point to the candidate point ($q_2(P)$): To ensure that the obtained path is the shortest, the distance between O_t and the candidate waypoint is considered.

(3) Yaw angle ($q_3(P)$): The offset angle between $O_t{\rightarrow}P$ and $S{\rightarrow}E$ is used to evaluate the path quality in the present study. A smaller yaw angle means that the degree of path detour is low.

Based on the above introductions, Eqs. (4), (5) and (6) are used to compute the weights of three performance indicators. Therefore, the quality of P_t and P'_t can be computed as follows:

$$F(P_t) = \omega_1 \times q_1(P_t) + \omega_2 \times q_2(P_t) + \omega_3 \times q_3(P_t) \tag{7}$$

$$F(P'_t) = \omega_1 \times q_1(P'_t) + \omega_2 \times q_2(P'_t) + \omega_3 \times q_3(P'_t) \tag{8}$$

Step 4: Judge whether the path O_tP is collision-free. If O_tP is collision-free, add P to the set PT; otherwise, add to the set CT. If CT is empty, execute Step 5 and the iterations T is output; otherwise, the algorithm continue to execute Step 2.

Step 5: Determine obstacles between the j-th waypoint and the $(j + 2)$-th waypoint. If no obstacles, delete the $(j + 1)$-th waypoint. Moreover, the trajectory point selection is finished until the path on the $P_j{\rightarrow}P_{j+2}$ is collision-free.

Step 6: The cubic B-spline curve method [21] is applied to smooth the trajectory points in the PT set.

Step 7: Output the path.

4   Experimental Results and Analyses

To verify the performance of the proposed algorithm, it is compared with other algorithms on E1-E26 environments. Moreover, to improve the performance of the elliptical tangent algorithm, the comprehensive performance indicators are used to evaluate the trajectory points, their effectiveness is verified in the experiments.

4.1   Environment Settings

To obtain reliable results regarding the proposed algorithm, five different types of obstacle layout scenarios are used in experiments. Five scenarios with different characteristics can be seen in Fig. 3. The scenario L1 contains randomly distributed circular obstacles of equal size, while the scenario L2 contains randomly distributed circular obstacles of different sizes. Randomly distributed elliptical obstacles in the scenario L3, which has a noticeable level of sparsity between obstacles. Scenarios L4 and L5 have randomly distributed irregular elliptical and circular obstacles, which increase the complexity of the scenarios. Moreover, the scenario L5 was simulated as the complicated maze scene.

Based on the five different types of scenarios, 26 experimental environments are established. Environments E1-E5 and E11-E15 are set up at 100 km × 100 km map size based on five different obstacle layout scenarios, each of which contains the different start and end points. Also, environments E6-E10 and E16-E26 are set up at 200 km × 200 km map size based on five different obstacle layout scenarios, each of which includes the different start and end points.

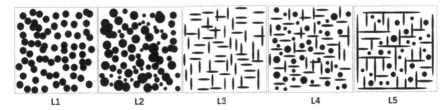

Fig. 3. Scenarios with different obstacle layouts

4.2 Comparison with Other Competitive Algorithms

In this section, the performance of the ETG-CPI is compared with that of other four competitive algorithms, which are A* [22], RRT [23], PRM [24], and SETG-TG [8]. The results of the experimental comparison are presented in Table 2.

It can be observed from Table 1 that the ETG-CPI algorithm outperforms selected competitors on all instances. This is because the proposed algorithm can use a comprehensive performance indicator to select promising candidate waypoints. The A* algorithm is able to obtain feasible paths based on the grid map, but Table 2 shows that that it cannot find optimal paths in complex scenes. Particularly, the A* algorithm is more susceptible to getting trapped in local optima when addressing with scenarios characterized as L5 obstacle layouts. Due to trajectory points generated by random sampling in the traditional PRM and RRT, they cannot guarantee the optimality of paths, which can easily lead the UAV to detour. Compared with the SETG-TG, the ETG-CPI utilizes the comprehensive performance indicators for evaluating and selecting candidate waypoints. Additionally, Table 2 also reveals that the computational time of the proposed algorithm is less than that of other compared algorithms.

Based on the above analyses, we can conclude that the proposed algorithm is an effective and efficient method to plan paths in various complex environments.

4.3 Effectiveness of the Comprehensive Performance Indicator

To further analyze the impact of the comprehensive performance indicator on the proposed algorithm, three performance indicators are deleted from the proposed algorithm, respectively. Namely, the ETG-CPI without $q1$, $q2$, and $q3$ are named as ETG-CPI-1, ETG-CPI-2, and ETG-CPI-3, respectively. Moreover, E1-E26 different environments are used in this experiment. Additionally, the results of all algorithms are analyzed by the Friedman's test [25].

From Fig. 4, it can be observed that three selected performance indicators can influence the performance of the proposed algorithm. Therefore, integrating these three ones into the comprehensive performance indicators is useful for finding high-quality candidate waypoints. Additionally, we can see from Fig. 4 that the yaw angle (i.e., ETG-CPI-3) has the greatest impact on the performance of the ETG-CPI algorithm among three performance indicators.

Table 2. Results of all compared algorithm on different environments

Environment	Layout	Path Length (km)					Run Time (sec)				
		A*	RRT	PRM	SETG-TG	ETG-CPI	A*	RRT	PRM	SETG-TG	ETG-CPI
E1	L1	196.0	175.3	171.7	171.6	**143.3**	0.576	0.077	0.081	0.021	0.017
E2	L2	200.3	173.9	175.2	172.9	**169.8**	0.621	0.089	0.062	0.032	0.021
E3	L3	198.2	175.2	176.7	**172.3**	**172.3**	0.541	0.079	0.064	0.027	0.026
E4	L4	189.2	172.8	175.3	169.2	**164.3**	0.456	0.065	0.065	0.021	0.013
E5	L5	186.3	178.3	170.3	162.1	**159.8**	0.547	0.068	0.055	**0.018**	**0.018**
E6	L1	317.3	304.2	312.5	303.2	**288.6**	0.443	0.057	0.077	0.024	0.022
E7	L2	304.6	305.3	322.5	301.1	**293.9**	0.216	0.023	0.027	0.013	0.012
E8	L3	358.3	340.3	345.2	339.9	**335.0**	0.457	0.022	0.019	0.017	0.015
E9	L4	412.3	336.7	330.2	321.7	**301.3**	0.981	0.048	**0.028**	**0.030**	**0.032**
E10	L5	522.6	498.7	514.3	494.8	**345.8**	0.812	0.095	0.105	0.023	0.022
E11	L1	172.3	153.2	150.2	145.4	**112.4**	0.495	0.044	0.042	0.024	0.021
E12	L2	147.3	145.8	145.2	**143.1**	**143.1**	0.231	0.021	0.017	**0.010**	**0.010**
E13	L3	159.3	145.3	146.2	144.9	**126.9**	0.332	0.025	0.035	0.025	0.022
E14	L4	200.6	186.3	192.3	181.3	**157.5**	0.223	0.056	0.066	0.027	0.023
E15	L5	288.6	275.3	251.7	250.4	**209.3**	0.378	0.078	0.042	0.025	0.024
E16	L1	291.2	278.6	298.3	270.5	**245.2**	0.451	0.074	0.089	0.037	0.012
E17	L2	316.5	292.7	296.3	290.4	**270.2**	0.512	0.041	0.066	0.045	0.022
E18	L3	223.3	208.7	210.5	207.9	**206.7**	0.365	**0.025**	0.033	**0.026**	0.033
E19	L4	325.2	332.4	321.6	318.9	**272.7**	0.612	0.033	0.025	0.021	**0.020**
E20	L5	390.1	372.2	369.3	353.7	**346.1**	0.891	0.057	0.026	0.023	**0.023**
E21	L5	388.1	366.2	367.3	358.7	**348.2**	0.992	0.065	0.056	0.021	**0.016**
E22	L5	492.3	463.5	462.6	462.6	**217.9**	0.856	0.078	0.074	0.035	**0.021**
E23	L5	352.6	340.2	342.3	337.1	**325.6**	0.768	0.072	0.075	0.019	**0.016**
E24	L1	322.3	311.4	305.2	293.1	**290.5**	0.664	0.056	0.045	0.022	**0.022**
E25	L3	258.3	230.2	235.4	223.9	**196.7**	0.582	0.035	0.032	0.016	**0.009**
E26	L4	351.4	327.8	325.2	318.6	**272.6**	0.685	0.041	0.038	0.021	**0.018**

Based on the above experimental analyses, it can be concluded that the comprehensive performance indicators can help the proposed algorithm find a high-quality trajectory in complex environments.

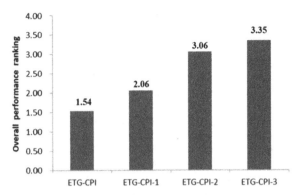

Fig. 4. Overall performance ranking obtained by Friedman's test on E1-E26 environments

5   Conclusions

In this study, an enhanced elliptical tangent graph algorithm (ETG-CPI) based on comprehensive performance indicator is proposed. In the ETG-CPI, we integrate three useful performance indicators to select promising trajectory points. To verify the performance of the proposed algorithm, the ETG-CPI is compared with four state-of-the-art algorithms on 26 complex environments. The experimental results demonstrate that the ETG-CPI performs better than other algorithms in complex environments. Moreover, the effectiveness of three performance indicators is demonstrated in the experiment, the results indicates that the comprehensive performance indicator can improve the performance of the proposed algorithm.

References

1. Wang, Y., Bai, P., Liang, X., Wang, W., Zhang, J., Fu, Q.: Reconnaissance mission conducted by UAV swarms based on distributed PSO path planning algorithms. IEEE Access **7**, 105086–105099 (2019)
2. Srivastava, A., Prakash, J.: Techniques, answers, and real-world UAV implementations for precision farming. Wirel. Pers. Commun. **131**, 2715–2746 (2023)
3. Li, S., Zhang, H., Li, Z., Liu, H.: An air route network planning model of logistics UAV terminal distribution in urban low altitude airspace. Sustainability **13**, 14 (2021)
4. Matlekovic, L., Schneider-Kamp, P.: Constraint programming approach to coverage-path planning for autonomous multi-UAV infrastructure inspection. Drones-Basel **7**, 25 (2023)
5. Batinovic, A., Ivanovic, A., Petrovic, T., Bogdan, S.: A shadowcasting-based next-best-view planner for autonomous 3D exploration. IEEE Robot. Autom. Lett. **7**, 2969–2976 (2022)
6. Zhang, H., Tao, Y., Zhu, W.: Global path planning of unmanned surface vehicle based on improved A-Star algorithm. Sensors **23**, 18 (2023)
7. Li, M., Qiao, L., Jiang, J.: A multigoal path-planning approach for explosive ordnance disposal robots based on bidirectional dynamic weighted-A* and learn memory-swap sequence PSO algorithm. Symmetry-Basel **15**, 37 (2023)
8. Liu, H., Li, X., Fan, M., Wu, G., Pedrycz, W., Suganthan, P.N.: An autonomous path planning method for unmanned aerial vehicle based on a tangent intersection and target guidance strategy. IEEE Trans. Intell. Transp. Syst. **23**, 3061–3073 (2022)

9. Cong, J., Hu, J., Wang, Y., He, Z., Han, L., Su, M.: FF-RRT*: a sampling-improved path planning algorithm for mobile robots against concave cavity obstacle. Complex Intell. Syst. **9**, 7249–7267 (2023)

10. Ye, L., Chen, J., Zhou, Y.: Real-time path planning for robot using OP-PRM in complex dynamic environment. Front. Neurorobot. **16**, 910859 (2022)

11. Liang, Y., Zhao, H.: CCPF-RRT*: an improved path planning algorithm with consideration of congestion. Expert Syst. Appl. **228**, 120403 (2023)

12. Chan, Y., Ng, K., Lee, C., Hsu, L., Keung, K.: Wind dynamic and energy-efficiency path planning for unmanned aerial vehicles in the lower-level airspace and urban air mobility context. Sustain. Energy Technol. Assess. **57**, 103202 (2023)

13. Miao, Z., Huang, W., Jiang, Q., Fan, Q.: A novel multimodal multi-objective optimization algorithm for multi-robot task allocation. Trans. Inst. Meas. Control **12** (2023)

14. Ma, Y., Li, B., Huang, W., Fan, Q.: An Improved NSGA-II based on multi-task optimization for Multi-UAV maritime search and rescue under severe weather. J. Mar. Sci. Eng. **11**(4), 781 (2023)

15. Akay, R., Yildirim, M.Y.: Multi-strategy and self-adaptive differential sine-cosine algorithm for multi-robot path planning. Expert Syst. Appl. **232**, 19 (2023)

16. Huang, P.-Q., Wang, Y., Wang, K., Yang, K.: Differential evolution with a variable population size for deployment optimization in a UAV-assisted IoT data collection system. IEEE Trans. Emerg. Top. Comput. Intell. **4**, 324–335 (2019)

17. Lin, J., Pan, L.: Multiobjective trajectory optimization with a cutting and padding encoding strategy for single-UAV-assisted mobile edge computing system. Swarm Evol. Comput. **75**, 101163 (2022)

18. Wu, X., Bai, W., Xie, Y., Sun, X., Deng, C., Cui, H.: A hybrid algorithm of particle swarm optimization, metropolis criterion and RTS smoother for path planning of UAVs. Appl. Soft Comput. **73**, 735–747 (2018)

19. Zheng, A., Li, B., Zheng, M., Zhong, H.: Multi-objective UAV trajectory planning in uncertain environment. Symmetry-Basel **13**, 28 (2021)

20. Huang, D., Han, M.: An optimization route selection method of urban oversize cargo transportation. Appl. Sci. **11**(5), 2213 (2021)

21. Berglund, T., Brodnik, A., Jonsson, H., Staffanson, M., Söderkvist, I.: Planning smooth and obstacle-avoiding b-spline paths for autonomous mining vehicles. IEEE Trans. Autom. Sci. Eng. **7**, 167–172 (2010)

22. Hart, P.E., Nilsson, N.J., Raphael, B.: A formal basis for the heuristic determination of minimum cost paths. IEEE Trans. Syst. Sci. Cybern. **4**, 100–107 (1968)

23. Devaurs, D., Siméon, T., Cortés, J.: Optimal path planning in complex cost spaces with sampling-based algorithms. IEEE Trans. Autom. Sci. Eng. **13**, 415–424 (2015)

24. Kavraki, L.E., Svestka, P., Latombe, J.-C., Overmars, M.H.: Probabilistic roadmaps for path planning in high-dimensional configuration spaces. IEEE Trans. Robot. Autom. **12**, 566–580 (1996)

25. Friedman, M.: The use of ranks to avoid the assumption of normality implicit in the analysis of variance. J. Am. Stat. Assoc. **32**, 675–701 (1937)

UUV Dynamic Path Planning Algorithm Based on A-Star and Dynamic Window

Fengyun Li[1](✉), Lihua Wu[1], Leixin Shi[1], Xu Cao[1], Xiangpeng Zhang[1], and Guanglong Zeng[2]

[1] Wuhan Second Ship Design and Research Institute, Wuhan 430025, China
lfynevermore@126.com
[2] Wuhan University of Technology, Wuhan 430070, China

Abstract. In order to enable the unmanned ship to have the ability of autonomous path planning in the complex marine environment, which can avoid obstacles and reach the destination accurately in the unknown environment. In this paper, a new unmanned ship dynamic path planning algorithm is proposed by combining the global optimal characteristic of A* algorithm with the real-time performance of dynamic window method. Simulation results in static and dynamic environments show that the algorithm can complete the task of path planning and obstacle avoidance and reach the destination, which shows the rationality and effectiveness of the algorithm. The path planning algorithm based on the combination of A* algorithm and dynamic window method performs well in the simulation experiment and meets the needs of the operation and use of the unmanned ship in the static or dynamic environment. This algorithm can be carried in the actual environment to realize the autonomous navigation of the unmanned ship.

Keywords: Unmanned Ship · Path Planning · A* Algorithm · Dynamic Window Approach

1 Introduction

Path planning is one of the key technologies to study mobile agents. Path planning means that the agent finds an optimal or sub-optimal path from the starting point to the target position in a complex environment containing obstacles under certain constraints (such as the shortest path, shortest time or shortest motion time, etc.) [1]. UUV path planning generally follows two steps: abstract modeling of the UUV environment; Then the path is calculated using the path planning algorithm [2–4].

Inspired by the concept of "field" in physics, Khatib et al. [5] proposed the artificial potential field method in 1986: the target point provides gravity to the agent, guiding the agent to move toward the target point, and the obstacle generates repulsion to the agent, guiding the agent to avoid the obstacle. In 1991, Dorigo M et al. [6] proposed ant colony algorithm in order to solve the travel agency problem. In view of the poor dynamic characteristics of traditional ant colony algorithm in dynamic path planning, Zhao Feng [7] combined with the idea of greedy algorithm, improved and optimized it to

L. Pan et al. (Eds.): BIC-TA 2023, CCIS 2062, pp. 159–170, 2024.
https://doi.org/10.1007/978-981-97-2275-4_13

meet the requirements of path planning in different dynamic environments. Chen Huiwei et al. [8] applied ant colony algorithm to UUV path planning by improving the heuristic function and updating the pheromone. Zadeh et al. [9] proposed the dynamic window method in 1997 and applied it to the local obstacle avoidance of robots. By changing the gravitational potential field and repulsive potential field models, Chen Huiwei et al. [10] applied the artificial potential field method to the path navigation of UUV and achieved good results.

Another commonly used path planning algorithm is A* algorithm [11]. A* algorithm has good performance in global planning, but poor adaptability in dynamic environment. In this paper, A new UUV dynamic path planning algorithm is proposed by combining the global optimal characteristics of A* algorithm with the real-time performance of dynamic window method. The heuristic function of A* algorithm is applied to dynamic window method and the weights of each index are adjusted.

2 Algorithm Design

The dynamic window method samples the linear and angular velocities of the intelligent agent in the velocity space (v, w), collecting multiple sets of motion data in the velocity space. Over a specific time period, the sampled data is used to simulate the motion trajectory of the intelligent agent. Finally, an evaluation function is applied to score the simulated motion trajectories, and the velocity associated with the highest score is used to drive the intelligent agent. The dynamic window method mainly includes the following three aspects:

(1) Motion Model

In the dynamic window method, simulating the motion model of the Unmanned Underwater Vehicle (UUV) is a prerequisite for predicting its motion trajectory. It is assumed that the UUV's velocity remains constant over a time interval, meaning zero acceleration within time Δt. Initially, only the motion along the x-axis of the body coordinate system is considered, with the velocity within the time interval Δt denoted as (v, w), as shown in Fig. 1. When predicting the motion trajectory of the UUV through velocity sampling, since the sampling time interval Δt is small, the trajectory between two adjacent points is approximated as a straight line. The UUV's motion distance along the x-axis in the body coordinate system is $v\Delta t$. Projecting this distance onto the x-axis and y-axis of the Earth coordinate system, the displacement (Δx and Δy) of the UUV's movement in the Earth coordinate system at time $t + 1$ relative to time t can be obtained:

$$\Delta x = v\Delta t \cdot \cos(\theta_t) \tag{1}$$

$$\Delta y = v\Delta t \cdot \sin(\theta_t) \tag{2}$$

Therefore, at time $t + 1$, the position of the UUV in the Earth coordinate system is given by Eqs. (3) and (4):

$$x_{t+1} = x_t + v_x\Delta t \cdot cos(\theta_t) \tag{3}$$

$$y_{t+1} = y_t + v_x \Delta t \cdot sin(\theta_t) \tag{4}$$

$$\theta_{t+1} = \theta_t + w\Delta t \tag{5}$$

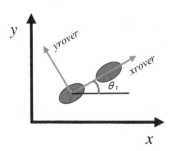

Fig. 1. Schematic Diagram of the UUV Coordinate System

Similarly, when the UUV is moving along the y-axis of the body coordinate system, the projection of the y-axis movement distance onto the Earth coordinate system is given by Eqs. (7) and (6):

$$\Delta x = v_y \Delta t \cdot cos\left(\theta_t + \frac{\pi}{2}\right) = -v_y \Delta t \cdot sin(\theta_t) \tag{6}$$

$$\Delta y = v_y \Delta t \cdot sin\left(\theta_t + \frac{\pi}{2}\right) = v_y \Delta t \cdot cos(\theta_t) \tag{7}$$

When the Unmanned Underwater Vehicle (UUV) is engaged in omnidirectional motion, having velocities along both the x and y axes of the body coordinate system, the motion model for the dynamic window method can be derived from Eqs. (3), (4), (6), and (7):

$$x_{t+1} = x_t + v_x \Delta t \cdot cos(\theta_t) - v_y \Delta t \cdot sin(\theta_t) \tag{8}$$

$$y_{t+1} = y_t + v_x \Delta t \cdot sin(\theta_t) - v_y \Delta t \cdot cos(\theta_t) \tag{9}$$

(2) Velocity Sampling

Velocity sampling for the Unmanned Underwater Vehicle (UUV) is one of the core components of the dynamic window method. After modeling the UUV motion model, the UUV's motion trajectory can be deduced from sampled velocities, predicting the UUV's next navigational position. Velocity sampling is performed along the time dimension to sample the UUV's linear and angular velocities, corresponding to its longitudinal and lateral speeds, respectively. Multiple sets of velocity combinations are obtained through velocity sampling, forming a dynamic window of velocities.

The evaluation function of the dynamic window method is then applied to assess the sampled velocities, determining the optimal velocity that is used to predict the

UUV's navigational trajectory. Velocity sampling is primarily influenced by factors such as maximum and minimum velocities, maximum acceleration, and safety avoidance considerations.

1) Maximum Velocity and Minimum Velocity

In the velocity space (v, w), whether the Unmanned Underwater Vehicle (UUV) is moving with linear velocity v or angular velocity w, there are constraints on the maximum and minimum values influenced by the actual environment and the UUV's own capabilities. Let V_s represent the set of maximum and minimum velocities for the UUV, then:

$$V_s = \{ (v, w) | v \in (v_{min}, v_{max}), w \in (w_{min}, w_{max}) \} \tag{10}$$

Here, v_{min}, v_{max} are the maximum and minimum linear velocities, respectively, and w_{max}, w_{min} are maximum and minimum angular velocities, respectively.

2) Maximum Acceleration

In the dynamic window method, there is a certain limit on the maximum acceleration due to constraints such as batteries and motors. Within the period of simulating the UUV's motion trajectory, there exists a dynamic window representing the set of velocities that the UUV can realistically achieve in the next cycle. Let. V_d be the set of attainable velocities for the UUV in the next window:

$$V_d = \{ (v, w) | v \in (v_a - \dot{v}t, v_a + \dot{v}t), w \in (w_a - \dot{w}t, w_a + \dot{w}t) \} \tag{11}$$

In the equation, v_a, w_a represent the linear velocity and angular velocity of the UUV, $\dot{v}$ represents the maximum acceleration of the UUV's linear velocity, $\dot{w}$ represents the maximum acceleration of the UUV's angular velocity, and V_a represents the sampling interval time.

3) Safety Obstacle Avoidance

When the Unmanned Underwater Vehicle (UUV) is close to obstacles, both its velocity and acceleration are subject to certain limitations. To ensure obstacle avoidance and bring the UUV to a speed of 0 just before encountering an obstacle, there needs to be a restriction on the maximum deceleration. Let the velocity of the UUV be represented as (v, w). Under the constraint of maximum acceleration, V_a denotes the safe velocity space for the UUV:

$$V_a = \left\{ (v, w) | v \le \sqrt{2\text{dist}(v, w)\dot{v}}, w \in \sqrt{2\text{dist}(v, w)\dot{w}} \right\} \tag{12}$$

In the equation, $\text{dist}(v, w)$ represents the distance between the UUV's motion trajectory and obstacles. $\dot{v}$ and $\dot{w}$ represent the deceleration and deceleration angular velocity of the UUV, respectively.

When sampling velocities for the Unmanned Underwater Vehicle (UUV), it is necessary to simultaneously satisfy the three conditions mentioned above. Let V_r be the velocity space for the UUV, then:

$$V_r = V_s \cap V_d \cap V_a \tag{13}$$

(3) Evaluation Function

After sampling linear and angular velocities for the Unmanned Underwater Vehicle (UUV), several sets of sampled data are obtained. An evaluation function is then employed to assess the sampled data and calculate the optimal solution. For the dynamic window method, the evaluation function needs to consider local objectives, cost maps, and global paths. The main evaluation functions include:

1) Heading Angle Evaluation Function: This function continuously adjusts the UUV's navigation trajectory based on the difference θ between the current heading angle and the target heading angle, represented by $180° - \theta$, as shown in Fig. 2.
2) Clearance Function: Implements obstacle avoidance for the UUV, ensuring that the UUV stays as far away from obstacles as possible.
3) Velocity Evaluation Function: In situations where the UUV is surrounded by obstacles or is in close proximity to obstacles, the UUV autonomously reduces its speed. The role of this velocity evaluation function is to maintain high-speed travel for the UUV while navigating towards the target point.
4) Goal Cost Function: During the UUV's navigation, the trajectory is divided into multiple segments, and for each calculation, one segment is selected. The function calculates the deviation between the current position and the expected endpoint, ensuring a short distance to the endpoint of the local path.
5) Oscillation Evaluation Function: Ensures that the UUV does not frequently move forward and backward or left and right when navigating in complex environments.
6) Path Evaluation Function: Guides the UUV to follow the expected trajectory, reducing the lateral distance deviation between the UUV and local target points.

The evaluation function $G(v, w)$ for the dynamic window method in this paper consists of three parts: the heading angle evaluation function, clearance evaluation function, and velocity evaluation function, as shown in Eq. (14):

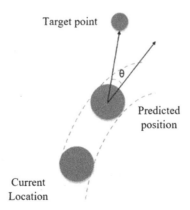

Fig. 2. Heading Angle Diagram

$$G(v, w) = \sigma(\alpha \cdot \text{heading}(v, w) + \beta \cdot \text{dist}(v, w) + \gamma \cdot \text{vel}(v, w)) \tag{14}$$

In the equation, $\text{heading}(v, w)$ represents the heading angle evaluation function, $\text{dist}(v, w)$ represents the clearance evaluation function, and $\text{vel}(v, w)$ represents the velocity evaluation function. α, β and γ are the weight coefficients for the heading angle evaluation function, clearance evaluation function, and velocity evaluation function, respectively. The purpose of these coefficients is to smooth the UUV's motion trajectory. Since the weighting of the heading angle function, clearance evaluation function, and velocity evaluation function may vary in different environments, it is necessary to normalize these three functions:

$$normalized\_heading(i) = \frac{heading(i)}{\sum_1^n heading(i)} \tag{15}$$

$$normalized\_dist(i) = \frac{dist(i)}{\sum_1^n dist(i)} \tag{16}$$

$$normalized\_vel(i) = \frac{vel(i)}{\sum_1^n vel(i)} \tag{17}$$

Under the constraints of the evaluation function, the Unmanned Underwater Vehicle (UUV) can navigate toward the endpoint as quickly and smoothly as possible, avoiding obstacles.

However, when the dynamic window method is used for dynamic path planning, differences in weights of various evaluation functions and navigation environments may lead to situations where the UUV deviates from the endpoint or cannot reach the endpoint.

To address this, the paper combines the dynamic window method with the A* algorithm to accomplish the UUV's path planning task. In the A* algorithm, the heuristic function remains constant after each iteration in the path planning process. In contrast, the dynamic window method divides the navigation path into numerous segments, and by sampling the UUV's linear and angular velocities and iterating the calculations based on known static and obstacle positions, the evaluation function is a constant value at each iteration. Therefore, the heuristic function from the A* algorithm can be added to the evaluation function of the dynamic window method. The heuristic function is given by Eq. (18):

$$h(pre) = \rho(pre, goal) \tag{18}$$

In the equation, pre represents the predicted position, and $goal$ represents the goal point. The heuristic function for the algorithm combining A* and the dynamic window method uses the Euclidean distance calculation method. The evaluation function is given by Eq. (19):

$$h(pre) = D \cdot \text{sqrt}\left((pre.x - goal.x)^2 + (pre.y - goal.y)^2\right) \tag{19}$$

In the equation, D represents the Euclidean coefficient.

By combining the evaluation function of the A* algorithm with the dynamic window method, the resulting evaluation function is given by Eq. (20):

$$G(v, w) = \sigma(\alpha \cdot heading(v, w) + \beta \cdot dist(v, w) + \gamma \cdot vel(v, w) + \lambda \cdot h(v, w)) \quad (20)$$

In the equation, λ represents the weight coefficient for the heuristic function.

3 Simulation

In this section, path planning and obstacle avoidance simulation experiments are conducted in a dynamic environment using the dynamic window approach. Subsequently, a simulation experiment is performed using the algorithm proposed in this paper, which combines the dynamic window approach with the A* algorithm. By comparing the simulation results obtained from these two methods, the rationality and effectiveness of the newly proposed algorithm are validated.

3.1 Simulation Experiment of Dynamic Window Approach

To better simulate the actual navigation environment of Unmanned Underwater Vehicles (UUVs), dynamic obstacle points are introduced in the simulation of this section. These dynamic obstacle points simulate the movements of other vessels in the real environment, while static obstacle points simulate islands on the surface of lakes or seas. An 11*11 grid map model is established, with the starting point set at A(0,0) and the destination point at D(10,10). Solid circles (○) represent static obstacle points, while dashed circles with asterisks (*○) represent randomly generated dynamic obstacle points. The radius of the obstacle points is uniformly set to 0.4 m. The dynamic obstacles move randomly at a speed of 0.1 m per second. During each iteration of the dynamic window approach, when calculating the expected values, the positions of dynamic obstacle points are known.

Assuming the Unmanned Underwater Vehicle (UUV) has a maximum linear velocity v of 4.0 m/s, a maximum angular velocity w of 5.0 rad/s, a maximum linear acceleration $\dot{v}$ of 2.0 m/s^2, and a maximum angular acceleration $\dot{w}$ of 4.0 rad/s^2. The weight coefficients (α, β, and γ) for the heading function, clearance function, and velocity function are set as 0.1, 0.5, and 2, respectively. The iteration is performed for 10,000 cycles. The UUV's navigation process is illustrated in Fig. 3.

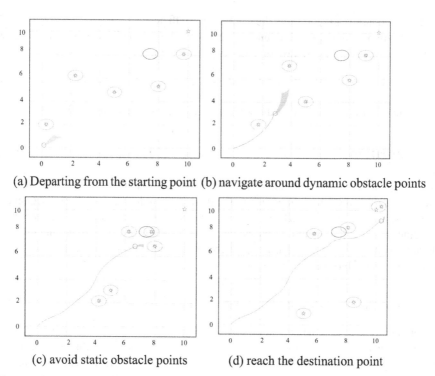

(a) Departing from the starting point (b) navigate around dynamic obstacle points

(c) avoid static obstacle points   (d) reach the destination point

Fig. 3. Diagram illustrating the dynamic obstacle avoidance and reaching the target point using the algorithm combining the dynamic window approach and A* algorithm.

From Fig. 3, it can be observed that in a dynamic navigation environment, utilizing the dynamic window approach for dynamic path planning leads to the Unmanned Underwater Vehicle (UUV) deviating from the target point and failing to reach the destination. In the simulated environment and with the given UUV navigation parameters, extensive simulation experiments indicate that, due to the absence of a target point evaluation function in the evaluation function of the dynamic window approach, the UUV is unable to reach the target point. In certain dynamic environments, the dynamic window approach proves inadequate for both path planning and obstacle avoidance.

3.2   Simulation Experiment of the Algorithm Combining A* Algorithm and Dynamic Window Approach

In a dynamic environment, the Unmanned Underwater Vehicle (UUV) utilizes the algorithm combining the dynamic window approach and A* algorithm for path planning and obstacle avoidance. The UUV navigation parameters, including simulation speed, acceleration, angular velocity, angular acceleration, and the weight coefficients of the evaluation function, remain the same as mentioned earlier. An additional heuristic function with a weight coefficient λ of 1.0 is introduced.

A simulation environment model is established with the same information for the starting point, endpoint, obstacle points, and other environmental details as mentioned before. Under identical external conditions and UUV parameters, the algorithm, which combines the dynamic window approach and A* algorithm, is employed for UUV path planning and obstacle avoidance in a dynamic environment. The trajectory of the UUV during the navigation process is illustrated in Fig. 4.

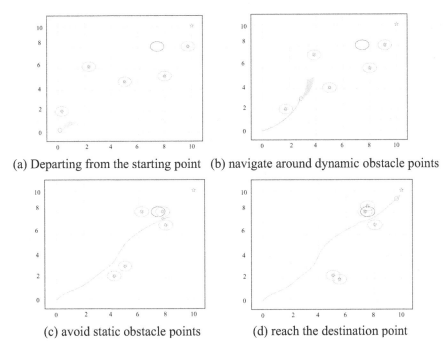

(a) Departing from the starting point   (b) navigate around dynamic obstacle points

(c) avoid static obstacle points   (d) reach the destination point

Fig. 4. Schematic Diagram of Dynamic Obstacle Avoidance and Arrival at the Destination Point Using the Algorithm Combining Dynamic Window Approach and A* Algorithm

From Fig. 4, it is evident that the Unmanned Underwater Vehicle (UUV) successfully reaches the destination point, and the path is relatively smooth. Through multiple experiments under the mentioned conditions, the UUV consistently manages to complete path planning, obstacle avoidance, and reach the target point. This demonstrates that the Autonomous Surface Vehicle (ASV) using the fused algorithm can also accomplish path planning and obstacle avoidance in a dynamic environment. Velocity and angular velocity comparison diagrams during the navigation process of the algorithm combining the dynamic window approach and the A* algorithm are shown in Fig. 5 and Fig. 6.

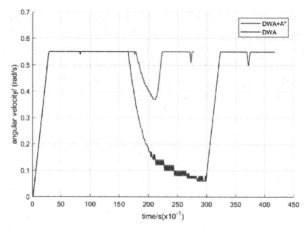

Fig. 5. Comparison graph of angular velocities between the dynamic window approach and the algorithm combining the dynamic window approach and A*.

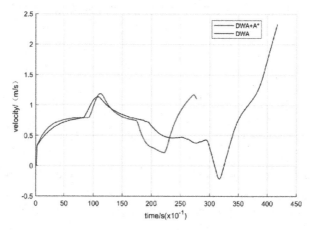

Fig. 6. Comparison graph of velocities between the dynamic window approach and the algorithm combining the dynamic window approach and A*.

From Fig. 5 and Fig. 6, it is evident that the angular velocity and velocity of the Unmanned Underwater Vehicle (UUV) decrease when approaching obstacles, indicating obstacle avoidance. With the addition of a heuristic function, the UUV can reach the destination smoothly, reduce the time spent, exhibit smoother velocity changes, and maintain stable navigation speed within a certain range, resulting in improved stability during navigation.

To further demonstrate the effectiveness of the algorithm combining the dynamic window approach and A* algorithm, the UUV is placed in a more complex environment. The algorithm is used for path planning and obstacle avoidance in the presence of multiple static and dynamic obstacle points, with simulation parameters and weight coefficients for the evaluation function being the same as before. The dynamic obstacle avoidance trajectory of the UUV is shown in Fig. 7.

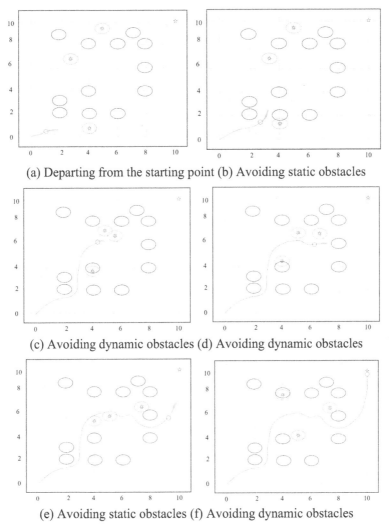

(a) Departing from the starting point (b) Avoiding static obstacles

(c) Avoiding dynamic obstacles (d) Avoiding dynamic obstacles

(e) Avoiding static obstacles (f) Avoiding dynamic obstacles

Fig. 7. Simulation Trajectory of UUV Combining Dynamic Window Approach and A* Algorithm.

From Fig. 7, it can be observed that the algorithm combining the dynamic window approach and A* algorithm successfully completes path planning, obstacle avoidance, and navigation to the destination point even in a complex environment. The navigation path is also relatively smooth, without jagged or stair-step patterns, validating the usability of the algorithm in dynamic environments.

4 Conclusion

In the real-world navigation environment for Unmanned Underwater Vehicles (UUVs), there exist both dynamic and static obstacles. This chapter focuses on the research of UUV path planning and obstacle avoidance in environments that combine both dynamic and static obstacles. Firstly, addressing the limitation of the dynamic window approach in handling real-time dynamic obstacle avoidance and path planning in complex environments, a new algorithm is proposed. This algorithm leverages the dynamic real-time planning characteristics of the dynamic window approach and the global optimality features of the A* algorithm for path planning. Subsequently, simulation comparative experiments are conducted in dynamic environments. Under the same experimental conditions, the new algorithm successfully completes the tasks of path planning and obstacle avoidance, reaching the destination point. This demonstrates the rationality and effectiveness of the proposed algorithm.

References

1. Wu, H., Tian, G.H., Li, Y., Zhou, F.Y., Duan, P.: Spatial semantic hybrid map building and application of mobile service robot. Robot. Auton. Syst. **62**(6), 923–941 (2014)
2. Wu, N.L., Wang, X.Y., Ge, T., Wu, C., Yang, R.: Parametric identification and structure searching for underwater vehicle model using symbolic regression. J. Mar. Sci. Technol. **22**, 51–60 (2017)
3. Manzanilla, A., Reyes, S., Garcia, M., Mercado, D., Lozano, R.: Autonomous navigation for unmanned underwater vehicles: real-time experiments using computer vision. IEEE Robot. Autom. Lett. **4**(2), 1351–1356 (2019)
4. Melo, J., Matos, A.: Survey on advances on terrain based navigation for autonomous underwater vehicles. Ocean Eng. **139**, 250–264 (2017)
5. Khatib, O.: Real-time obstacle avoidance for manipulators and mobile robots. In: Proceedings. 1985 IEEE International Conference on Robotics and Automation, vol. 2, pp. 500–505. IEEE (1985)
6. Dorigo, M., Di Caro, G., Gambardella, L.M.: Ant algorithms for discrete optimization. Artif. Life **5**(2), 137–172 (1999)
7. Zhao, F.: Dynamic path planning based on ant colony algorithm and its simulation application in formation. Master Thesis, Kunming University of Science and Technology (2017)
8. Chen, H.W., Liu, P.X., Liu, S.M.: Research on UUV path planning method based on improved ant colony algorithm. Eng. Technol. Dev. **1**(8), 118–120 (2019)
9. Zadeh, S.M., Powers, D.M., Sammut, K.: An autonomous reactive architecture for efficient AUV mission time management in realistic dynamic ocean environment. Robot. Auton. Syst. **87**, 81–103 (2017)
10. Chen, H.W., Chen, Y.J., Feng, F.: Research status analysis of UUV track planning based on artificial potential field method. Sci. Technol. Innov. **17**, 23–25 (2020)
11. Liu, H., Bao, Y.L.: Application of A* algorithm in vector map optimal path search. Comput. Simul. **4**, 253–257 (2008)

Cuckoo Search Algorithm with Balanced Learning to Solve Logistics Distribution Problem

Juan Li[1] and Han-xia Liu[2(✉)]

[1] School of Information Engineering, Wuhan Business University, Wuhan 430058, China
[2] College of International Business and Economics, Wuhan Textile University, Wuhan 430202, China
178172060@qq.com

Abstract. Cuckoo search (CS) has been successfully applied to solve various optimization problems. Despite its simplicity and efficiency, the CS is easy to suffer from the premature convergence and fall into local optimum. In this study, a differential CS extension with balanced learning namely DFCS is proposed. In DFCS, two sets including the better fitness set (FSL) and the better diversity set (DSL) are produced in the iterative process. Two excellent individuals are selected from two sets to participate, which improved the search ability by learning their beneficial behaviors in search process. The performance of DFCS algorithm is evaluated on the logistics distribution problem. The results show that DFCS algorithm has stronger competitiveness in solving logistics distribution problem than CS algorithm.

Keywords: Cuckoo search algorithm · Balanced-Learning · Optimization algorithm · Diversity

1 Introduction

Lots of real-world problems can be converted to optimization problems, such as multi-robot path planning, economic load dispatch, wireless sensor networks, and image segmentation. Optimization algorithms [1] select the best alternative in a given objective function based on nature-inspired ideas. In general, the optimization algorithms can be either a heuristic or a metaheuristic approach.

Metaheuristic algorithms [2] have proved to be a viable solution to alternative optimization problems. Some of the well-known methods in this arena are genetic algorithms (GAs) [3, 4], particle swarm optimization (PSO) [5, 6], differential evolution (DE) [7, 8], monarch butterfly optimization (MBO) [9, 10], earthworm optimization algorithm (EWA) [11], chicken swarm optimization (CSO) [12], krill herd (KH) [13, 14], firefly algorithm (FA) [15, 16], simulated annealing (SA) [17], intelligent water drop (IWD) [18], moth search (MS) [19], monkey algorithm (MA) [20], evolutionary strategy (ES) [21], free search (FS) [22], and cuckoo search (CS) [23–27].

Yang and Deb [23] applied successfully the CS algorithm to diverse fields. A number of CS variants can be generally divided into five categories to improve the performance of

L. Pan et al. (Eds.): BIC-TA 2023, CCIS 2062, pp. 171–181, 2024.
https://doi.org/10.1007/978-981-97-2275-4_14

the CS algorithm. Knowledge learning strategy is adopted to enhanced the ability of the cuckoo search algorithm. Ammar et al. [28] proposed a novel enhanced cuckoo search (ECS) algorithm which proposed a new range of search space for the parameters of the local/global enhancement (LGE) transformation that need to be optimized. Majumder [29] proposed a hybrid discrete cuckoo search (HDCS) algorithm, in which a modified lévy flight was proposed to transform a continuous position into a discrete schedule.

In this paper, we proposed an improved CS algorithm namely DFCS that adopts balanced learning strategies. In DFCS, the better fitness set (FSL) and the better diversity set (DSL) are produced in the iterative process. The FSL and DSL learning factors improve the global search ability and search accuracy of the algorithm and effectively balance the contradiction between exploitation and exploration. To verify the effectiveness of DFCS, we conducted comprehensive experiments on the logistics distribution center location problem. The results show that DFCS algorithm has stronger competitiveness in solving logistics distribution problem than CS algorithm.

The remainder of this paper is organized as follows. Section 2 reviews the basic characteristics of CS, and then Sect. 3 describes balanced-learning model and DFCS algorithm steps. Benchmark problems and corresponding experimental results are given in Sect. 4. Finally, Sect. 5 concludes this paper and points out some future research directions.

2 Cuckoo Search

The cuckoo search algorithm (CS) is a nature-inspired evolutionary algorithm, which inspired by parasitism behavior of cuckoo species that lay eggs in other host birds. The algorithm is based on Lévy flights which is a type of random walk with a heavy tail. CS is based on three idealized rules:

1) Each cuckoo lays one egg at a time, and places it in a randomly chosen nest.
2) The best nests with the highest quality eggs (solutions) will be carried over to the next generations.
3) The number of available host nests is fixed, and a host can discover an alien egg with the probability P_a. If the alien egg is discovered, the nest is abandoned and a new nest is built in a new location.

The position of the number i nest are indicated by using D-dimensional vector, the offspring are produced by using *Lévy* flights (based on random walks). *Lévy* flight is performed as follows:

$$X_i^{t+1} = x_i^t + a \otimes levy(\lambda) \quad (i = 1, 2, ..., n), \tag{1}$$

$$a = a_0 \otimes (x_j^t - x_i^t) \tag{2}$$

where x_i^t and x_j^t are two different solutions selected randomly. α is the step size with the range of the random search. Step size can be computed by using Eq. (2). After partial solutions are discarded, a new solution with the same number of cuckoos is generated by using Eq. (3).

$$X_i^{t+1} = x_i^t + r(X_m^t - X_n^t) \tag{3}$$

where X_m^t and X_n^t are random solutions for the t-th generation, r generates a random number between -1 and 1.

$$a = a_0 + (a_1 - a_0) \cdot d_i \tag{4}$$

$$d_i = \frac{||x_i - x_{best}||}{d_{\max}} \tag{5}$$

where α_0 and α_1 represent the minimum and maximum step size, respectively. X_{best} represents the optimal nest position, x_i represents the i-th nest position, $d_{\max}$ is the maximum distance between the optimal nest and all other nests. The structure of CS algorithm is described in Algorithm 1.

Algorithm 1: *CS algorithm*

---

(1) randomly initialize population of n host nests
(2) calculate fitness value for each solution in each nest
*(3) **while** (stopping criterion is not meet do)*
*(4) **for** I = 1 to n*
(5)   generate as new solution by using lévy flights;
(6)   choose candidate solution ;
*(7)   **if** $f(x_i^t) > f(x_i^{t+1})$*
(8)   replace with new solution ;
*(9)   **end if***
*(10) **end for***
(11) throw out a fraction (p_a) of worst nests
(12) for each abandoned nest $k \in c$ do
(13)   for each $i \in n$ do
(14)   generate solution k_i^{t+1} using Eq. (3)
*(15)   **if** $f(x_i^t) > f(x_i^{t+1})$*
(16)   replace with new solution ;
*(17)   **end if***
*(18)   **end for***
*(19) **end for***
(20)   rank the solution and find the current best
*(21) **end while***

---

3 Improved Cuckoo Search Algorithm with Balanced-Learning Model

For optimization algorithm, the balance between exploitation and exploration is an important goal. In this study, fitness sorting learning mechanism (FSL) and diversity sorting learning mechanism (DSL) are introduced into the CS algorithm. According to the fitness value of individuals, population is sorted in descending order in FSL. The new population after sorted by fitness, the first individual is the worst, and the n-th individual is the best. Diversity sorting learning mechanism (DSL) is also introduced into individual updates. In DSL, Xt is the current i-th individual. The individuals are sorted in descending order according to the diversity of individuals.

$$\bar{x}_i = k(a_i + b_i) - x_i \tag{6}$$

$$d_{i,j} = d(X_i, X_j) = \sqrt{\sum_{k=1}^{D}(X_{i,k} - X_{j,k})^2} \tag{7}$$

$$d = \begin{bmatrix} d_{1,1},\ d_{1,2}\ ...,\ d_{1,n} \\ d_{2,1},\ d_{2,2}\ ...,\ d_{2,n} \\ ... \\ d_{n,1},\ d_{n,2}\ ...,\ d_{n,n} \end{bmatrix} \tag{8}$$

$$\bar{d}(X_i) = \sum_{j=1, j \neq i}^{D} d_{i,j}/(n-1) \tag{9}$$

The diversity matrix d can be calculated by Eq. (8). The diversity of each individual is denoted by the mean diversity $\bar{d}(X_i)$. The distance-based diversity and fitness-based diversity are investigated in some researches. The individual with the biggest diversity value is the best. X_i is the current i-th individual. An individual is randomly selected from $\{X_{q+1}, ..., X_n\}$ and used for the updating process of the current individual.

$$x_{g+1,i} = x_{g,i} + a \oplus levy(\beta) + R_1(x_{FSL,g} - x_{g,i}) + R_2(x_{DSL,g} - x_{g,i}) \tag{10}$$

where $x_{FSL,g}$ and $x_{dSL,g}$ are randomly chosen from $\{X_{t+1}, ..., X_n\}$ with better fitness and better diversity than the current individual, respectively. R_1 and R_2 are learning factor from $x_{FSL,g}$ fitness learns behaviors and $x_{DSL,g}$ diversity learns behaviors, respectively. R_1 and R_2 can be computed by using Eqs. (11) and (12), respectively.

$$R_1 = (f(X_i) - f_{min})/(f_{mean} - f_{min}) \tag{11}$$

$$R_2 = (T/t)^2 \tag{12}$$

where f_{min} is the optimal solution at current generation, $f(X_i)$ is the solution of the current, and f_{mean} is the mean solution. The procedure pseudo-code of DFCS algorithm can be described in Algorithm 2.

***Algorithm 2**: DFCS algorithm*

---

(1) initialize population in terms of opposition-based learning model

(2) calculate fitness value for each solution in each nest.

*(3) **for** t=1 to T do*

(4)   calculate the fitness values of individuals and sort according to fitness values.

(5)   calculate the diversity of individuals and sort according to diversity.

(6)   update the learning factor R_2 of DSL according to Eq. (12).

*(7)   **for** i=1 to NP do*

(8)   update the learning factor R_1 of FSL according to Eq. (12).

(9)   randomly choose the better individual from the FSL set.

(10)   randomly choose the better individual from the DSL set.

(11)   generate x_i^{t+1} as new solution by using Eq. (12).

(12)   choose candidate solution .

*(13)   **if** $f(x_i^t) > f(x_i^{t+1})$*

(14)   replace x_i^t with new solution x_i^{t+1};

*(15)   **end if***

*(16)   **end for***

(17)   throw out a fraction (p_a) of worst nests.

*(18)   **for** each abandoned nest $k \in c$ do*

*(19)   **for** each $i \in n$ do*

(20)   generate solution k_i^{t+1} using Eq. (3).

*(21)   **if** $f(x_i^t) > f(x_i^{t+1})$*

(22)   replace x_i^{t+1} with new solution x_i^{t+1}.

*(23)   **end if***

*(34)   **end for***

*(25)   **end for***

(26)   rank the solution and find the current best.

*(27) **end for***

---

4 Results

Logistics distribution center location problem belongs to the research problem of the logistics management strategy level. In this paper, multiple distribution center location is discussed. The logistics distribution center location selects certain number of locations in a number of known sites. This type of problem is a nonlinear model with non-smooth characteristics and more complex constraints. The problem can be described as: m cargo distribution center are searched in n demand point. The distance between m and n is the shortest. The constraint conditions are as follows:

$$\min(\cos t) = \sum_{i=1}^{m} \sum_{j=1}^{n} (need_j \cdot dist_{i,j} \cdot \mu_{i,j}) \tag{13}$$

$$s.t. \quad \sum_{i=1}^{m} \mu_{i,j} = 1, i \in M, j \in N \tag{14}$$

$$\mu_{i,j} \le h_j, i \in M, j \in N \tag{15}$$

$$\sum_{t=1}^{m} h_i \le p, i \in M \tag{16}$$

$$h_j \in \{0, 1\}, i \in M \tag{17}$$

$$\mu_{i,j} \in \{0, 1\}, i \in M, j \in N \tag{18}$$

$$M = \{j | j = 1, 2, ..., m\} \quad N = \{j | j = 1, 2, ..., n\} \tag{19}$$

where Eq. (13) represents the objective function. Equations (14)–(19) are the constraints. Equation (14) indicates that a demand point of goods can only be distributed by a distribution center, Eq. (15) defines that each demand point of goods must have a distribution center to distribute goods, and Eq. (16) indicates the number of goods demand points for a distribution center.

In this section, there is a logistics network with 40 demand points. The geographical position coordinates and demands were shown in Table 1. The maximum number of iterations $T = 30,000$, population size $NP = 15$.

Table 1. The geographical position coordinates and demands

No	coordinates		demand	No	coordinates		demand	No	coordinates		demand	No	coordinates		demand
	x	y			x	y			x	y			x	y	
1	97	28	94	11	91	96	85	21	111	117	92	31	125	66	45
2	100	56	11	12	39	90	54	22	63	42	99	32	169	49	98
3	45	67	50	13	50	101	25	23	67	105	98	33	31	188	31
4	150	197	88	14	67	66	87	24	160	156	88	34	86	42	91
5	105	48	80	15	157	54	66	25	100	125	47	35	90	21	79
6	24	158	29	16	104	35	82	26	35	48	47	36	46	53	47
7	88	61	93	17	169	95	48	27	143	172	34	37	62	30	84
8	55	105	10	18	48	39	78	28	94	56	33	38	163	176	52
9	120	120	18	19	115	61	16	29	57	73	43	39	190	141	10
10	43	105	38	20	154	174	49	30	25	127	100	40	170	30	77

For the first set of experiments, the effectiveness of the DFCS is verified by comparing CS algorithms. 4 and 6 points were compared in this experiment. When the number of iterations T = 500, the optimal convergence curve, average convergence curve, and the optimal route found by CS are shown in Fig. 1(a). The optimal convergence curve, average convergence curve, and the optimal route found by CS are shown in Fig. 1(b).

It can be seen from Fig. 1(a), the optimal convergence curve of CS in 4 distribution centers for 500 iteratings can converge at 50 iterations, the average convergence curve can converge at 200 iterations. The optimal distribution, average distribution and the worst distribution cost obtained in 4 distribution centers for 500 iteratings are 6.5268E04,

7.4188E04, and 7.7992E04, respectively. The optimal distribution, average distribution and the worst distribution cost obtained in 6 distribution centers for 500 iterations are 4.7128E04, 5.399E04, and 5.6927E04, respectively. The optimal distribution center points in 4 and 6 distribution centers for 500 iterations are (8, 22, 27, 15) and (30, 23, 20, 18, 1, 15), respectively (Tables 2 and 3).

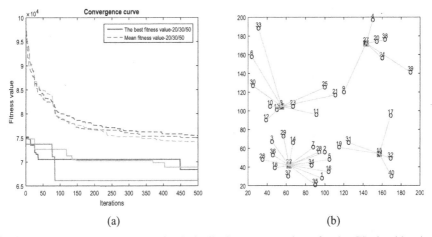

(a)   (b)

Fig. 1. Convergence curves and optimal distribution centers schemefor the CS algorithm in 4 distribution center ($T = 500$). Figure 1(a) indicated the average convergence curve, optimal convergence curve, and the optimal route found by CS for running 20, 30, and 50 times in 4 distribution centers. Figure 1(b) showed distribution range 100 iterations for 4 distribution.

Table 2. The distribution scheme for the CS algorithm in 4 distribution center ($T = 500$)

Distribution center	Distribution scope
8	33, 6, 30, 10, 12, 13, 23, 25, 21, 11
22	26, 36, 3, 29, 14, 18, 37, 7, 28, 2, 34, 5, 16, 1, 35
27	4, 9, 20, 24, 38, 39
15	19, 31, 17, 32, 40

For the second set of experiments. When the number of iterations $T = 500$, the optimal convergence curve, average convergence curve, and the optimal route found by DFCS are shown in Fig. 3 for 4 distribution centers. When the number of iterations $T = 500$, the optimal convergence curve, average convergence curve, and the optimal route found by DFCS are shown in Fig. 2 for 6 distribution centers. Table 4–5 show distribution ranges 500 iterations for 4 and 6 distribution centers, respectively.

From Fig. 2(a) shows that the optimal convergence curve of DFCS in 4 distribution centers for 500 iteratings can converge at 20 iterations, the average convergence curve can converge at 15 iterations. The optimal distribution, average distribution and the worst distribution cost obtained in 4 distribution centers for 500 iterations are 6.4218E04,

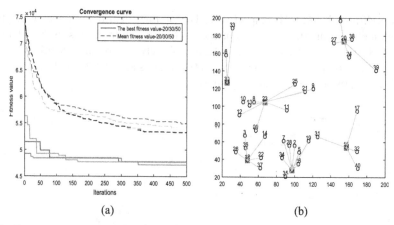

Fig. 2. Convergence curves and optimal distribution centers scheme for the CS algorithm in 6 distribution centers ($T = 500$). Figure 2(a) indicated the average convergence curve, optimal convergence curve, and the optimal route found by CS for running 20, 30, and 50 times in 6 distribution centers. Figure 2(b) showed distribution range 100 iterations for 6 distribution.

Table 3. The distribution scheme for the CS algorithm in 6 distribution centers ($T = 500$)

Distribution center	Distribution scope
30	33, 6
23	11, 25, 9, 21, 10, 13, 12, 8, 29
20	4, 27, 38, 24, 39
18	14, 3, 36, 26, 37, 22
1	28, 7, 34, 2, 5, 16, 35
15	31, 17, 32, 40

6.5624E04, and 6.5990E04, respectively. The optimal distribution, average distribution and the worst distribution cost obtained in 6 distribution centers for 500 iterations are 4.5825E04, 4.6154E04, and 4.6561E04, respectively. The optimal distribution center points in 4 and 6 distribution centers are (23, 22, 27, 15) and (12, 21, 20, 22, 16, 15) for 500 iteratings, respectively.

It can be seen from Fig. 1, 2, 3 and 4, the average convergence curve of DFCS in 4 distribution centers and 6 distributions can converge at 500 iterations. The average convergence curve of CS can converge at 250 iterations, which in include DFCS is far superior to CS for terms of convergence speed. It indicates that DFCS has fast speed and high solution accuracy, which effectively reduces the cost of logistics distribution. At the same time, we can say that the DFCS outperforms CS in terms of convergence rate and robustness.

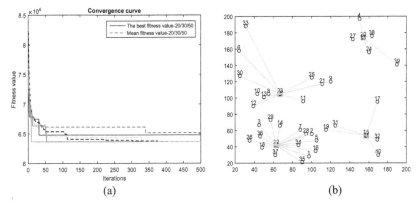

(a)   (b)

Fig. 3. Convergence curves and optimal distribution centers scheme for the DFCS algorithm in 4 distribution centers ($T = 500$). Figure 3(a) indicated the average convergence curve, optimal convergence curve, and the optimal route found by DFCS for running 20, 30, and 50 times in 4 distribution centers. Figure 3(b) showed distribution range 100 iterations for 4 distribution.

Table 4. The distribution scheme for the DFCS algorithm in 4 distribution centers ($T = 500$)

Distribution center	Distribution scope
23	33, 6, 30, 10, 12, 13, 8, 25, 21, 11, 9
22	26, 36, 3, 29, 14, 18, 37, 7, 28, 2, 34, 5, 16, 1, 35
27	4, 9, 20, 24, 38, 39
15	19, 31, 17, 32, 40

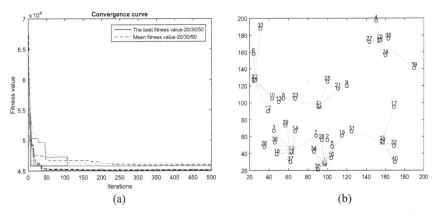

(a)   (b)

Fig. 4. Convergence curves and optimal distribution centers scheme for the DFCS algorithm in 6 distribution centers ($T = 500$). Figure 4(a) indicated the average convergence curve, optimal convergence curve, and the optimal route found by DFCS for running 20, 30, and 50 times in 6 distribution centers. Figure 4(b) showed distribution range 100 iterations for 6 distribution.

Table 5. The distribution scheme for the DFCS algorithm in 6 distribution centers ($T = 500$)

Distribution center	Distribution scope
30	33, 6, 12, 10, 8, 13
11	23, 25, 21,9
20	4, 27, 38, 24, 39
22	14, 29, 3, 25, 26, 37, 18,7
16	28, 34, 2, 5, 19, 35, 16
15	31, 17, 32, 40

5 Conclusions

This paper proposed an improved CS algorithm with balanced-learning scheme namely DFCS, in which two excellent individuals are selected from two sets to participate in search process. The two learning factors are adjusted at each generation to improve the global search ability. In order to verify the performance of DFCS, this algorithm is applied to solve the logistics distribution problem. The effectiveness of the DFCS method is verified in both 6 distribution centers and 10 distribution centers.

Acknowledgement. This work was supported by the fundamental research funds for Ministry of University-Industry Collaborative Education Program (220905181091456).

References

1. Li, J., Lei, H., Alavi, A.H., Wang, G.G.: Elephant herding optimization: variants, hybrids, and applications. Mathematics **8**(9), 1415 (2020)
2. Wang, G.G., Tan, Y.: Improving metaheuristic algorithms with information feedback models. IEEE Trans. Cybern. **49**(2), 542–555 (2019)
3. Goldberg, D.E., Holland, J.H.: Genetic algorithms and machine learning. Mach. Learn. **3**, 95–99 (1988)
4. Kennedy, J., Eberhart, R.: Particle swarm optimization. In: Proceeding of the IEEE International Conference on Neural Networks, Perth, Australia, 27 November–1 December, pp 1942–1948. IEEE (1995)
5. Wang, G.G., Gandomi, A.H., Yang, X.-S., Alavi, A.H.: A novel improved accelerated particle swarm optimization algorithm for global numerical optimization. Eng. Comput. **31**(7), 1198–1220 (2014)
6. Storn, R., Price, K.: Differential evolution-a simple and efficient heuristic for global optimization over continuous spaces. J. Glob. Optim. **11**, 341–359 (1997)
7. Xu, Z., Unveren, A., Acan, A.: Probability collectives hybridised with differential evolution for global optimisation. Int. J. Bio-Inspired Comput. **8**(3), 133–153 (2016)
8. Wang, G.G., Zhao, X., Deb, S.: A novel monarch butterfly optimization with greedy strategy and self-adaptive crossover operator. In: 2015 Second International Conference on Soft Computing and Machine Intelligence (ISCMI), Hong Kong, 23–24 November 2015, pp 45–50. IEEE (2015)

9. Wang, G.G., Deb, S., Zhao, X., Cui, Z.: A new monarch butterfly optimization with an improved crossover operator. Oper. Res. **18**, 731–755 (2018)

10. Wang, G.G., Deb, S., Dos Santos Coelho, L.: Earthworm optimization algorithm: a bio-inspired metaheuristic algorithm for global optimization problems. Int. J. Bio-Inspired Comput. **12**(1), 1–22 (2018)

11. Meng, X., Liu, Yu., Gao, X., Zhang, H.: A new bio-inspired algorithm: chicken swarm optimization. In: Tan, Y., Shi, Y., Coello, C.A., Coello, (eds.) ICSI 2014. LNCS, vol. 8794, pp. 86–94. Springer, Cham (2014). https://doi.org/10.1007/978-3-319-11857-4_10

12. Gandomi, A.H., Alavi, A.H.: Krill herd: a new bio-inspired optimization algorithm. Commun. Nonlinear Sci. Numer. Simul. **17**(12), 4831–4845 (2012)

13. Wang, G.G., Gandomi, A.H., Alavi, A.H., Gong, D.: A comprehensive review of krill herd algorithm: variants, hybrids and applications. Artif. Intell. Rev. **51**, 119–148 (2019)

14. Gálvez, A., Iglesias, A.: New memetic self-adaptive firefly algorithm for continuous optimisation. Int. J. Bio-Inspired Comput. **8**, 300–317 (2016)

15. Nasiri, B., Meybodi, M.R.: History-driven firefly algorithm for optimisation in dynamic and uncertain environments. Int. J. Bio-Inspired Comput. **8**, 326–339 (2016)

16. Kirkpatrick, S., Gelatt, C.D., Jr., Vecchi, M.P.: Optimization by simulated annealing. Science **220**, 671–680 (1983)

17. Shah-Hosseini, H.: The intelligent water drops algorithm: a nature-inspired swarm-based optimization algorithm. Int. J. Bio-Inspired Comput. **1**, 71–79 (2009)

18. Wang, G.G.: Moth search algorithm: a bio-inspired metaheuristic algorithm for global optimization problems. Memetic Comput. **10**, 151–164 (2018)

19. Zhao, R., Tang, W.: Monkey algorithm for global numerical optimization. J. Uncertain Syst. **2**, 165–176 (2008)

20. Beyer, H.: The Theory of Evolution Strategies. Springer, New York (2001). https://doi.org/10.1007/978-3-662-04378-3

21. Penev, K., Littlefair, G.: Free search-a comparative analysis. Inf. Sci. **172**, 173–193 (2005)

22. Yang, X.-S., Deb, S.: Cuckoo search via Lévy flights. In: Proceeding of World Congress on Nature & Biologically Inspired Computing (NaBIC 2009), Coimbatore, India, December 2009, pp. 210–214. IEEE (2009)

23. Li, J., Li, Y.-X., Tian, S.-S., Zou, J.: Dynamic cuckoo search algorithm based on Taguchi opposition-based search. Int. J. Bio-Inspired Comput. **13**, 59–69 (2019)

24. Li, J., Xiao, D.-D., Lei, H., Zhang, T., Tian, T.: Using cuckoo search algorithm with Q-learning and genetic operation to solve the problem of logistics distribution center location. Mathematics **8**(149), 56–77 (2020)

25. Li, J., Xiao, D.-D., Zhang, T., Liu, C., Li, Y.-X., Wang, G.-G.: Multi-swarm cuckoo search algorithm with Q-learning model. Comput. J. **12**, 156–167 (2020)

26. Li, J., Li, Y.-X., Tian, S.-S., Xia, J.: An improved cuckoo search algorithm with self-adaptive knowledge learning. Neural Comput. Appl. **32**, 31 (2020)

27. Kamoona, A.M., Patra, J.C.: A novel enhanced cuckoo search algorithm for contrast enhancement of gray scale images. Appl. Soft Comput. **85**, 15–19 (2019)

28. Majumder, A., Laha, D., Suganthan, P.N.: A hybrid cuckoo search algorithm in parallel batch processing machines with unequal job ready times. Comput. Ind. Eng. **124**, 65–76 (2018)

Badminton Detection Using Lightweight Neural Networks for Service Fault Judgement

Tiandong Li[1], Jianqing Lin[1], Linqiang Pan[1], and Zhenxing Wang[2(✉)]

[1] Key Laboratory of Image Information Processing and Intelligent Control
of Education Ministry of China, School of Artificial Intelligence and Automation,
Huazhong University of Science and Technology, Wuhan 430074, China
[2] School of Physical Education, Huazhong University of Science and Technology,
Wuhan 430074, China
wzx0605@hust.edu.cn

Abstract. The Badminton World Federation stipulates that the height of the badminton during the serving shall not exceed 1.15 m. In current official matches, enforcement still relies on the referee's eye judgment. The Hawk-Eye system in matches mainly judges shuttlecock landings, which is not suited for service fault judgement. Recently, some algorithms based on computer vision have been proposed for badminton rules judgement. However, it is a challenging task for badminton service fault detection. Firstly, the ball is usually small and moves fast within the imaging field of view, complicating the detection process. Furthermore, accurately locating the service point and verifying its height against standards is technically challenging. Additionally, the scarcity of public datasets for badminton detection poses further challenges to developing effective detection algorithms. To address these deficiencies, an annotated dataset is collected for badminton detection. A deep learning network based on the concept of CenterNet is proposed to locate badminton in the imaging field of view accurately and in real time. Experiments are conducted on the collected dataset compared with four state-of-the-art methods. The experimental results demonstrate that our method used for badminton detection can achieve high detection accuracy in real-time on ordinary computing devices such as personal computers or smartphones.

Keywords: Object detection · Badminton dataset · Lightweight Neural network

1 Introduction

The Badminton World Federation has changed the badminton serving rules in 2018. Specifically, the height of the serve was changed from no more than the waist to no more than 1.15 m [1]. In current official matches, referee uses his visual judgment assisted by a fixed height service tool to enforce this rule. However, this measure does not provide evidence to substantiate the referee's decisions, and there is no publicly available data about its accuracy. In addition, long hours

L. Pan et al. (Eds.): BIC-TA 2023, CCIS 2062, pp. 182–194, 2024.
https://doi.org/10.1007/978-981-97-2275-4_15

of work for referee can lead to fatigue, increasing the likelihood of incorrect or missed judgments. The commonly used Hawk-Eye system in official badminton matches is mainly employed for judging whether the badminton lands outside the boundaries of the court, but it lacks the capability to locate the service point and verify its height against standards. Thus, the Hawk-Eye system can not be directly used for badminton service fault judgement. Additionally, the Hawk-Eye system, composed of multiple high-speed cameras, servers, display units, and advanced analysis software, incurs a weekly usage cost of several tens of thousands of dollars. Recently, some algorithms based on computer vision have been proposed for badminton rules judgement. However, it is a challenging task for badminton service fault detection. The ball is usually small and moves fast within the imaging field of view, complicating the detection process. In the collected badminton dataset, the pixel values for the width and height of the badminton range between 30 and 50, and the dimensions of the entire image are 3840×2160, which indicates that the badminton occupies a very small proportion of pixels in the field of view. In addition, the badminton can achieve exceptionally high velocities, with speeds of up to $137\,\mathrm{m/s}$ [2], so that the badminton is blurred, residual, or even invisible in the imaging field of view. Thus, it is more difficult to detect badminton and judge service faults. The two major contributions of this work are as follows.

1. A high-resolution and annotated dataset is collected for badminton detection. Specifically, many live videos are recorded from the perspective of a referee in badminton matches with the resolution of 3840×2160 pixels at 30 frames per second. After that, about 1,000 RGB images of badminton are extracted from these videos to obtain the detection dataset.
2. A neural network based on the concept of CenterNet [3] is proposed to locate badminton in the imaging field of view accurately and in real time. The proposed network is composed of a convolutional neural network followed by a fractionally-strided convolutional neural network.

The rest of this paper is organized as follows. The next section provides an overview of related work in the field of object detection. Section 3 details the algorithm proposed in this work. Section 4 presents our experimental results and analysis. Finally, Sect. 5 provides a detailed elaboration of our conclusions.

2 Related Work

In recent years, advancements in camera and sensor technology are employed to maintain fairness in badminton competitions. The Badminton World Federation first introduced the Hawk-Eye technology in 2013 [4], also known as the Instant Review System. Initially deployed in tennis, this system's mechanism involves capturing images through multiple high-definition cameras and processing these images with high-performance computers to reconstruct and replay the trajectory of the ball using virtual technology, thereby aiding in decision-making.

Different from its application in tennis, the instant replay technology in badminton does not track the shuttlecock's movement. Instead, it concentrates on high-definition replays of the baseline or sidelines. The primary purpose is to ascertain whether the shuttlecock lands in or out of bounds. But this system is unable to detect violations in players' actions and cannot be used for judging service faults in badminton. Beyond the instant replay system, there are additional support systems for badminton. Waghmare and team develop a system, which utilizes multiple scanning devices to detect and predict the trajectory of a shuttlecock, presenting a more affordable option compared to the Hawk-Eye [5]. In a separate initiative, ultrasonic sensors are tested for tracking service height in badminton. However, this approach encountered limitations, including a narrow scanning range and difficulties in identifying service events [6].

Concurrently, the use of computer vision and object detection methods in sports technology has expanded, encompassing areas like match analysis [7–10], player performance evaluation [11], and the development of electronic umpiring systems [12,13]. In the realm of badminton, object detection and machine learning technologies have been applied to facilitate automatic recognition of badminton actions for post-match analysis that isn't conducted in real-time [14–16]. YOLO and its subsequent variants are classic examples of one-stage algorithms in the field of object detection [17–19]. This series of algorithms can directly predict the class probabilities and positions of objects based on regression. Cao et al. introduce two novel variants of Tiny YOLOv2 designed for shuttlecock detection to assist in the navigation of robots [20]. Vrajesh et al. improve upon YOLOv3 to detect the shuttlecock and predict its landing spot [21]. Menon and colleagues developed a machine learning-based framework for badminton service fault judgement [22]. However, the framework faces a challenge due to its operation at a relatively low sample rate of 25.8 frames per second.

3   Network Model

In this work, the proposed network utilizes a fully convolutional encoder-decoder architecture, which can quickly and accurately identify the location of badminton on a high-resolution imaging field of view. For badminton detection, ResNet-18 [23] is streamlined to act as the encoder. Then, a fractionally-strided convolution neural network [24] is used as the decoder. Afterwards, the decoder output is passed through the convolution layer to generate a feature map. The badminton position in the map is calculated based on the feature map generated by the neural network.

3.1   Network Architecture

The proposed network is composed of a convolutional neural network based on ResNet-18 followed by a fractionally-strided convolutional neural network. It takes single frame as input to generate a heatmap indicating the object position. The overall structure of the network is illustrated in Fig. 1.

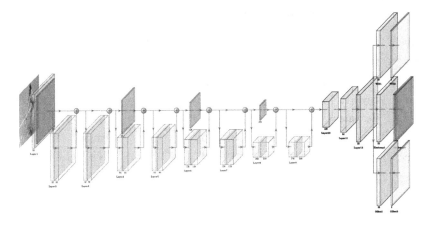

Fig. 1. The architecture of the proposed network.

In Fig. 1, the first 9 layers are designed as the encoder with reference to ResNet-18 [23]. The 10–12 layers are designed as the decoder based on DCGAN [24]. The specific parameters of the encoder and decoder are described in detail in Table 1 and Table 2, respectively. Furthermore, the last 2 layers, including Heatmap, WH, and Offset, refer to CenterNet [3]. To achieve pixel-level prediction, transpose convolution is used to recover the information loss from downsampling layers.

Table 1. Architecture of the encoder.

Name	Module	Filters	Stride	Padding	Activation
Layer1	Convolution	7×7	2	3	BN + ReLU
	MaxPool	3×3	2	1	
Layer2	Convolution	3×3	1	1	BN + ReLU
	Convolution	3×3	1	1	BN
	Add				ReLU
Layer3	Same Parameters as Layer2				
Layer4	Convolution	3×3	2	1	BN + ReLU
	Convolution	3×3	1	1	BN
	DownSample	1×1	2	0	BN
	Add				ReLU
Layer5	Same Parameters as Layer2				
Layer6	Same Parameters as Layer3				
Layer7	Same Parameters as Layer2				
Layer8	Same Parameters as Layer3				
Layer9	Same Parameters as Layer2				

Table 2. Architecture of the decoder.

Name	Module	Filters	Stride	Padding	Activation
Layer10	TransConv	4×4	2	1	BN + ReLU
Layer11	Same Parameters as Layer10				
Layer12	Same Parameters as Layer10				·
Heatmap1	Convolution	3×3	1	1	BN + ReLU
Heatmap2	Convolution	1×1	1	0	Sigmoid
WH1	Convolution	3×3	1	1	BN + ReLU
WH2	Convolution	1×1	1	0	
Offset1	Convolution	3×3	1	1	BN + ReLU
Offset2	Convolution	1×1	1	0	

3.2 Network Output

The network generates three feature maps, comprising a heatmap of the badminton center point, a local offset for each center point, and a regression of the object size (including width W and height H). Below is a description of their respective implications.

An input image $I \in R^{W \times H \times 3}$ has a width of W and height of H, where the bounding box of the badminton object is denoted with (x_1, y_1, x_2, y_2), and its center point is positioned at $c = ((x_1 + x_2)/2, (y_1 + y_2)/2)$. The network produces a heatmap $\hat{Y} \in [0, 1]^{\frac{W}{R} \times \frac{W}{R}}$, where $R = 4$ is the output stride because of our network parameters. $\hat{Y}$ is utilized to predict the center point of badminton, meaning that $\hat{Y}_{x,y} = 1$ corresponds to finding out object, while $\hat{Y}_{x,y} = 0$ is background. Additionally, a size prediction $\hat{S} \in R^{\frac{W}{R} \times \frac{H}{R} \times 2}$ is used to regress to the object size $(x_2 - x_1, y_2 - y_1)$ for badminton. Furthermore, to recover the discretization error caused by the output stride, a local offset $\hat{O} \in R^{\frac{W}{R} \times \frac{W}{R} \times 2}$ is predicted for center point.

3.3 Loss Function

For the ground truth of the heatmap, first a low-resolution equivalent $\tilde{c} = \lfloor \frac{c}{R} \rfloor$ is computed. Then the center coordinates available in the labeled dataset are splatted onto a heatmap $Y \in [0, 1]^{\frac{W}{R} \times \frac{W}{R}}$ using a Gaussian kernel. The kernel is expressed as

$$Y_{x,y} = \exp\left(-\frac{(x - \tilde{c}_x)^2 + (y - \tilde{c}_y)^2}{2\sigma_p^2}\right), \tag{1}$$

where σ_p is an object size-adaptive standard deviation [25]. The training objective involves performing pixel-level logistic regression with a focus on reducing penalties through the implementation of focal loss [26].

$$L_{hm} = \begin{cases} -\frac{1}{N}\sum_{x,y}\left(1 - \hat{Y}_{x,y}\right)^{\alpha}\log\left(\hat{Y}_{x,y}\right), & \text{if } Y_{x,y} = 1, \\ -\frac{1}{N}\sum_{x,y}\left(1 - Y_{x,y}\right)^{\beta}\left(\hat{Y}_{x,y}\right)^{\alpha}\log\left(1 - \hat{Y}_{x,y}\right), & \text{otherwise,} \end{cases} \tag{2}$$

where α and β are hyper-parameters of the focal loss [26], and N is the number of badminton in image I.

In addition, L1 loss is utilized at the regression of the object size and the offset following Zhou and Wang [3], as shown below:

$$L_{size} = \frac{1}{N}\sum_{k=1}^{N}\left|\hat{S}_k - s_k\right|, \tag{3}$$

$$L_{off} = \frac{1}{N}\sum_{c}\left|\hat{O}_{\tilde{c}} - \left(\frac{c}{R} - \tilde{c}\right)\right|. \tag{4}$$

In summary, the overall loss function is

$$L = L_k + \lambda_{size}L_{size} + \lambda_{off}L_{off}. \tag{5}$$

In all our experiments, $\alpha = 2$ and $\beta = 4$ are set following Law and Deng [25], and set $\lambda_{size} = 0.1$ and $\lambda_{off} = 1$ following Zhou and Wang [3].

4 Experimental Studies

To evaluate the proposed network, we mimic the referee's view to fix the camera, shoot many live videos of badminton sports with the resolution of 3840×2160 pixels at 30 FPS and use it to create a badminton dataset. After extensive testing on the NVIDIA GeForce RTX 3060 Laptop GPU hardware environment, our method is capable of processing high-resolution images of 1024×576 pixels at a speed of 31 frames per second. Furthermore, our approach achieves a mean average precision (mAP) of 92.5% on the dataset.

4.1 Dataset Collection and Processing

There is no well-annotated public dataset available, for the badminton detection task studied. Therefore, an annotated dataset is collected for badminton classification and detection according to the following procedure. First, a series of badminton match videos are recorded from different angles of the badminton court with the resolution of 3840×2160 pixels at 30 FPS. Next, about 1,000 RGB images of badminton from these videos are extracted to obtain the detection dataset. A few examples of these contexts are shown in Fig. 2. 80% of the data are randomly selected as the training set (about 800 images) and the remaining 20% of the data is set as the test set (about 200 images). In addition, the following methods are used for data enhancement, including cropping, rotation, oversampling images and copy-pasting methods [27].

4.2   Comparison and Evaluation Metrics

The proposed method is compared to several detectors, including SSD [28], Faster R-CNN [31], Tiny YOLOv2 [18], and YOLOv3 [19]. The comparison builds on the dataset described above, based on metrics like pression, recall, mean Average Precision (mAP), and FPS.

In the area of target detection, precision and recall are the most widely used metrics [29]. Precision is defined as the percentage of accurately identified positive samples, whereas recall measures the fraction of actual positive samples that are correctly detected, which are shown as follows:

$$\text{Precision} = \frac{N_{tp}}{N_{tp} + N_{fp}}, \tag{6}$$

$$\text{Recall} = \frac{N_{tp}}{N_{tp} + N_{fn}}, \tag{7}$$

where N_{tp}, N_{fp}, and N_{fn} are the numbers of true positives, false positives, and false negatives, respectively.

Fig. 2. Different badminton scenarios for training and validation.

Similarly, mAP is one of the evaluation metrics commonly used in target detection tasks, and its calculation involves several steps. Initially, for each category, the detection results are divided into positive and negative samples according to the real labeled objects, and their confidence scores are recorded. Subsequently, all detection results are sorted in descending order according to their confidence scores. Starting from the highest confidence level, the detection results are thresholded as positive samples one by one, and the corresponding precision and recall are calculated. Based on the Precision-Recall curve, the precision at each recall point is calculated and averaged as the AP for that category. Ultimately, mAP is obtained by averaging the APs of all categories. Different intersection over union (IOU) thresholds can be used for evaluating correct detections for different object classes, computed as follows:

$$IOU = \frac{Prediction \cap GroundTruth}{Prediction \cup GroundTruth}. \tag{8}$$

The IOU value is between 0 and 1. A prediction bounding box is defined as positive when its IOU value is greater than the threshold value and negative otherwise. In our experiments, the threshold is set to 0.5 to balance the number of true positives and false negatives.

FPS is the number of images processed every second to evaluate the computational efficiency of object detection.

4.3   Implementation Details

All training and inference processes are performed on the Pytorch framework. The hardware setup includes a laptop equipped with an i5-12450H CPU, an NVIDIA GeForce RTX 3060 Laptop GPU, 16 GB of RAM, and runs on a Linux OS. Below, a detailed description is provided on how to implement our proposed network.

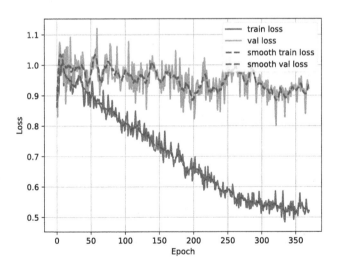

Fig. 3. The loss curve of our proposed model training.

All images are resized from 3840×2160 to 1024×576, when being imported into the model. The adaptive moment estimation (Adam) optimizer is used to optimize the network weights. The initial learning rate is set to 5×10^{-4}, the decay strategy for the learning rate is cosine annealing [30], and the minimum value of the learning rate is set to 0.01 times the initial value. During training, a batch size of 24 is used due to hardware limitations. The appropriate number of epochs can effectively prevent underfitting and overfitting, thus 380 epochs are chosen by repeating the experiment. Loss and mAP versus the number of epochs is shown in Fig. 3 and Fig. 4, respectively.

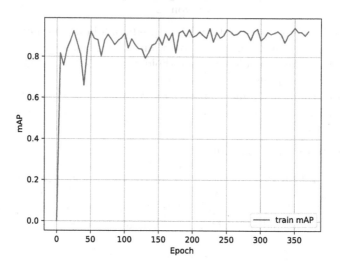

Fig. 4. The mAP curve of our proposed model training.

4.4   Results and Discussions

In this subsection, experiments are conducted to gain insights into how detection performance is improved by the proposed network, compared to other detectors such as SSD, Faster R-CNN, Tiny YOLOv2, and YOLOv3.

Experiments are conducted to evaluate our method, ensuring that the detector can handle every frame in the video pipeline. The test set described in Sect. 4.1 is selected for evaluation. Table 3 presents the experimental results. The results show that the proposed method achieves a running speed of 31 FPS (using MP4 video format with H.265 compression), which exceeds the typical camera frame rate of 30 FPS and meets the real-time requirements for badminton detection. Figure 5 shows the effect of prediction with a few examples. Compared to Faster R-CNN, Tiny YOLOv2, and YOLOv3, the proposed network demonstrates outstanding performance in terms of both detection accuracy and speed. Although Faster R-CNN performs best in terms of detection accuracy and recall rate, it is inefficient in computational efficiency, making it unsuitable for real-time detection. The proposed method achieves a frame rate of 31 FPS, which is sufficient to support real-time detection. This highlights the limitations of Faster R-CNN in most real-time applications. Compared to Tiny YOLOv2 and YOLOv3, our approach slightly lags behind in terms of computational efficiency but exhibits significant improvements in detection accuracy and recall rate. This can be attributed to the design of heatmaps in the network architecture.

Table 3. Evaluation results on the test set.

Method	Precision (%)	Recall (%)	FPS
Faster R-CNN	**97.3**	**99.2**	4
SSD	68.1	65.8	18
Tiny YOLOv2	89.3	75.4	**36**
YOLOv3	74.2	88.0	33
Our Method	95.4	96.2	31

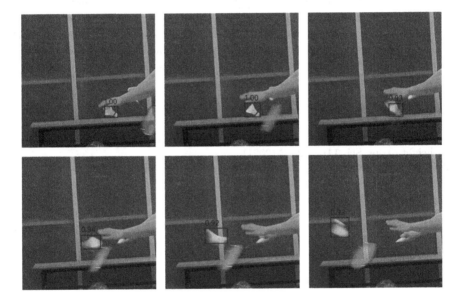

Fig. 5. Evaluation results on the test set.

However, when the model is applied to a computer vision system, the frame rate drops to 24 FPS, which is slightly slower than the previous frame rate of processing video streams (31 FPS). This decrease in frame rate is due to limitations in camera frame rate (FPS maximum) and additional overhead from image processing tasks such as encoding conversion, image reading, object detection, and image display. Consequently, the detector is forced to skip some frames, leading to a decrease in detection speed. However, this issue can be mitigated by optimizing the program architecture, such as transitioning from single-threaded to multi-threaded processing. This optimization can significantly improve the overall performance and enable the system to handle the processing of video streams more efficiently.

5   Conclusion

A set of the badminton service fault detection algorithms offers significant commercial potential. Within this set, a lightweight algorithm for detecting the position of badminton is the most crucial component. However, this task presents significant challenges due to the small size and fast movement of badminton object. Additionally, the limited availability of well-annotated public datasets further hampers progress in this field. To address these challenges, a deep learning network, specifically tailored for badminton detection, is developed. It comprises a convolutional neural network and a fractionally-strided convolutional neural network. Furthermore, a high-resolution dataset is collected that encompasses badminton movements in competition scenes from various angles. To evaluate the effectiveness of the network, comprehensive experimental comparisons are undertaken with other detectors. The results demonstrate that our network achieves high accuracy with efficient computation. In addition, this work provides potential application value for detecting other high-speed and small objects.

References

1. Rasmussen, J., de Zee, M.: A simulation of the effects of badminton serve release height. Appl. Sci. **11**(7), 2903 (2021)
2. Cohen, C., Texier, B.D., Quéré, D., Clanet, C.: The physics of badminton. New J. Phys. **17**(6), 063001 (2015)
3. Zhou, X., Wang, D., Krähenbühl, P.: Objects as points. arXiv preprint arXiv:1904.07850 (2019)
4. Leong, L.H., Zulkifley, M.A., Hussain, A.B.: Computer vision approach to automatic linesman. In: 2014 IEEE 10th International Colloquium on Signal Processing and its Applications, pp. 212–215 (2014). https://doi.org/10.1109/CSPA.2014.6805750
5. Waghmare, G., Borkar, S., Saley, V., Chinchore, H., Wabale, S.: Badminton shuttlecock detection and prediction of trajectory using multiple 2 dimensional scanners. In: 2016 IEEE First International Conference on Control, Measurement and Instrumentation (CMI), pp. 234–238 (2016).https://doi.org/10.1109/CMI.2016.7413746
6. Syafani, A., Subarkah, A., Marani, I.N.: Making service level measuring equipment (service detector) in badminton sports. Competitor **13**(1), 61–70 (2021)
7. Pers, J., Kovacic, S.: Computer vision system for tracking players in sports games. In: IWISPA 2000. Proceedings of the First International Workshop on Image and Signal Processing and Analysis. in conjunction with 22nd International Conference on Information Technology Interfaces IEEE, pp. 177–182. IEEE (2000)
8. Moon, S., Lee, J., Nam, D., Yoo, W., Kim, W.: A comparative study on preprocessing methods for object tracking in sports events. In: 2018 20th International Conference on Advanced Communication Technology (ICACT), pp. 460–462. IEEE (2018)

9. Host, K., Ivašić-Kos, M.: An overview of human action recognition in sports based on computer vision. Heliyon (2022)
10. Voeikov, R., Falaleev, N., Baikulov, R.: TTNet: real-time temporal and spatial video analysis of table tennis. In: Proceedings of the IEEE/CVF Conference on Computer Vision and Pattern Recognition Workshops, pp. 884–885 (2020)
11. Tahan, O., Rady, M., Sleiman, N., Ghantous, M., Merhi, Z.: A computer vision driven squash players tracking system. In: 2018 19th IEEE Mediterranean Electrotechnical Conference (MELECON), pp. 155–159. IEEE (2018)
12. Mendes-Neves, T., Meireles, L., Mendes-Moreira, J.: A survey of advanced computer vision techniques for sports. arXiv preprint arXiv:2301.07583 (2023)
13. Leong, L.H., Zulkifley, M.A., Hussain, A.B.: Computer vision approach to automatic linesman. In: 2014 IEEE 10th International Colloquium on Signal Processing and its Applications, pp. 212–215. IEEE (2014)
14. Rahmad, N.A., As'ari, M.A.: The new convolutional neural network (CNN) local feature extractor for automated badminton action recognition on vision based data. In: Journal of Physics: Conference Series. vol. 1529, p. 022021. IOP Publishing (2020)
15. Rahmad, N., As'Ari, M., Soeed, K., Zulkapri, I.: Automated badminton smash recognition using convolutional neural network on the vision based data. In: IOP Conference Series: Materials Science and Engineering, vol. 884, p. 012009. IOP Publishing (2020)
16. Luo, J., et al.: Vision-based movement recognition reveals badminton player footwork using deep learning and binocular positioning. Heliyon **8**(8) (2022)
17. Redmon, J., Divvala, S., Girshick, R., Farhadi, A.: You only look once: unified, real-time object detection. In: Proceedings of the IEEE Conference on Computer Vision and Pattern Recognition, pp. 779–788 (2016)
18. Redmon, J., Farhadi, A.: Yolo9000: better, faster, stronger. In: Proceedings of the IEEE Conference on Computer Vision and Pattern Recognition, pp. 7263–7271 (2017)
19. Redmon, J., Farhadi, A.: Yolov3: an incremental improvement. arXiv preprint arXiv:1804.02767 (2018)
20. Cao, Z., Liao, T., Song, W., Chen, Z., Li, C.: Detecting the shuttlecock for a badminton robot: a yolo based approach. Expert Syst. Appl. **164**, 113833 (2021). https://doi.org/10.1016/j.eswa.2020.113833
21. Vrajesh, S.R., Amudhan, A., Lijiya, A., Sudheer, A.: Shuttlecock detection and fall point prediction using neural networks. In: 2020 International Conference for Emerging Technology (INCET), pp. 1–6. IEEE (2020)
22. Menon, A., Siddig, A., Muntean, C.H., Pathak, P., Jilani, M., Stynes, P.: A machine learning framework for shuttlecock tracking and player service fault detection. In: Conte, D., Fred, A., Gusikhin, O., Sansone, C. (eds.) DeLTA 2023. Communications in Computer and Information Science, vol. 1875, pp. 71–83. Springer, Cham (2023). https://doi.org/10.1007/978-3-031-39059-3_5
23. He, K., Zhang, X., Ren, S., Sun, J.: Deep residual learning for image recognition. In: Proceedings of the IEEE Conference on Computer Vision and Pattern Recognition, pp. 770–778 (2016)
24. Radford, A., Metz, L., Chintala, S.: Unsupervised representation learning with deep convolutional generative adversarial networks. arXiv preprint arXiv:1511.06434 (2015)
25. Law, H., Deng, J.: CornerNet: detecting objects as paired keypoints. In: Proceedings of the European Conference on Computer Vision (ECCV), pp. 734–750 (2018)

26. Lin, T.Y., Goyal, P., Girshick, R., He, K., Dollár, P.: Focal loss for dense object detection. In: Proceedings of the IEEE International Conference on Computer Vision, pp. 2980–2988 (2017)
27. Kisantal, M., Wojna, Z., Murawski, J., Naruniec, J., Cho, K.: Augmentation for small object detection. arXiv preprint arXiv:1902.07296 (2019)
28. Liu, W., et al.: SSD: single shot multibox detector. In: Leibe, B., Matas, J., Sebe, N., Welling, M. (eds.) ECCV 2016. LNCS, vol. 9905, pp. 21–37. Springer, Cham (2016). https://doi.org/10.1007/978-3-319-46448-0_2
29. Hosang, J., Benenson, R., Dollár, P., Schiele, B.: What makes for effective detection proposals? IEEE Trans. Pattern Anal. Mach. Intell. **38**(4), 814–830 (2015)
30. Loshchilov, I., Hutter, F.: SGDR: stochastic gradient descent with warm restarts. arXiv preprint arXiv:1608.03983 (2016)
31. Ren, S., He, K., Girshick, R., Sun, J.: Faster r-cnn: Towards real-time object detection with region proposal networks. Adv. Neural Inf. Proc. Syst. **28** (2015)

Intelligent Control and Application

Controllability of Windmill Networks

Pengcheng Guo[1], Pengchao Lv[2], Junjie Huang[2], and Bo Liu[1(✉)]

[1] School of Information Engineering, Key Laboratory of Ethnic Language Intelligent Analysis and Security Governance, Minzu University of China, Beijing 100081, China
`22301984@muc.edu.cn, boliu2020@muc.edu.cn`
[2] School of Mathematical Sciences, Inner Mongolia University, Hohhot 010021, China
`0131122584@mail.imu.edu.cn, huangjunjie@imu.edu.cn`

Abstract. This paper mainly studies the controllability of the windmill networks. Through analysing the adjacency matrix of the windmill graph, the eigenvalue and eigenvector of this matrix are obtained, then some sufficient and necessary conditions for the controllability of the windmill networks are obtained. Moreover, some examples and simulations are given to verify the correctness of the result obtained.

Keywords: Windmill networks · Controllability · Multi-agent system

1 Introduction

The multi-agent network (MAN) is a collection of multiple agents and each of which has the ability to communicate and process information. In recent years, since the coordination and cooperation in MANs can improve the performance of the system and complete the complex tasks, such networks have been applied in many fields such as intelligent management of transportation systems, cooperation in scientific research projects and so on [1–4]. These applications make the MAN become an important research direction in academia [5–8]. In [5], the authors presented a distributed algorithm based on event-triggered strategy to achieve elastic consensus of MANs under spoofing attacks. A security control scheme against attack event triggering for a class of nonlinear MANs with input quantization was proposed in [7].

Since Tanner [9] first studied the controllability of MANs in fixed topology, as a basic problem of the coordinative control of MANs, controllability has been widely used in many fields [10,11] and received more and more scholars' attention [12–18]. Liu et al. studied the controllability and switching controllability

This work was supported in part by the National Natural Science Foundation of China under Grant 62173355 and Grant 11961052; in part by the Program for Innovative Research Team in Universities of Inner Mongolia Autonomous Region under Grant NMGIRT2317; in part by the Natural Science Foundation of Inner Mongolia under Grant 2021MS01006.

L. Pan et al. (Eds.): BIC-TA 2023, CCIS 2062, pp. 197–212, 2024.
https://doi.org/10.1007/978-981-97-2275-4_16

of discrete-time MANs in [12,13]. The authors in [15] considered the structural controllability of MANs with switching topology under sampled data settings. In [16], a fractional-order model with time delay was established, and the fractional-order controllability of MANs in fixed topology and switching topology was discussed. From the view of graph theory, [17] studied the controllability of MANs by using almost equitable partitions method. The authors in [18] investigated the controllability of matrix-weighted discrete-time leader-follower MANs by dividing the Laplacian matrix. At the same time, there are other results about the observability, such as observability for MANs with switching topology [19], observability for heterogeneous MANs [20], second-order observability for matrix-weighted networks [21], etc. Besides, the observability and controllability of the MANs were considered simultaneously in [22–25].

As a special MAN, the windmill network has become a popular topic recently. In [26], the authors determined the total constraint domination number and the fine fraction of the windmill diagram. In [27], the strong metric dimension of the windmill graph was obtained. The authors in [28] studied the topological index of the windmill graph, and a series of indexes such as the Wiener index and hyper Wiener index of the windmill graph were calculated. Estrada [29] investigated the spectra of adjacency matrix and Laplacian matrix of the windmill graph, on this basis, some properties of the generalized windmill graph were studied in [30]. The metric dimension and side metric dimension of two kinds of windmill graphs: the French windmill graph and the Dutch windmill graph were calculated in [31], and some generalizations of these two kinds of windmill graphs were given. It is worth noting that the research on the controllability of windmill graph is still in its fancy. This paper focus on the controllability of the windmill graph, the main contributions of this paper are as follows: (i) This paper studied the controllability of windmill graph. (ii) The eigenvalues and eigenvectors of the adjacency matrix are obtained. (iii) Some necessary and sufficient conditions for the controllability of windmill graph are proposed.

The content of this paper is arranged as follows: The model and some preliminaries are given in Sect. 2. Section 3 exhibits the main theoretical results. Examples and simulations are given in Sect. 4. Finally, Sect. 5 offers a conclusion.

2   Problem Statement

Consider a MAN consisting of $3n+1$ agents with an undirected and un-weighted topology G in the following form:

$$\dot{x}_i(t) = \sum_{j=1}^{3n+1} a_{ij}x_j(t) + \delta_i u_i, \quad i = 1, 2, \ldots, 3n+1, \tag{1}$$

where x_i, $u_i \in R$ are respectively the state and control input of agent i, $a_{ij} \in R$ is the coupling strength between agents i and j. $\delta_i = 1$ if agent i has information with the leader, otherwise $\delta_i = 0$. Let $x = [x_1, x_2, \ldots, x_{3n+1}]^T$ and

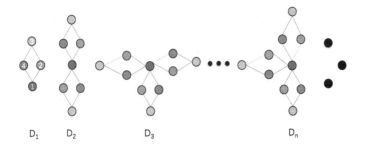

Fig. 1. Windmill diagram topology

$u = [u_1, u_2, \ldots, u_{3n+1}]^T$. Then, the network (1) can be rewritten in a compact form as

$$\dot{x}(t) = A(G)x(t) + Bu(t), \tag{2}$$

where $A(G) = [a_{ij}] \in R^{(3n+1)\times(3n+1)}$ and $B = diag\{\delta_1, \delta_2, \ldots, \delta_{3n+1}\}$. In this work, let D_n be a windmill graph with $3n+1$ vertices and $4n$ edges, which is formed by merging n circle graph D_1 with a common vertex and shown in Fig. 1.

Without loss of generality, we label the agents of the cycle graph D_1 as 1, 2, 3, and 4 shown in Fig. 1, where agent 1 is the common vertex, and the weights between interaction edges are denotes as a, b, c, d, which are all non-zero. For the sake of convenience in discussion, we rearrange the order of the agents. Let $X_1 = [x_1, \underbrace{x_2, x_3, x_4}] \triangleq [x_1, \overline{x}]^T$ and $U_1 = [u_1, \underbrace{u_2, u_3, u_4}] \triangleq [u_1, \overline{u}]^T$ be the whole state and the total external control input of the network D_1. respectively, where $\overline{x} = [x_2, x_3, x_4]$ and $\overline{u} = [u_2, u_3, u_4]$; and $X_n = [x_1, x_2, x_3, x_4, x_2, x_3, x_4, \ldots, x_2, x_3, x_4]^T = [x_1, \overline{x}, \overline{x}, \ldots, \overline{x}]$ and $U_n = [u_1, u_2, u_3, u_4, u_2, u_3, u_4, \ldots, u_2, u_3, u_4]^T = [u_1, \underbrace{\overline{u}, \overline{u}, \ldots, \overline{u}}]$ be the whole state and the total external control input of the network D_n for n $\geqslant$ 2. Then the network (2) can be rewritten in a compact form as

$$\dot{X}_n = A_n X_n + B_n U_n, \tag{3}$$

where

$$A_1 = \begin{bmatrix} 0 & a & 0 & d \\ a & 0 & b & 0 \\ 0 & b & 0 & c \\ d & 0 & c & 0 \end{bmatrix} \triangleq \begin{bmatrix} 0 & \alpha^T \\ \alpha & \mathcal{A} \end{bmatrix}, \tag{4}$$

$$A_n = \begin{bmatrix} 0 & \mathbf{1}_n^T \otimes \alpha^T \\ \mathbf{1}_n \otimes \alpha & I_n \otimes \mathcal{A} \end{bmatrix}_{(3n+1)\times(3n+1)}, \tag{5}$$

$$B_1 = \begin{bmatrix} \delta_1 & 0 & 0 & 0 \\ 0 & \delta_2 & 0 & 0 \\ 0 & 0 & \delta_3 & 0 \\ 0 & 0 & 0 & \delta_4 \end{bmatrix} \triangleq \begin{bmatrix} \delta_1 & \mathbf{0}_3^T \\ \mathbf{0}_3 & \mathcal{B} \end{bmatrix}, \tag{6}$$

$$B_n = \begin{bmatrix} \delta_1 & \mathbf{0}_{3n}^T \\ \mathbf{0}_{3n} & I_n \otimes \mathcal{B} \end{bmatrix}_{(3n+1)\times(3n+1)}, \tag{7}$$

$\alpha = \begin{bmatrix} a & 0 & d \end{bmatrix}^T$, $\mathbf{0}_n$ and $\mathbf{1}_n$ are respectively the all-0 and all-1 column vector, I_n is the $n \times n$ identity matrix.

The characteristic polynomial of D_n is $\Phi_{D_n}(A_n : \lambda) \triangleq det(\lambda I_{3n+1} - A_n) = \lambda^{n-1} [\lambda^4 - n\lambda^2(a^2 + d^2) - \lambda^2(b^2 + c^2) + n(ac - bd)^2](\lambda^2 - b^2 - c^2)^{n-1}$. Moreover, the spectra of D_n can be given by:

(i) 0, repeated $n - 1$ times;
(ii) $\sqrt{b^2 + c^2}$ repeated $n - 1$ times;
(iii) $-\sqrt{b^2 + c^2}$ repeated $n - 1$ times;
(iv) the four roots of the equation

$$\lambda^4 - n\lambda^2(a^2 + d^2) - \lambda^2(b^2 + c^2) + n(ac - bd)^2 = 0. \tag{8}$$

Denote $e_1, e_2, \ldots, e_n$ as the standard basis vectors. For the parameters a, b, c, d, we will discuss the two cases as follows.

Case (I). As $ac = bd$, the linear-independent left eigenvectors corresponding to the eigenvalues $\lambda_i = 0$ (repeated $n + 1$ times), $\lambda_{j+} = \sqrt{b^2 + c^2}$, $\lambda_{j-} = -\sqrt{b^2 + c^2}$, $\lambda_{k+} = \sqrt{na^2 + nd^2 + b^2 + c^2}$, $\lambda_{k-} = -\sqrt{na^2 + nd^2 + b^2 + c^2}$ can be expressed as

$$\xi_i = \begin{bmatrix} e_i \otimes \begin{bmatrix} 0 \\ c \\ 0 \\ -b \end{bmatrix} \end{bmatrix}^T, (e_i \in R^n, \ i = 1, \ldots, n), \tag{9}$$

$$\xi_i = \begin{bmatrix} \mathbf{1}_n \otimes \begin{bmatrix} b \\ 0 \\ -a \\ 0 \end{bmatrix} \end{bmatrix}^T, (i = n + 1), \tag{10}$$

$$\xi_{j+} = \begin{bmatrix} \begin{bmatrix} -1 \\ e_j \end{bmatrix} \otimes \begin{bmatrix} 0 \\ b \\ \sqrt{b^2 + c^2} \\ c \end{bmatrix} \end{bmatrix}^T, \tag{11}$$

$$\xi_{j-} = \begin{bmatrix} \begin{bmatrix} -1 \\ e_j \end{bmatrix} \otimes \begin{bmatrix} 0 \\ b \\ -\sqrt{b^2 + c^2} \\ c \end{bmatrix} \end{bmatrix}^T, (e_j \in R^{n-1}, \ j = 1, \ldots, n - 1), \tag{12}$$

$$\xi_{k+} = \begin{bmatrix} \mathbf{1}_n \otimes \begin{bmatrix} \sqrt{na^2 + nd^2 + b^2 + c^2} - \frac{b^2 + c^2}{\sqrt{na^2 + nd^2 + b^2 + c^2}} \\ a \\ \frac{ab + cd}{\sqrt{na^2 + nd^2 + b^2 + c^2}} \\ d \end{bmatrix} \end{bmatrix}^T, \tag{13}$$

$$\xi_{k-} = \left[\begin{array}{c} \frac{b^2+c^2}{\sqrt{na^2+nd^2+b^2+c^2}} - \sqrt{na^2+nd^2+b^2+c^2} \\ \mathbf{1}_n \otimes \left[\begin{array}{c} a \\ -\frac{ab+cd}{\sqrt{na^2+nd^2+b^2+c^2}} \\ d \end{array} \right] \end{array} \right]^T , \quad (14)$$

respectively. Let

$$P_1 \triangleq [\underbrace{\dots \xi_i \dots,}_{n+1} \underbrace{\dots \xi_{j+} \dots,}_{n-1} \underbrace{\dots \xi_{j-} \dots,}_{n-1} \xi_{k+}, \xi_{k-}], \quad (15)$$

$$\Lambda_1 \triangleq diag\{\underbrace{0,0,\dots,0,}_{n+1} \underbrace{\dots, \sqrt{b^2+c^2}, \dots,}_{n-1} \underbrace{\dots, -\sqrt{b^2+c^2}, \dots,}_{n-1} \lambda_{k+}, \lambda_{k-}\}, \quad (16)$$

we can know that P_1 is invertible.

Case (II). As $ac \neq bd$, for the eigenvalues $\lambda_i = 0$, $\lambda_{j+} = \sqrt{b^2+c^2}$, $\lambda_{j-} = -\sqrt{b^2+c^2}$,

$$\lambda_{k+} = \sqrt{\frac{na^2+nd^2+b^2+c^2 + \sqrt{(na^2+nd^2+b^2+c^2)^2 - 4n(ac-bd)^2}}{2}}, \quad (17)$$

$$\lambda_{k-} = -\sqrt{\frac{na^2+nd^2+b^2+c^2 + \sqrt{(na^2+nd^2+b^2+c^2)^2 - 4n(ac-bd)^2}}{2}}, \quad (18)$$

$$\lambda_{l+} = \sqrt{\frac{na^2+nd^2+b^2+c^2 - \sqrt{(na^2+nd^2+b^2+c^2)^2 - 4n(ac-bd)^2}}{2}}, \quad (19)$$

$$\lambda_{l-} = -\sqrt{\frac{na^2+nd^2+b^2+c^2 - \sqrt{(na^2+nd^2+b^2+c^2)^2 - 4n(ac-bd)^2}}{2}}, \quad (20)$$

the linear-independent left eigenvectors corresponding to the eigenvalues mentioned above can be expressed as

$$\xi_i = \left[\begin{bmatrix} -1 \\ e_i \end{bmatrix} \otimes \begin{bmatrix} 0 \\ c \\ 0 \\ -b \end{bmatrix} \right]^T , (e_i \in R^{n-1}, i = 1, \dots, n-1), \quad (21)$$

$$\xi_{j+} = \left[\begin{bmatrix} -1 \\ e_j \end{bmatrix} \otimes \begin{bmatrix} 0 \\ b \\ \sqrt{b^2+c^2} \\ c \end{bmatrix} \right]^T , \quad (22)$$

$$\xi_{j-} = \left[\begin{bmatrix} -1 \\ e_j \end{bmatrix} \otimes \begin{bmatrix} 0 \\ b \\ -\sqrt{b^2 + c^2} \\ c \end{bmatrix} \right]^T, (e_j \in R^{n-1}, \; j = 1, \ldots, n-1), \qquad (23)$$

$$\xi_{k+} = \left[1_n \otimes \begin{bmatrix} n(a\lambda_{k+}^2 - ac^2 + bcd) \\ \lambda_{k+}(\lambda_{k+}^2 - nd^2 - c^2) \\ b\lambda_{k+}^2 + nacd - nbd^2 \\ \lambda_{k+}(nad + bc) \end{bmatrix} \right]^T, \qquad (24)$$

$$\xi_{k-} = \left[1_n \otimes \begin{bmatrix} n(a\lambda_{k-}^2 - ac^2 + bcd) \\ \lambda_{k-}(\lambda_{k+}^2 - nd^2 - c^2) \\ b\lambda_{k-}^2 + nacd - nbd^2 \\ \lambda_{k-}(nad + bc) \end{bmatrix} \right]^T, \qquad (25)$$

$$\xi_{l+} = \left[1_n \otimes \begin{bmatrix} n(a\lambda_{l+}^2 - ac^2 + bcd) \\ \lambda_{l+}(\lambda_{k+}^2 - nd^2 - c^2) \\ b\lambda_{l+}^2 + nacd - nbd^2 \\ \lambda_{l+}(nad + bc) \end{bmatrix} \right]^T, \qquad (26)$$

$$\xi_{l-} = \left[1_n \otimes \begin{bmatrix} n(a\lambda_{l-}^2 - ac^2 + bcd) \\ \lambda_{l-}(\lambda_{k+}^2 - nd^2 - c^2) \\ b\lambda_{l-}^2 + nacd - nbd^2 \\ \lambda_{l-}(nad + bc) \end{bmatrix} \right]^T, \qquad (27)$$

respectively. Let

$$P_2 \triangleq [\underbrace{\ldots \xi_i \ldots}_{n-1}, \underbrace{\ldots \xi_{j+} \ldots}_{n-1}, \underbrace{\ldots \xi_{j-} \ldots}_{n-1}, \xi_{k+}, \xi_{k-}, \xi_{l+}, \xi_{l-}], \qquad (28)$$

$$\Lambda_1 \triangleq diag\{\underbrace{0, 0, \ldots, 0}_{n-1}, \underbrace{\ldots, \sqrt{b^2 + c^2}, \ldots}_{n-1}, \underbrace{\ldots, -\sqrt{b^2 + c^2}, \ldots}_{n-1}, \lambda_{k+}, \lambda_{k-}, \lambda_{l+}, \lambda_{l-}\}, \qquad (29)$$

we can know that P_2 is invertible.

3   Controllability Analysis

Proposition 1. *MAN* (3) *is controllable if and only if* $\alpha^T A_n = \lambda_i \alpha^T$ & $\alpha^T B_n = 0_{3n+1}^T \Rightarrow \alpha = 0_{3n+1}$, *where* λ_i *is the eigenvalue of* A_n *for* $\forall i \in \{1, \ldots, 3n+1\}$.

Theorem 1. *If* $ac = bd$, *then system* (3) *is controllable if and only if* $1 \in \{\delta_1, \delta_3\} \cap \{\delta_2, \delta_4\}$, *that is, at least one of* δ_1 *and* δ_3 *is non-zero, at least one of* δ_2 *and* δ_4 *is non-zero.*

Proof. Sufficiency. If $1 \in \{\delta_1, \delta_3\} \cap \{\delta_2, \delta_4\}$, we can have

$$
\xi_i B_n = \left[e_i \otimes \begin{bmatrix} 0 \\ \begin{bmatrix} c \\ 0 \\ -b \end{bmatrix} \end{bmatrix} \right]^T \begin{bmatrix} \delta_1 & \mathbf{0}_{3n}^T \\ \mathbf{0}_{3n} & I_n \otimes \mathcal{B} \end{bmatrix} = \left[e_i \otimes \begin{bmatrix} 0 \\ \begin{bmatrix} c \\ 0 \\ -b \end{bmatrix} \end{bmatrix} \right]^T \begin{bmatrix} \delta_1 & \mathbf{0}_{3n}^T \\ \mathbf{0}_{3n} & I_n \otimes \mathcal{B}^T \end{bmatrix}^T
$$

$$
= \left[\begin{bmatrix} \delta_1 & \mathbf{0}_{3n}^T \\ \mathbf{0}_{3n} & I_n \otimes \mathcal{B}^T \end{bmatrix} \left[e_i \otimes \begin{bmatrix} 0 \\ \begin{bmatrix} c \\ 0 \\ -b \end{bmatrix} \end{bmatrix} \right] \right]^T = \left[e_i \otimes \begin{bmatrix} 0 \\ \begin{bmatrix} c\delta_2 \\ 0 \\ -b\delta_4 \end{bmatrix} \end{bmatrix} \right]^T \neq \mathbf{0}_{3n+1}^T, \ i = 1, 2, \ldots, n;
$$

$$(30)$$

$$
\xi_i B_n = \left[\mathbf{1}_n \otimes \begin{bmatrix} b \\ \begin{bmatrix} 0 \\ -a \\ 0 \end{bmatrix} \end{bmatrix} \right]^T \begin{bmatrix} \delta_1 & \mathbf{0}_{3n}^T \\ \mathbf{0}_{3n} & I_n \otimes \mathcal{B} \end{bmatrix} = \left[\mathbf{1}_n \otimes \begin{bmatrix} b \\ \begin{bmatrix} 0 \\ -a \\ 0 \end{bmatrix} \end{bmatrix} \right]^T \begin{bmatrix} \delta_1 & \mathbf{0}_{3n}^T \\ \mathbf{0}_{3n} & I_n \otimes \mathcal{B}^T \end{bmatrix}^T
$$

$$
= \left[\begin{bmatrix} \delta_1 & \mathbf{0}_{3n}^T \\ \mathbf{0}_{3n} & I_n \otimes \mathcal{B}^T \end{bmatrix} \left[\mathbf{1}_n \otimes \begin{bmatrix} b \\ \begin{bmatrix} 0 \\ -a \\ 0 \end{bmatrix} \end{bmatrix} \right] \right]^T = \left[\mathbf{1}_n \otimes \begin{bmatrix} b\delta_1 \\ \begin{bmatrix} 0 \\ -a\delta_3 \\ 0 \end{bmatrix} \end{bmatrix} \right]^T \neq \mathbf{0}_{3n+1}^T, i = n+1;
$$

$$(31)$$

$$
\xi_{j+} B_n = \left[\begin{bmatrix} -1 \\ e_j \end{bmatrix} \otimes \begin{bmatrix} 0 \\ \begin{bmatrix} b \\ \sqrt{b^2+c^2} \\ c \end{bmatrix} \end{bmatrix} \right]^T \begin{bmatrix} \delta_1 & \mathbf{0}_{3n}^T \\ \mathbf{0}_{3n} & I_n \otimes \mathcal{B} \end{bmatrix} = \left[\begin{bmatrix} -1 \\ e_j \end{bmatrix} \otimes \begin{bmatrix} 0 \\ \begin{bmatrix} b\delta_2 \\ \sqrt{b^2+c^2}\delta_3 \\ c\delta_4 \end{bmatrix} \end{bmatrix} \right]^T
$$

$$
\neq \mathbf{0}_{3n+1}^T, \ j = 1, 2, \cdots, n-1;
$$

$$(32)$$

$$
\xi_{j-} B_n = \left[\begin{bmatrix} -1 \\ e_j \end{bmatrix} \otimes \begin{bmatrix} 0 \\ \begin{bmatrix} b \\ -\sqrt{b^2+c^2} \\ c \end{bmatrix} \end{bmatrix} \right]^T \begin{bmatrix} \delta_1 & \mathbf{0}_{3n}^T \\ \mathbf{0}_{3n} & I_n \otimes \mathcal{B} \end{bmatrix}
$$

$$(33)$$

$$
= \left[\begin{bmatrix} -1 \\ e_j \end{bmatrix} \otimes \begin{bmatrix} 0 \\ \begin{bmatrix} b\delta_2 \\ -\sqrt{b^2+c^2}\delta_3 \\ c\delta_4 \end{bmatrix} \end{bmatrix} \right]^T \neq \mathbf{0}_{3n+1}^T, \ j = 1, 2, \cdots, n-1;
$$

$$\xi_{k+}B_n = \begin{bmatrix} \sqrt{na^2 + nd^2 + b^2 + c^2} - \frac{b^2+c^2}{\sqrt{na^2+nd^2+b^2+c^2}} \\ \mathbf{1}_n \otimes \begin{bmatrix} a \\ \frac{ab+cd}{\sqrt{na^2+nd^2+b^2+c^2}} \\ d \end{bmatrix} \end{bmatrix}^T \begin{bmatrix} \delta_1 & \mathbf{0}_{3n}^T \\ \mathbf{0}_{3n} & I_n \otimes \mathcal{B} \end{bmatrix}$$

$$= \begin{bmatrix} (\sqrt{na^2 + nd^2 + b^2 + c^2} - \frac{b^2+c^2}{\sqrt{na^2+nd^2+b^2+c^2}})\delta_1 \\ \mathbf{1}_n \otimes \begin{bmatrix} a\delta_2 \\ \frac{ab+cd}{\sqrt{na^2+nd^2+b^2+c^2}}\delta_3 \\ d\delta_4 \end{bmatrix} \end{bmatrix}^T \neq \mathbf{0}_{3n+1}^T; \tag{34}$$

$$\xi_{k-}B_n = \begin{bmatrix} \frac{b^2+c^2}{\sqrt{na^2+nd^2+b^2+c^2}} - \sqrt{na^2 + nd^2 + b^2 + c^2} \\ \mathbf{1}_n \otimes \begin{bmatrix} a \\ -\frac{ab+cd}{\sqrt{na^2+nd^2+b^2+c^2}} \\ d \end{bmatrix} \end{bmatrix}^T \begin{bmatrix} \delta_1 & \mathbf{0}_{3n}^T \\ \mathbf{0}_{3n} & I_n \otimes \mathcal{B} \end{bmatrix}$$

$$= \begin{bmatrix} (\frac{b^2+c^2}{\sqrt{na^2+nd^2+b^2+c^2}} - \sqrt{na^2 + nd^2 + b^2 + c^2})\delta_1 \\ \mathbf{1}_n \otimes \begin{bmatrix} a\delta_2 \\ -\frac{ab+cd}{\sqrt{na^2+nd^2+b^2+c^2}}\delta_3 \\ d\delta_4 \end{bmatrix} \end{bmatrix}^T \neq \mathbf{0}_{3n+1}^T. \tag{35}$$

Therefore, system (3) is controllable if $1 \in \{\delta_1, \delta_3\} \cap \{\delta_2, \delta_4\}$ from Proposition 1.

Necessity. By contradiction, if $1 \notin \{\delta_1, \delta_3\} \cap \{\delta_2, \delta_4\}$, we can have $\delta_1 = \delta_3 = 0$ or $\delta_2 = \delta_4 = 0$. According to the calculation mentioned in the sufficiency, there must be a left eigenvector ξ of A_n such that $\xi B_n = \mathbf{0}_{3n+1}^T$, it means system (3) is not controllable from Proposition 1, which is a contradiction.

Theorem 2. *If $ac \neq bd$ and $nad + bc \neq 0$, then system (3) is controllable if and only if at least one of δ_2 and δ_4 is non-zero.*

Proof. If $ac \neq bd$, the eigenvalues λ_{k+}, λ_{k-}, λ_{l+} and λ_{l-} are all non-zero. Besides, let $\lambda \in \{\lambda_{k+}, \lambda_{k-}, \lambda_{l+}, \lambda_{l-}\}$, then

$$\lambda^2 - nd^2 - c^2 = \frac{na^2 - nd^2 + b^2 - c^2 \pm \sqrt{(na^2 - nd^2 + b^2 - c^2)^2 + 4(nad + bc)^2}}{2} \tag{36}$$

is non-zero since $nad + bc \neq 0$.

Sufficiency. If at least one of δ_2 and δ_4 is non-zero, we can have

$$
\xi_i B_n = \left[\begin{bmatrix} -1 \\ e_i \end{bmatrix} \otimes \begin{bmatrix} 0 \\ c \\ 0 \\ -b \end{bmatrix} \right]^T \begin{bmatrix} \delta_1 & \mathbf{0}_{3n}^T \\ \mathbf{0}_{3n} & I_n \otimes \mathcal{B} \end{bmatrix} = \left[\begin{bmatrix} \delta_1 & \mathbf{0}_{3n}^T \\ \mathbf{0}_{3n} & I_n \otimes \mathcal{B}^T \end{bmatrix} \left[\begin{bmatrix} -1 \\ e_i \end{bmatrix} \otimes \begin{bmatrix} 0 \\ c \\ 0 \\ -b \end{bmatrix} \right] \right]^T
$$
$$
= \left[\begin{bmatrix} -1 \\ e_i \end{bmatrix} \otimes \begin{bmatrix} 0 \\ c\delta_2 \\ 0 \\ -b\delta_4 \end{bmatrix} \right]^T \neq \mathbf{0}_{3n+1}^T, \ i = 1, 2, \ldots, n-1;
$$
$$
\tag{37}
$$

$$
\xi_{j+} B_n = \left[\begin{bmatrix} -1 \\ e_j \end{bmatrix} \otimes \begin{bmatrix} 0 \\ b \\ \sqrt{b^2 + c^2} \\ c \end{bmatrix} \right]^T \begin{bmatrix} \delta_1 & \mathbf{0}_{3n}^T \\ \mathbf{0}_{3n} & I_n \otimes \mathcal{B} \end{bmatrix} = \left[\begin{bmatrix} -1 \\ e_j \end{bmatrix} \otimes \begin{bmatrix} 0 \\ b\delta_2 \\ \sqrt{b^2 + c^2}\delta_3 \\ c\delta_4 \end{bmatrix} \right]^T
$$
$$
\neq \mathbf{0}_{3n+1}^T, \ j = 1, 2, \cdots, n-1;
$$
$$
\tag{38}
$$

$$
\xi_{j-} B_n = \left[\begin{bmatrix} -1 \\ e_j \end{bmatrix} \otimes \begin{bmatrix} 0 \\ b \\ -\sqrt{b^2 + c^2} \\ c \end{bmatrix} \right]^T \begin{bmatrix} \delta_1 & \mathbf{0}_{3n}^T \\ \mathbf{0}_{3n} & I_n \otimes \mathcal{B} \end{bmatrix} = \left[\begin{bmatrix} -1 \\ e_j \end{bmatrix} \otimes \begin{bmatrix} 0 \\ b\delta_2 \\ -\sqrt{b^2 + c^2}\delta_3 \\ c\delta_4 \end{bmatrix} \right]^T
$$
$$
\neq \mathbf{0}_{3n+1}^T, \ j = 1, 2, \cdots, n-1;
$$
$$
\tag{39}
$$

$$
\xi_{k+} B_n = \left[\mathbf{1}_n \otimes \begin{bmatrix} n(a\lambda_{k+}^2 - ac^2 + bcd)\delta_1 \\ \lambda_{k+}(\lambda_{k+}^2 - nd^2 - c^2)\delta_2 \\ (b\lambda_{k+}^2 + nacd - nbd^2)\delta_3 \\ \lambda_{k+}(nad + bc)\delta_4 \end{bmatrix} \right]^T \neq \mathbf{0}_{3n+1}^T; \tag{40}
$$

$$
\xi_{k-} B_n = \left[\mathbf{1}_n \otimes \begin{bmatrix} n(a\lambda_{k-}^2 - ac^2 + bcd)\delta_1 \\ \lambda_{k-}(\lambda_{k-}^2 - nd^2 - c^2)\delta_2 \\ (b\lambda_{k-}^2 + nacd - nbd^2)\delta_3 \\ \lambda_{k-}(nad + bc)\delta_4 \end{bmatrix} \right]^T \neq \mathbf{0}_{3n+1}^T; \tag{41}
$$

$$
\xi_{l+} B_n = \left[\mathbf{1}_n \otimes \begin{bmatrix} n(a\lambda_{l+}^2 - ac^2 + bcd)\delta_1 \\ \lambda_{l+}(\lambda_{k+}^2 - nd^2 - c^2)\delta_2 \\ (b\lambda_{l+}^2 + nacd - nbd^2)\delta_3 \\ \lambda_{l+}(nad + bc)\delta_4 \end{bmatrix} \right]^T \neq \mathbf{0}_{3n+1}^T; \tag{42}
$$

$$\xi_{l-}B_n = \left[\mathbf{1}_n \otimes \begin{bmatrix} n(a\lambda_{l-}^2 - ac^2 + bcd)\delta_1 \\ \lambda_{l-}(\lambda_{k+}^2 - nd^2 - c^2)\delta_2 \\ (b\lambda_{l-}^2 + nacd - nbd^2)\delta_3 \\ \lambda_{l-}(nad + bc)\delta_4 \end{bmatrix} \right]^T \neq \mathbf{0}_{3n+1}^T. \tag{43}$$

Therefore, system (3) is controllable from Proposition 1.

Necessity. By contradiction, if $\delta_2 = \delta_4 = 0$. According to the calculation mentioned in the sufficiency, there must be a left eigenvector ξ of A_n such that $\xi B_n = \mathbf{0}_{3n+1}^T$, it means system (3) is not controllable from Proposition 1, which is a contradiction.

4   Examples and Simulations

In this section, we will use some examples to verify the correctness of the results obtained.

Example 1. A seven-node windmill network is described by graph D_2 in Fig. 1, where $a = b = c = d = 1$, $\delta_1 = 1$, $\mathcal{B} = \begin{bmatrix} 1\,0\,0 \\ 0\,0\,0 \\ 0\,0\,0 \end{bmatrix}$, let A_2 and B_2 be

$$A_2 = \begin{bmatrix} 0\,1\,0\,1\,1\,0\,1 \\ 1\,0\,1\,0\,0\,0\,0 \\ 0\,1\,0\,1\,0\,0\,0 \\ 1\,0\,1\,0\,0\,0\,0 \\ 1\,0\,0\,0\,0\,1\,0 \\ 0\,0\,0\,0\,1\,0\,1 \\ 1\,0\,0\,0\,0\,1\,0 \end{bmatrix}, \tag{44}$$

$$B_2 = \begin{bmatrix} 1\,0\,0\,0\,0\,0\,0 \\ 0\,1\,0\,0\,0\,0\,0 \\ 0\,0\,0\,0\,0\,0\,0 \\ 0\,0\,0\,0\,0\,0\,0 \\ 0\,0\,0\,0\,1\,0\,0 \\ 0\,0\,0\,0\,0\,0\,0 \\ 0\,0\,0\,0\,0\,0\,0 \end{bmatrix}. \tag{45}$$

Since $ac = bd$ and $\delta_1 = \delta_2 = 1$, the controllability of the windmill network (3) can be obtained from Theorem 1.

Besides, the eigenvalues and matrix $\xi_{A_2} = \begin{bmatrix} \xi_1^T & \xi_2^T & \xi_3^T & \xi_4^T & \xi_5^T & \xi_6^T & \xi_7^T \end{bmatrix}^T$ consisting of the left eigenvectors of A_2 can be obtained respectively:

$$\lambda(A_2) = \{-2.4495, -1.4142, 0, 0, 0, 1.4142, 2.4495\}, \tag{46}$$

$$\xi_{A_2} = \begin{bmatrix} -0.5774 & 0.3536 & -0.2887 & 0.3536 & 0.3536 & -0.2887 & 0.3536 \\ 0 & -0.3536 & 0.5000 & -0.3536 & 0.3536 & -0.5000 & 0.3536 \\ 0.5033 & 0.3392 & -0.5033 & -0.3392 & -0.0702 & -0.5033 & 0.0702 \\ 0.2610 & -0.5020 & -0.2610 & 0.5020 & 0.3818 & -0.2610 & -0.3818 \\ -0.1089 & 0.3646 & 0.1089 & -0.3646 & 0.5910 & 0.1089 & -0.5910 \\ 0 & -0.3536 & -0.5000 & -0.3536 & 0.3536 & 0.5000 & 0.3536 \\ -0.5774 & -0.3536 & -0.2887 & -0.3536 & -0.3536 & -0.2887 & -0.3536 \end{bmatrix} . \quad (47)$$

Since

$$\xi_{A_2 B_2} = \begin{bmatrix} -0.5774 & 0.3536 & 0\,0 & 0.3536 & 0\,0 \\ 0 & -0.3536 & 0\,0 & 0.3536 & 0\,0 \\ 0.5033 & 0.3392 & 0\,0 & -0.0702 & 0\,0 \\ 0.2610 & -0.5020 & 0\,0 & 0.3818 & 0\,0 \\ -0.1089 & 0.3646 & 0\,0 & 0.5910 & 0\,0 \\ 0 & -0.3536 & 0\,0 & 0.3536 & 0\,0 \\ -0.5774 & -0.3536 & 0\,0 & -0.3536 & 0\,0 \end{bmatrix} , \quad (48)$$

from the PBH criterion, the network is controllable.

Figure 2 shows that the seven agents in the system can reach the desired target state (denoted by "△") from any given initial state (denoted by "★"), and forms a triangle. This figure shows that the system is controllable.

Both the PBH criterion and simulation show that the windmill network is controllable, which can verify the effectiveness of Theorem 1.

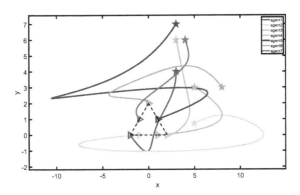

Fig. 2. State trajectories of 7 agents for MAN (3).

Example 2. A four-node windmill network is described by graph D_1 in Fig. 1, where $a = 1$, $b = 2$, $c = 3$, $d = 4$, $\delta_1 = 1$, $\mathcal{B} = \begin{bmatrix} 0\,0\,0 \\ 0\,1\,0 \\ 0\,0\,1 \end{bmatrix}$, let A_1 and B_1 be

$$A_1 = \begin{bmatrix} 0 & 1 & 0 & 4 \\ 1 & 0 & 2 & 0 \\ 0 & 2 & 0 & 3 \\ 4 & 0 & 3 & 0 \end{bmatrix} , \quad (49)$$

$$B_1 = \begin{bmatrix} 0\,0\,0\,0 \\ 0\,0\,0\,0 \\ 0\,0\,1\,0 \\ 0\,0\,0\,1 \end{bmatrix}. \tag{50}$$

Since $ac \neq bd$, $ad + bc \neq 0$ and $\delta_4 = 1$, the controllability of the windmill network (3) can be obtained from Theorem 2.

By calculation, we can get eigenvalues and matrix $\xi_{A_1} = \begin{bmatrix} \xi_1^T & \xi_2^T & \xi_3^T & \xi_4^T \end{bmatrix}^T$ consisting of the left eigenvectors of A_1 as follows:

$$\lambda(A_1) = \{-5.3983, -0.9262, 0.9262, 5.3983\}, \tag{51}$$

$$\xi_{A_1} = \begin{bmatrix} 0.5342 & -0.2706 & 0.4633 & -0.6533 \\ 0.4633 & 0.6533 & -0.5342 & -0.2706 \\ 0.4633 & -0.6533 & -0.5342 & 0.2706 \\ -0.5342 & -0.2706 & -0.4633 & -0.6533 \end{bmatrix}. \tag{52}$$

Since

$$\xi_{A_1} B_1 = \begin{bmatrix} 0\,0 & 0.4633 & -0.6533 \\ 0\,0 & -0.5342 & -0.2706 \\ 0\,0 & -0.5342 & 0.2706 \\ 0\,0 & -0.4633 & -0.6533 \end{bmatrix}, \tag{53}$$

the system is controllable from the PBH criterion.

Figure 3 shows that the seven agents in the system can reach the desired target state (denoted by "$\triangle$") from any given initial state (denoted by "$\bigstar$"), and forms a square. This figure shows that the system is controllable.

Both the PBH criterion and simulation show that the windmill network is controllable, which can verify the effectiveness of Theorem 2.

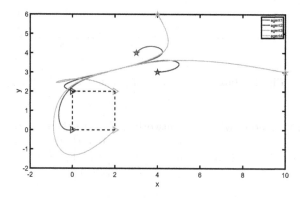

Fig. 3. State trajectories of 4 agents for MAN (3)

Example 3. A seven-node windmill network is described by graph D_2 in Fig. 1, where $a = 1$, $b = 2$, $c = 3$, $d = 4$, $\delta_1 = 1$, $\mathcal{B} = \begin{bmatrix} 1 & 0 & 0 \\ 0 & 0 & 0 \\ 0 & 0 & 0 \end{bmatrix}$, let A_2 and B_2 be

$$A_2 = \begin{bmatrix} 0 & 1 & 0 & 4 & 1 & 0 & 4 \\ 1 & 0 & 2 & 0 & 0 & 0 & 0 \\ 0 & 2 & 0 & 3 & 0 & 0 & 0 \\ 4 & 0 & 3 & 0 & 0 & 0 & 0 \\ 1 & 0 & 0 & 0 & 0 & 2 & 0 \\ 0 & 0 & 0 & 0 & 2 & 0 & 3 \\ 4 & 0 & 0 & 0 & 0 & 3 & 0 \end{bmatrix}, \tag{54}$$

$$B_2 = \begin{bmatrix} 1 & 0 & 0 & 0 & 0 & 0 & 0 \\ 0 & 1 & 0 & 0 & 0 & 0 & 0 \\ 0 & 0 & 0 & 0 & 0 & 0 & 0 \\ 0 & 0 & 0 & 0 & 0 & 0 & 0 \\ 0 & 0 & 0 & 0 & 1 & 0 & 0 \\ 0 & 0 & 0 & 0 & 0 & 0 & 0 \\ 0 & 0 & 0 & 0 & 0 & 0 & 0 \end{bmatrix}. \tag{55}$$

Since $ac \neq bd$, $2ad + bc \neq 0$ and $\delta_2 = 1$, the controllability of the windmill network (3) can be obtained from Theorem 2.

Besides, the eigenvalues and matrix $\xi_{A_2} = \begin{bmatrix} \xi_1^T & \xi_2^T & \xi_3^T & \xi_4^T & \xi_5^T & \xi_6^T & \xi_7^T \end{bmatrix}^T$ consisting of the left eigenvectors of A_2 can be obtained respectively:

$$\lambda(A_2) = \{-6.7758, -3.6056, -1.0436, 0, 1.0436, 3.6056, 6.7758\}, \tag{56}$$

$$\xi_{A_2} = \begin{bmatrix} 0.6059 & -0.1655 & 0.2577 & -0.4718 & -0.1655 & 0.2577 & -0.4718 \\ -0.0000 & 0.2774 & -0.5000 & 0.4160 & -0.2774 & 0.5000 & -0.4160 \\ 0.3645 & 0.4718 & -0.4284 & -0.1655 & 0.4718 & -0.4284 & -0.1655 \\ -0.0000 & 0.5883 & 0.0000 & -0.3922 & -0.5883 & 0.0000 & 0.3922 \\ -0.3645 & 0.4718 & 0.4284 & -0.1655 & 0.4718 & 0.4284 & -0.1655 \\ -0.0000 & -0.2774 & -0.5000 & -0.4160 & 0.2774 & 0.5000 & 0.4160 \\ -0.6059 & -0.1655 & -0.2577 & -0.4718 & -0.1655 & -0.2577 & -0.4718 \end{bmatrix}. \tag{57}$$

Since

$$\xi_{A_2} B_2 = \begin{bmatrix} 0.6059 & -0.1655 & 0 & 0 & -0.1655 & 0 & 0 \\ -0.0000 & 0.2774 & 0 & 0 & -0.2774 & 0 & 0 \\ 0.3645 & 0.4718 & 0 & 0 & 0.4718 & 0 & 0 \\ -0.0000 & 0.5883 & 0 & 0 & -0.5883 & 0 & 0 \\ -0.3645 & 0.4718 & 0 & 0 & 0.4718 & 0 & 0 \\ -0.0000 & -0.2774 & 0 & 0 & 0.2774 & 0 & 0 \\ -0.6059 & -0.1655 & 0 & 0 & -0.1655 & 0 & 0 \end{bmatrix}, \tag{58}$$

from the PBH criterion, the network (3) is controllable.

Figure 4 shows that the seven agents in the system can reach the desired target state (denoted by "$\triangle$") from any given initial state (denoted by "$\bigstar$"), and forms a heptagon. This figure shows that the system is controllable.

Both the PBH criterion and simulation show that the windmill network is controllable, which can verify the effectiveness of Theorem 2.

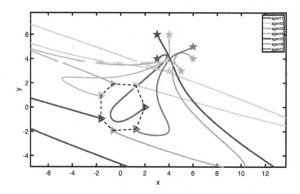

Fig. 4. State trajectories of 7 agents for MAN (3)

5   Conclusion

This paper has considered the controllability of the windmill networks. By computing the eigenvalues and eigenvectors of the adjacency matrix of the system, the controllability of windmill networks can by obtained if some agents are controlled, which is easier to check. In our future work, the observability issue of discrete-time windmill networks will be explored.

References

1. Lu, J.: From single-layer network to multi-layer network structure, dynamics and function. Mod. Phys. Knowl. **27**(4), 3–8 (2015)
2. Guo, K.X., Li, X.X., Xie, L.H.: Simultaneous cooperative relative localization and distributed formation control for multiple UAVs. Sci. China Inf. Sci. **63**, 119201 (2020)
3. Angulo, M.T., Aparicio, A., Moog, C.H.: Structural accessibility and structural observability of nonlinear networked systems. IEEE Trans. Netw. Sci. Eng. **7**(3), 1656–1666 (2020)
4. Wu, W., Peng, Z., Liu, L., Wang, D.: A general safety-certified cooperative control architecture for interconnected intelligent surface vehicles with applications to vessel train. IEEE Trans. Intell. Veh. **7**(3), 627–637 (2022)
5. Wu, Y., Xu, M., Zheng, N., et al.: Event-triggered resilient consensus for multi-agent networks under deception attacks. IEEE Access **8**, 78121–78129 (2020)
6. Zhang, W., Mao, S., Huang, J., et al.: Data-driven resilient control for linear discrete-time multi-agent networks under unconfined cyber-attacks. IEEE Trans. Circ. Syst. I Regul. Pap. **68**(2), 776–785 (2021)

7. Xu, Y., Li, T., Yang, Y., Shan, Q., et al.: Anti-attack event-triggered control for nonlinear multi-agent systems with input quantization. IEEE Trans. Neural Netw. Learn. Syst. **34**(12), 10105–10115 (2022). https://doi.org/10.1109/TNNLS.2022.3164881

8. Peng, H., Shen, X.: Multi-agent reinforcement learning based resource management in MEC- and UAV-assisted vehicular networks. IEEE J. Sel. Areas Commun. **39**(1), 131–141 (2021)

9. Tanner, H.G.: On the controllability of nearest neighbor interconnections. In: 43rd IEEE Conference on Decision and Control, Nassau, The Bahamas, pp. 2467–2472 (2004)

10. Gu, S., Pasqualetti, F., Cieslak, M., et al.: Controllability of structural brain networks. Nat. Commun. **6**, 1–10 (2015). Article number: 8414

11. Pirani, M., Taylor, J.A.: Controllability of AC power networks with DC lines. IEEE Trans. Power Syst. **36**(2), 1649–1651 (2021)

12. Liu, B., Hu, W., Zhang, J., Su, H.: Controllability of discrete-time multi-agent systems with multiple leaders on fixed networks. Commun. Theor. Phys. **58**(6), 856–862 (2012)

13. Liu, B., Su, H., Li, R., et al.: Switching controllability of discrete-time multi-agent systems with multiple leaders and time-delays. Appl. Math. Comput. **228**, 571–588 (2014)

14. Wang, L., Chen, G., Wang, X., Tang, W.K.S.: Controllability of networked MIMO systems. Automatica **69**, 405–409 (2016)

15. Lu, Z., Ji, Z., Zhang, Z.: Sampled-data based structural controllability of multi-agent systems with switching topology. J. Franklin Inst. **357**(15), 10886–10899 (2020)

16. Liu, B., Su, H., Wu, L., et al.: Fractional-order controllability of multi-agent systems with time-delay. Neurocomputing **424**, 268–277 (2021)

17. Zhang, X.F., Sun, J.: Almost equitable partitions and controllability of leader-follower multi-agent systems. Automatica **131**, 109740 (2021)

18. Miao, S., Su, H., Liu, B.: Controllability of discrete-time multi-agent systems with matrix-weighted networks. IEEE Trans. Circ. Syst. II Express Briefs **70**(8), 2984–2988 (2023)

19. Lu, Z., Zhang, L., Wang, L.: Observability of multi-agent systems with switching topology. IEEE Trans. Circ. Syst. II Express Briefs **64**(11), 1317–1321 (2017)

20. Liu, B., Shen, X.: Observability of heterogeneous multi-agent systems. IEEE Trans. Netw. Sci. Eng. **8**(2), 1828–1841 (2021)

21. Lv, P., Huang, J.J., Liu, B., et al.: Second-order observability of matrix-weight-based networks. J. Franklin Inst. **360**(4), 2769–2793 (2023)

22. Franceschelli, M., Martini, S., Egerstedt, M.: Observability and controllability verification in multi-agent systems through decentralized Laplacian spectrum estimation. In: 49th IEEE Conference on Decision and Control, Atlanta, GA, USA, pp. 5775–5780 (2010)

23. Sundaram, S., Hadjicostis, C.N.: Structural controllability and observability of linear systems over finite fields with applications to multi-agent systems. IEEE Trans. Autom. Control **58**(1), 60–73 (2013)

24. Tian, L., Guan, Y., Wang, L.: Controllability and observability of switched multi-agent systems. Int. J. Control **92**(8), 1742–1752 (2017)

25. Tian, L., Guan, Y., Wang, L.: Controllability and observability of multi-agent systems with heterogeneous and switching topologies. Int. J. Control **93**(3), 437–448 (2018)

26. Jeyanthi, P., Hemalatha, G., Davvaz, B.: Results on total restrained domination number and subdivision number for certain graphs. J. Discrete Math. Sci. Cryptogr. **18**(4), 363–369 (2014)
27. Widyaningrum, M., Atmojo Kusmayadi, T.: On the strong metric dimension of sun graph, windmill graph, and Mäbius ladder graph. J. Phys: Conf. Ser. **1008**, 12–32 (2018)
28. Kang, S.: Distance and eccentricity based invariants of windmill graph. J. Discrete Math. Sci. Cryptogr. **22**(7), 1323–1334 (2019)
29. Estrada, E.: When local and global clustering of networks diverge. Linear Algebra Appl. **488**, 249–263 (2016)
30. Kooij, R.: On generalized windmill graphs. Linear Algebra Appl. **565**, 25–46 (2019)
31. Singh, P.: Metric dimension and edge metric dimension of windmill graphs. AIMS Math. **6**(9), 9138–9153 (2021)

Incremental Learning with Maximum Dissimilarity Sampling Based Fault Diagnosis for Rolling Bearings

Yue Fu[1,2] , Juanjuan He[1,2(✉)] , Liuyan Yang[1,2] , and Zilin Luo[1,2]

[1] College of Computer Science and Technology, Wuhan University of Science and Technology, Wuhan 430081, China
{fymm,hejuanjuan}@wust.edu.cn
[2] Hubei Province Key Laboratory of Intelligent Information Processing and Real-Time Industrial System, Wuhan, China

Abstract. Traditional fault diagnosis methods for rolling bearings require retraining the model from scratch when faced with new fault signals, consuming significant computational resources and exhibiting lower efficiency. Class incremental learning can be applied to online fault diagnosis, effectively reducing resource consumption. However, the model often experiences catastrophic forgetting during the training process, a problem that can be addressed by replaying data. Nonetheless, the subsets obtained by the current sampling strategy often fail to cover the distribution of classes, leading to localized information gaps and subsequent degradation in the performance of the fault diagnosis model. To address this, this paper introduces Maximum Dissimilarity Sampling (MDS). Dissimilarity refers to the Euclidean distance between feature vectors of different samples. MDS iteratively selects samples with the maximum dissimilarity to the already chosen subset, aiming to maximize the dissimilarity between samples within the subset and comprehensively cover the class distribution. Additionally, the model tends to favor new classes in decision-making due to the imbalance in class quantities. This paper recommends using Logit Balanced Cross-Entropy Loss (Laloss) to mitigate this issue. Experimental results on the CWRU dataset demonstrate that MDS can comprehensively cover the class distribution and better retain the model's ability to recognize previous tasks. Compared to classical algorithms, our method exhibits superior performance on both the CWRU and MFPT datasets.

Keywords: Fault Diagnosis · Incremental Learning · Dissimilarity

1 Introduction

Rolling bearings play a crucial role in the transmission processes of industrial equipment. Operating in harsh and complex environments, ensuring bearings'

L. Pan et al. (Eds.): BIC-TA 2023, CCIS 2062, pp. 213–226, 2024.
https://doi.org/10.1007/978-981-97-2275-4_17

reliability and safety is paramount. Swiftly diagnosing operational faults in bearings holds significant research importance [1–3]. Online fault diagnosis is a critical factor in enhancing the reliability of industrial systems [4], enabling prompt identification and resolution of faults. Traditional fault diagnosis often employs data-driven approaches, necessitating substantial training data to establish reliable models [5]. This approach incurs high computational and time costs. Moreover, in practical industrial settings, fault data is limited, compromising the model's reliability. Therefore, effectively utilizing limited data to avoid excessive resource wastage has become an urgent concern. Currently, incremental learning methods are used for online fault diagnosis, which can improve the efficiency of fault diagnosis.

Incremental learning, also called continual learning [6] or lifelong learning [7,8], faces one of its most prominent challenges: catastrophic forgetting. Unable to access previous data, incremental learning endeavors to acquire new knowledge from new data while preserving the old knowledge. However, in practical scenarios, the class distribution of each task within a given task stream continually evolves. Consequently, acquiring new knowledge inevitably interferes with maintaining the old knowledge, leading to catastrophic forgetting.

Various advanced methods have been proposed to mitigate catastrophic forgetting. LwF [9] employs a new model to fit the output of the old model. EWC [10] introduces a penalty term to constrain the model's optimization direction. Packnet [11] divides the network into multiple modules, each dedicated to training individual tasks. Rpsnet [12] randomly selects pathways for training the current task. However, these methods are primarily designed to address task-incremental learning, where task increments [11,12] involve multi-headed testing, requiring knowledge of the task ID for each test, making them unsuitable for fault diagnosis in complex industrial environments. In contrast, class-incremental learning [13–17] involves single-headed testing without needing a task ID. It allows testing against all known classes, making it better suited for online fault diagnosis in industrial environments and presenting a more challenging scenario. Replay is one of the effective strategies to mitigate catastrophic forgetting. Icarl [13] proposes retaining a portion of old data and replaying old data while learning new tasks to achieve an effect akin to joint training. Nevertheless, limited memory capacity results in an imbalance in the number of new and old categories and a long-tailed distribution, causing the fully connected (FC) layer to bias towards categorizing instances into new categories [14,15], leading to misclassification of new and old classes [16]. Bic [17] utilizes a balanced subset to fine-tune the model after training. WA [15] corrects the weights of the fc layer to mitigate bias. Lucir [17] introduces cosine normalization to balance new and old classes.

This paper focuses on class-incremental learning based on replaying old data. Many existing articles have made significant contributions to data replay methods, but these methods primarily select subsets that represent the entire dataset, often overlooking the uniform distribution of these subsets. Such non-uniform distribution of subsets can lead to information loss. To maximize the preservation of global information from old data and ensure the data is as evenly dispersed as possible, this paper introduces the Maximum Dissimilarity Sampling (MDS) method. MDS employs the Euclidean distance to measure the dissimilarity between two samples, where greater distance indicates a higher dissimilarity [18]. The essential algorithm implementation proceeds: it starts with selecting a central point as the initial point, forming the selected sample set A. Subsequently, it calculates the distance between the remaining points and set A, selecting the sample with the farthest distance to add to set A, repeating this selection process until the desired number is reached. This ensures that the distances between all selected samples are as far apart as possible, maximizing the dissimilarity between the samples. Meanwhile, approximately 30% of the space is allocated to store samples near the center of the sample set to maintain the stability of the distribution of old data. Experimental results demonstrate that MDS is reliable and enhances the model's ability to recognize old tasks. Due to the imbalance in the number of new and old classes, there is a bias in the fc layer's classification. It is recommended to use the Laloss [19] as a replacement for the standard cross-entropy loss function. Additionally, using Laloss eliminates the need for additional balance fine-tuning work, resulting in a reduction in training costs.

In summary, there are three main points of contribution to this paper:

- This paper introduces a novel diversity sampling strategy, Maximum Dissimilarity Sampling (MDS), which maximizes the coverage of the distribution of old samples and preserves global information to the greatest extent.
- To mitigate the classification bias resulting from the imbalance between the number of new and old classes, we recommend replacing the standard cross-entropy loss function with Laloss.
- This paper compared MDS with classical sampling strategies on the CWRU dataset, confirming its effectiveness and reliability. Compared to other classical algorithms, our approach performs better on CWRU and MFPT.

2   Related Work

2.1   Incremental Learning

Incremental learning encompasses various scenarios and settings, primarily task-incremental [11,12] and class-incremental [13–17]. Our work predominantly focuses on class-incremental learning. Multiple methods for class-incremental learning are primarily classified into those based on replay, regularization, and model extension.

Replay. Replay methods primarily involve the storage of a portion of old data [13,14,20,37,39], the generation of synthetic images [21,22], or artificial features [23,24]. When training for a new task, this stored data is replayed to achieve results similar to joint training. Replay-based methods often perform better in class-incremental learning (CIL) [25]. How to sample old data so that the sampled subset effectively covers the old data distribution is a topic worthy of discussion.

Regularization. Regularization methods can be primarily categorized into two types. One type involves knowledge distillation to fit the output of new and old models [9,26,38]. For instance, LwF [9] matches the logit output, and Podnet [26] compares the feature output of both new and old models. The other type focuses on parameter constraints by introducing penalty terms to restrict the network's update direction. EWC [10] calculates the importance of parameters within the network and constrains the variations of crucial parameters. However, this approach can result in a substantial computational burden. To address this, oEwc [36] proposes online estimation of importance weights, reducing the computational load.

Model Extension. Model Extension allocates additional models [27] or parameters [28] for each new task. This increases model parameters as the number of tasks grows, resulting in more memory consumption.

2.2   Sampling Strategy

In replay methods, a range of different data sampling techniques contribute significantly to mitigating catastrophic forgetting.

Herding. Icarl [13] proposed this approach. It retains the samples closest to the center. It begins by computing the sample set's class mean to represent the sample set's center. Subsequently, the distance from each sample to the center is calculated, and examples closest to the center are selected in succession until the desired quantity is reached. The ones stored first are also the closest to the center. However, this method will forget information about boundary data during the training process.

Distance. They are proposed in [29]; this method stores boundary samples. It approximates the distance from a given sample to the decision boundary by calculating the inverse distance to the class mean. A smaller distance indicates a greater distance from the decision boundary. However, this approach tends to forget information about central data during training. In [16], the inverse form of distance is proposed. It iteratively selects a single sample closest to the class mean. This method is akin to herding and tends to forget information about boundary samples during training.

Random. Random sampling is the simplest sample selection method. It involves generating random numbers to select samples, with each sample having an equal probability of being chosen. However, as the number of samples decreases, randomness introduces uncertainty, and therefore, the remaining samples cannot guarantee whether they adequately represent the characteristics of the original data class.

3 Method

Class-incremental learning aims to learn a unified classifier from a series of training tasks, including classes that have not been encountered before. In each training step, the model can also access a small buffer containing samples from the classes located in previous incremental steps. Our objective is to utilize subsets to replace the global information of samples and address the class imbalance issue due to the limited buffer capacity.

3.1 Maximum Dissimilarity Sampling

Algorithm 1: Maximum Dissimilarity Sampling

Input: $X_s, ..., X_t$ // training examples;
Input: N // memory size;
Output: $M = (M_1, M2, , , , M_t)$ //new diversity exemplar set;
1   $n = 0.7 * N/(t * 2)$ //number of exemplars per class
2   Initialize m //initialize a sample
3   Initialize $distance$ //Initialize a distance matrix distance;
4   **for** y in $s, ..., t$ **do**
5   **for** k in $1, ..., n$ **do**
6   M_y =add_data(m)
7   F =Extract_features($X_s, ..., X_t, \theta$)
8       $dist(i) = $ Distance_calculation(X_i, F)
9       $distance(i) = Min(distance(i - 1), dist(i))$
10      $m = Max(distance)$
11   **end**
12 **end**

We believe that samples near the distribution center are representative, while samples near the classification boundaries are discriminative. The selected stored classes should represent their corresponding classes and possess discriminative characteristics. To satisfy both of these attributes, we recommend uniformly sampling the dataset.

We introduce a novel sample sampling strategy, Maximum Dissimilarity Sampling (MDS), to maximize the dissimilarity between the selected samples and ensure sample diversity. MDS iteratively selects samples with the maximum dissimilarity to the already chosen set, aiming to increase the dissimilarity between samples and preserve their diversity. Additionally, we reserve 30% of the space to store samples closest to the sample set's center, creating a representative sample set. The divariety set and representative set together form the buffer we need, ensuring the stability of old data distribution while preserving the global information from the original samples to the maximum extent.

We utilize the Euclidean distance to calculate the dissimilarity between samples, where greater distance indicates more significant dissimilarity between samples. We choose the point nearest to the distribution center as the initial point to obtain the initially selected sample set, denoted as set $A = (A_1)$. Next, we calculate the distance of the remaining samples to set A and select the sample with the farthest distance to add to set A.

We will describe how to compute the distance of samples to the set. For any remaining sample, we compute the Euclidean distance between that sample and all samples in set $A = (A_1, A_2...A_i)$, resulting in a distance set, $dist(m)$:

$$d(F_i, F_j) = \sqrt{\sum_{k=1}^{p}(x_{ik} - x_{jk})^2} \tag{1}$$

$$F_i = (x_{i1}, x_{i2}, ..., x_{ik}) \tag{2}$$

$$dist(m) = (d_{(F_m, F_1)}, d_{(F_m, F_2)}, ...d_{(F_m, F_i)}) \tag{3}$$

where $d(F_i, F_j)$ is the Euclidean distance from each data to the other data in the same class, p is the feature dimension, and F_i is the feature vector of the current sample.

Then, the minimum value is taken as the distance from that sample to the set, denoted as $distance_m = min(dist(m))$. After calculating the set of distance for all remaining points, the sample with the maximum distance is selected and added to set $A = (A_1, A_2...A_{i+1})$. This process is repeated until the desired number of samples is reached.

Furthermore, we employ a herding sampling strategy to reserve 30% of the space to store central samples, forming a representative sample set. The representative sample set and the diverse sample set together constitute the buffer required for training.

3.2 Logit Adjusted Softmax Cross-Entropy

The model can access a small buffer for replaying old data in each incremental step. However, there is a noticeable difference in the number of samples between the buffer and the training set. This results in the fc layer being more biased towards the most recently learned new classes [14,15], causing the model to predict old classes as new ones. One of the most direct solutions to this issue is continuously increasing the buffer's capacity, which is impractical in natural industrial environments.

Therefore, we recommend using the Laloss [19] during training to replace the standard cross-entropy loss function. It assigns different weights to each class based on the sample proportion. When new classes are added, the proportion of old classes decreases, and readjusting the weights can enhance the management of old classes. The Logit Adjusted Softmax function is defined as follows:

$$q_k(x) = \frac{e^{f_y(x) + \tau \log \pi_y}}{\sum_{j=1}^{N_t} e^{f_j(x) + \tau \log \pi_j}} \tag{4}$$

$$\pi_y = \frac{n_i}{N} \tag{5}$$

where x is the input, $f(x) = [f_1(x), ..., f_{N_t}(x)]$ is the output logits of the current model. Where τ is a hyperparameter, N represents the total number of classes, and n_i is the sample number of each class.

The new classification loss Lc is then defined as the Cross-Entropy loss using the logit adjusted Softmax instead of the Softmax:

$$L_{clf}(x) = \sum_{k=1}^{N_t} -\delta_{k=y} \log(q_k(x)) \tag{6}$$

the new cross-entropy loss function can replace the standard cross-entropy loss function and can be integrated with other incremental learning methods. This method does not introduce excessive computational complexity and is easy to incorporate. Therefore, the final loss function is as follows:

$$L = L_{clf} + \gamma L_{kd} \tag{7}$$

4   Experiments

In this section, we compared our sampling method with various other sampling strategies on a common standard rolling bearing dataset. Additionally, we compared it with other classic incremental learning methods.

4.1   Experimental Setup

Datasets. The experiments were conducted on the renowned public experimental dataset from the Case Western Reserve University (CWRU) Bearing Data Center [30] and the Mechanical Fault Prognosis Technology (MFPT) [31] rolling bearing fault diagnosis dataset to validate the effectiveness of the proposed method. In this paper, we selected 10 types of vibration signals from the CWRU dataset, including one normal signal and nine different fault vibration signals. These signals were collected under a sampling rate of 12 kHz, with a drive motor speed of 1797 RPMs and no load (0 hp). There are a total of ten bearing conditions, and the images are saved at a size of 32×32 pixels. We randomly selected 2000 samples for training and 400 samples for testing. MFPT selected 17 fault vibration signals and one normal rolling bearing signal. These 17 fault vibration signals include 10 outer and 7 inner race faults. Among the outer race faults, there are 3 fault signals collected at a sampling rate of 97,656 Hz for 6 s under a load of 270 pounds, and there are 7 inner race faults sampled at 48,828 Hz for 3 s under loads of 0, 50, 100, 150, 200, 250, and 300 pounds. In total, there are 18 bearing conditions. We randomly selected 3480 training samples and 838 test samples.

Experimental Details. All of our methods were implemented using PyTorch [32]. We used the standard ResNet18 [33] as the network feature extractor. The training images were randomly horizontally flipped with no additional augmentation applied. During the experiments, class-incremental training was performed with a batch size set to 64. The learning rate started at 0.1 and gradually decayed to 0 using the SGD optimizer [34]. (The CWRU was trained for 30 epochs, while the MFPT was trained for 50 epochs).

4.2   Comparison to the State-Of-The-Art Methods

In this section, we conducted experiments on the rolling bearing fault diagnosis datasets CWRU and MFPT. We compared MDS with various existing sampling strategies. Additionally, we combined it with Laloss and compared it with other classic incremental learning methods.

Sampling Strategy Comparison. We compared MDS with standard sampling strategies, including Herding, Random, Distance, and Inverse Distance. The experiments in the Table 1 demonstrate that MDS exhibits better resistance to forgetting. By effectively capturing diverse samples, it preserves the global information of the sample set. The Distance strategy keeps boundary samples, which are difficult to train and pose challenges. Difficulty samples have a relatively weaker ability to resist forgetting than other methods. In contrast, Inverse Distance and Herding select samples close to the sample center and are more representative. Representative samples better assist in model training and resistance to forgetting. However, both difficulty and representative samples only select samples from a specific distribution within the sample set and cannot represent the global distribution of the sample set. Random samples introduce randomness, and when k is large, random samples can effectively cover the class distribution, yielding better results. But as k decreases, the effect worsens because of the randomness of the random samples and, thus, the result.

Table 1. Average accuracy of different sampling strategies on CWRU

	K = 100	K = 200	K = 500
Herding	68.44 ± 1.17	75.87 ± 2.58	80.69 ± 2.51
Distance	60.67 ± 0.83	67.11 ± 1.38	79.14 ± 1.91
Inverse Distance	66.67 ± 2.52	69.77 ± 2.14	82.62 ± 0.62
Random	68.22 ± 1.08	78.27 ± 1.21	81.92 ± 0.83
Ours	**69.3 ± 0.91**	**79.26 ± 1.92**	**84.03 ± 0.92**

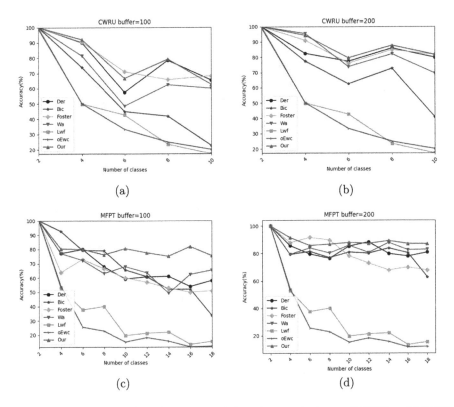

Fig. 1. Comparison results of incremental learning methods on CWRU and MFPT with buffers of 100 and 200, respectively.

Classic Algorithms Comparison. We replaced the standard cross-entropy loss function with Laloss and combined it with MDS. We compared our approach to several classic incremental learning methods on the CWRU and MFPT rolling bearing fault diagnosis datasets. These methods included two regularization methods (Lwf [9], oEwc [36]), two model expansion methods(Der [27], Foster [35]), and three replay methods (Icarl [13], BIC [14], WA [15], where WA and BIC primarily addressing classification bias caused by class imbalance). The Table 2 showed that our method outperforms the others in resisting forgetting. This is because we preserve global information of old data by retaining diverse and representative samples while mitigating classification bias due to class imbalance using Laloss. oEwc and Lwf do not use replay data, but their resistance to forgetting is poor compared to replay-based methods. WA mitigates classification bias caused by class imbalance by correcting the weights of the FC layer, while BIC adjusts the output logits using a balanced subset, incurring additional computation. Der and Foster are model expansion methods that offer better resistance to forgetting but come with more model parameters and higher costs. Figure 1 also demonstrates that the performance gap between our

Table 2. Comparison with some current classical methods on CWRU and MFPT

Memory	Method	CWRU			MFPT		
		Paras (M)	last	Avg	Paras (M)	last	Avg
–	Lwf [9]	11.2	17.17	46.6	11.2	15.51	35.81
	oEwc [36]	11.2	20.00	45.4	11.2	12.17	30.62
100	Icarl [13]	11.2	62.18	64.04	11.2	43.91	59.58
	Bic [14]	11.2	22.62	56.61	11.2	33.53	68.31
	WA [15]	11.2	59.98	70.40	11.2	66.54	66.59
	Foster [35]	22.4	67.05	77.40	22.4	50.84	63.78
	Der [27]	55.9	**73.95**	78.87	112.3	67.42	63.80
	Ours	23.2	62.65	**79.56**	23.2	**71.24**	**79.21**
200	Icarl	11.2	76.42	75.10	11.2	74.70	74.40
	Bic	11.2	25.42	68.30	11.2	62.77	80.33
	WA	11.2	69.50	84.10	11.2	82.94	84.82
	Foster	22.4	81.92	84.99	22.4	61.93	78.27
	Der	55.9	80.82	82.44	112.3	80.79	83.74
	Ours	11.2	**82.35**	**86.49**	11.2	**86.99**	**89.15**

method and others widens as the incremental steps progress. This indicates that as incremental learning advances, the superiority of our method becomes more pronounced, showcasing its ability to learn longer sequences of new classes.

4.3   Ablation Study

To gain further insight into the working mechanism of our method, we conducted additional experiments on CWRU and MFPT datasets.

Ablation of Components. To analyze why our method has shown improvement, we established three hybrid mechanisms: the first one where none of the methods are used, the second one using only MDS, and the third one replacing the standard cross-entropy loss with Laloss while using MDS. The Table 3 explains our method's performance, indicating that the synergy of these components leads to a more effective resistance to forgetting. MDS can maximize and preserve global information from old data, better retaining old knowledge. Laloss is effective in mitigating classification bias due to imbalanced class numbers. The experimental results highlight the effectiveness of each component in our method.

Table 3. Ablation per component on CWRU and MFPT (average accuracy)

Method	CWRU		MFPT	
	K = 100	K = 200	K = 100	K = 200
Finetune	45.63 ± 0.17	45.63 ± 0.17	29.64 ± 0.47	29.64 ± 0.47
MDS	69.3 ± 0.91	78.27 ± 1.21	72.63 ± 1.63	88.92 ± 1.24
+Laloss	$\mathbf{79.56 \pm 0.52}$	$\mathbf{86.49 \pm 1.58}$	$\mathbf{79.21 \pm 1.56}$	$\mathbf{89.15 \pm 2.81}$

Confusion Matrix. To better assess the effectiveness of Laloss in mitigating classification bias, we compared the confusion matrices of finetuning, WA, and our method. The Fig. 2 show that the confusion matrices based on finetuning are heavily weighted towards the new classes, with old classes often being misclassified as new classes. This indicates that finetuning lacks the ability to resist forgetting. WA, on the other hand, corrects classification bias by adjusting the weights of the FC layer. The confusion matrices based on WA show that most weights are along the diagonal, but old classes are still misclassified as new classes. This suggests that WA possesses some resistance to forgetting but still exhibits classification bias. Our method mitigates classification bias by replacing the standard cross-entropy loss with Laloss. As observed from the confusion matrices, our method has fewer instances of old classes being misclassified as new classes than the other methods. Our method achieves good accuracy on both new and old classes.

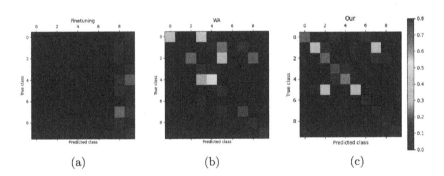

(a)   (b)   (c)

Fig. 2. Comparison of the confusion matrix of Finetuning, WA [15] and Our.

5   Conclusion

This paper introduces a novel sampling strategy to preserve global information of old data by retaining diverse samples. MDS iteratively selects samples with maximum dissimilarity to the already chosen set, ensuring sufficient dissimilarity among the subset's samples. At the same time, we reserve 30% of the space to store representative samples near the center, enhancing stability while preserving information about the old classes as much as possible. Furthermore, we recommend using Laloss to replace the standard cross-entropy loss function to alleviate the classification bias arising from class imbalance. Experimental results on the CWRU and MFPT rolling bearing fault diagnosis datasets demonstrate the effectiveness and reliability of MDS in preserving knowledge from old data. It also confirms that Laloss effectively mitigates classification bias. Combining Laloss and MDS achieves better resistance to forgetting on the rolling bearing fault diagnosis dataset.

Acknowledgements. This research is supported by the National Natural Science Foundation of China under Grant 62272355, 61702383, and 62176191.

References

1. Chu, W., Liu, T., Wang, Z., Liu, C., Zhou, J.: Research on the sparse optimization method of periodic weights and its application in bearing fault diagnosis. Mech. Mach. Theory **177**, 105063 (2022)
2. Zhuang, D., et al.: The IBA-ISMO method for rolling bearing fault diagnosis based on VMD-sample entropy. Sensors **23**(2), 991 (2023)
3. Gu, H., Liu, W., Zhang, Y., Jiang, X.: A novel fault diagnosis method of wind turbine bearings based on compressed sensing and AlexNet. Measur. Sci. Technol. **33**(11), 115011 (2022)
4. Zou, W., Xia, Y., Li, H.: Fault diagnosis of Tennessee-Eastman process using orthogonal incremental extreme learning machine based on driving amount. IEEE Trans. Cybern. **48**(12), 3403–3410 (2018)
5. Peng, P., et al.: Progressively balanced supervised contrastive representation learning for long-tailed fault diagnosis. IEEE Trans. Instrum. Meas. **71**, 1–12 (2022)
6. Lao, Q., Mortazavi, M., Tahaei, M., Dutil, F., Fevens, T., Havaei, M.: FoCL: feature-oriented continual learning for generative models. Pattern Recogn. **120**, 108127 (2021)
7. Lan, C., et al.: Towards lifelong object recognition: a dataset and benchmark. Pattern Recogn. **130**, 108819 (2022)
8. Dong, J., Cong, Y., Sun, G., Zhang, T.: Lifelong robotic visual-tactile perception learning. Pattern Recogn. **121**, 108176 (2022)
9. Li, Z., Hoiem, D.: Learning without forgetting. IEEE Trans. Pattern Anal. Mach. Intell. **40**(12), 2935–2947 (2017)
10. Kirkpatrick, J., et al.: Overcoming catastrophic forgetting in neural networks. Proc. Natl. Acad. Sci. **114**(13), 3521–3526 (2017)
11. Hung, C.Y., Tu, C.H., Wu, C.E., Chen, C.H., Chan, Y.M., Chen, C.S.: Compacting, picking and growing for unforgetting continual learning. In: Advances in Neural Information Processing Systems, vol. 32 (2019)
12. Rajasegaran, J., Hayat, M., Khan, S.H., Khan, F.S., Shao, L.: Random path selection for continual learning. In: Advances in Neural Information Processing Systems, vol. 32 (2019)
13. Rebuffi, S.A., Kolesnikov, A., Sperl, G., Lampert, C.H.: iCaRL: incremental classifier and representation learning. In: Proceedings of the IEEE Conference on Computer Vision and Pattern Recognition, pp. 2001–2010 (2017)

14. Wu, Y., et al.: Large scale incremental learning. In: Proceedings of the IEEE/CVF Conference on Computer Vision and Pattern Recognition, pp. 374–382 (2019)
15. Zhao, B., Xiao, X., Gan, G., Zhang, B., Xia, S.T.: Maintaining discrimination and fairness in class incremental learning. In: Proceedings of the IEEE/CVF Conference on Computer Vision and Pattern Recognition, pp. 13208–13217 (2020)
16. Masana, M., Liu, X., Twardowski, B., Menta, M., Bagdanov, A.D., Van De Weijer, J.: Class-incremental learning: survey and performance evaluation on image classification. IEEE Trans. Pattern Anal. Mach. Intell. **45**(5), 5513–5533 (2022)
17. Hou, S., Pan, X., Loy, C.C., Wang, Z., Lin, D.: Learning a unified classifier incrementally via rebalancing. In: Proceedings of the IEEE/CVF Conference on Computer Vision and Pattern Recognition, pp. 831–839 (2019)
18. Lin, Y.S., Jiang, J.Y., Lee, S.J.: A similarity measure for text classification and clustering. IEEE Trans. Knowl. Data Eng. **26**(7), 1575–1590 (2013)
19. Menon, A.K., Jayasumana, S., Rawat, A.S., Jain, H., Veit, A., Kumar, S.: Long-tail learning via logit adjustment. arXiv preprint arXiv:2007.07314) (2020
20. Castro, F.M., Marín-Jiménez, M.J., Guil, N., Schmid, C., Alahari, K.: End-to-end incremental learning. In: Proceedings of the European Conference on Computer Vision (ECCV), pp. 233–248 (2018)
21. Shin, H., Lee, J.K., Kim, J., Kim, J.: Continual learning with deep generative replay. In: Advances in Neural Information Processing Systems, vol. 30 (2017)
22. Ostapenko, O., Puscas, M., Klein, T., Jahnichen, P., Nabi, M.: Learning to remember: a synaptic plasticity driven framework for continual learning. In: Proceedings of the IEEE/CVF Conference on Computer Vision and Pattern Recognition, pp. 11321–11329 (2019)
23. Xiang, Y., Fu, Y., Ji, P., Huang, H.: Incremental learning using conditional adversarial networks. In: Proceedings of the IEEE/CVF International Conference on Computer Vision, pp. 6619–6628 (2019)
24. Kemker, R., Kanan, C.: FearNet: brain-inspired model for incremental learning. arXiv preprint arXiv:1711.10563 (2017)
25. Prabhu, A., Torr, P.H., Dokania, P.K.: GDumb: a simple approach that questions our progress in continual learning. In: Vedaldi, A., Bischof, H., Brox, T., Frahm, J.M. (eds.) ECCV 2020. LNCS, vol. 12347, pp. 524–540. Springer, Cham (2020). https://doi.org/10.1007/978-3-030-58536-5_31
26. Douillard, A., Cord, M., Ollion, C., Robert, T., Valle, E.: PODNet: pooled outputs distillation for small-tasks incremental learning. In: Vedaldi, A., Bischof, H., Brox, T., Frahm, J.M. (eds.) ECCV 2020. LNCS, vol. 12365, pp. 86–102. Springer, Cham (2020). https://doi.org/10.1007/978-3-030-58565-5_6
27. Yan, S., Xie, J., He, X.: DER: dynamically expandable representation for class incremental learning. In: Proceedings of the IEEE/CVF Conference on Computer Vision and Pattern Recognition, pp. 3014–3023 (2021)
28. Xu, J., Zhu, Z.: Reinforced continual learning. In: Advances in Neural Information Processing Systems, vol. 31 (2018)
29. Chaudhry, A., Dokania, P.K., Ajanthan, T., Torr, P.H.: Riemannian walk for incremental learning: understanding forgetting and intransigence. In: Proceedings of the European Conference on Computer Vision (ECCV), pp. 532–547 (2018)
30. Case Western Reserve University Bearing Data Center. http://csegroups.case.edu/bearingdatacenter/home. Accessed 22 Dec 2019
31. Bellini, A., Filippetti, F., Tassoni, C., Capolino, G.A.: Advances in diagnostic techniques for induction machines. IEEE Trans. Ind. Electron. **55**(12), 4109–4126 (2008)

32. Paszke, A., et al.: Automatic differentiation in PyTorch (2017)

33. He, K., Zhang, X., Ren, S., Sun, J.: Deep residual learning for image recognition. In: Proceedings of the IEEE Conference on Computer Vision and Pattern Recognition, pp. 770–778 (2016)

34. Loshchilov, I., Hutter, F.: SGDR: stochastic gradient descent with warm restarts. arXiv preprint arXiv:1608.03983 (2016)

35. Wang, F.Y., Zhou, D.W., Ye, H.J., Zhan, D.C.: FOSTER: feature boosting and compression for class-incremental learning. In: Avidan, S., Brostow, G., Cissé, M, Farinella, G.M., Hassner, T. (eds.) ECCV 2022. LNCS, vol. 13685, pp. 398–414. Springer, Cham (2022). https://doi.org/10.1007/978-3-031-19806-9_23

36. Schwarz, J., et al.: Progress & compress: a scalable framework for continual learning. In: International Conference on Machine Learning, pp. 4528–4537. PMLR (2018)

37. Min, Q., He, J., Yang, L., Fu, Y.: Continual learning with a memory of non-similar samples. In: Pan, L., Zhao, D., Li, L., Lin, J. (eds.) BIC-TA 2022. CCIS, vol. 1801, pp. 316–328. Springer, Singapore (2022). https://doi.org/10.1007/978-981-99-1549-1_25

38. Yu, P., He, J., Min, Q., Zhu, Q.: Metric learning with distillation for overcoming catastrophic forgetting. In: Pan, L., Cui, Z., Cai, J., Li, L. (eds.) BIC-TA 2021. CCIS, vol. 1566, pp. 232–243. Springer, Singapore (2021). https://doi.org/10.1007/978-981-19-1253-5_17

39. Min, Q., He, J., Yu, P., Fu, Y.: Incremental fault diagnosis method based on metric feature distillation and improved sample memory. IEEE Access 11, 46015–46025 (2023)

Application and Prospect of Knowledge Graph in Unmanned Vehicle Field

Yi-ting Shen[1,2] and Jun-tao Li[1,2(✉)]

[1] School of Information, Beijing Wuzi University, Beijing 101149, China
[2] Beijing Key Laboratory of Intelligent Logistics System, Beijing 101149, China
`Ljtletter@126.com`

Abstract. With the rapid development of unmanned vehicle technology, unmanned vehicle plays an increasingly important role in related fields, which is of great significance to the future development of intelligent transportation. In order to break through the bottleneck of the intelligent degree of existing unmanned vehicles, unmanned vehicle driving technology based on knowledge graph has become one of the development trends. Firstly, this paper gives a brief overview of the decision-making process of unmanned vehicles, and introduces the basic principles and related concepts of unmanned vehicle technology based on knowledge graph. Then the application of knowledge graph in intelligent decision-making of unmanned vehicle is introduced, including target recognition, semantic segmentation, target trajectory prediction and scene understanding. Then an automatic driving decision system based on knowledge graph is proposed. Finally, the future development of intelligent decision-making of unmanned vehicles based on knowledge graph is prospected, and various possibilities of its future development are discussed.

Keywords: Unmanned vehicles · Knowledge graph · Autonomous driving

1 Introduction

Unmanned vehicle is an intelligent vehicle that can automatically perceive the surrounding environment and navigate without human intervention. With the continuous progress of unmanned vehicle technology, it has been widely used in public transportation, environmental protection, logistics and transportation. Unmanned vehicles not only become a typical paradigm of the integration and innovation of information technology and automobile technology, but also open a new track for smart transportation, and will become an important infrastructure of smart cities in the future. Among the technologies of unmanned vehicles, intelligent decision-making is one of the most critical technologies, which plays a decisive role in the driving safety of vehicles. The intelligent decision-making system of unmanned vehicle mainly includes three aspects: perception, decision-making and control. Perception mainly uses cameras and other sensors to collect environmental information around unmanned vehicles to realize object recognition

© The Author(s), under exclusive license to Springer Nature Singapore Pte Ltd. 2024
L. Pan et al. (Eds.): BIC-TA 2023, CCIS 2062, pp. 227–241, 2024.
https://doi.org/10.1007/978-981-97-2275-4_18

and tracking. Decision-making is mainly based on the perceived information for vehicle behavior prediction and path planning, which is equivalent to the "brain" of unmanned vehicles. Control is mainly to control the vehicle behavior according to the plan given by the decision-making part, and the development of this module is relatively mature at present. In recent years, the intelligent decision-making technology of unmanned vehicles has achieved a major breakthrough, but these methods are all based on the existing data for modeling and analysis, and can not integrate some well-known information (such as weather conditions, road signs, obstacles, drivers' driving intentions, etc.) to achieve a more advanced semantic understanding. The decision-making system of unmanned vehicle based on knowledge graph combines data-driven and knowledge-driven, which can not only express the semantic relationship between entities, but also show some heterogeneous knowledge in the form of graphs. With the deepening of information technology research, the research on knowledge graphping technology has become a hot spot at present, and has achieved fruitful results in the fields of intelligent question answering, language understanding, recommendation calculation, big data analysis and so on. At present, there is little research on unmanned vehicle technology based on knowledge graph. This paper aims to summarize the application of knowledge graph in unmanned vehicle field, and proposes an automatic driving decision system based on knowledge graph.

2   Related Work

At present, automatic driving based on knowledge graph is a new research field. Knowledge graph technology can integrate data from multiple sources, better integrate heterogeneous data, and realize more advanced semantics so as to have a deeper and comprehensive understanding of the situation of automatic driving. Its construction process involves many aspects, including knowledge representation technology, ontology technology, data mining technology, automatic reasoning and so on. To construct the knowledge graph of autonomous driving, we must first extract entities, relationships and attributes from various databases, and then store the extracted data into the databases respectively. Each database can generate corresponding local knowledge graphs, and then the local knowledge graphs are fused into a global knowledge graph through knowledge fusion technology. Finally, based on the generated global knowledge graph, reasoning is carried out to realize a series of automatic driving applications and finally complete the decision. The construction process of automatic driving knowledge graph is shown in "Fig. 1".

2.1   Automatic Driving Knowledge Modeling

Knowledge graph is a knowledge base with directed graph [1], which can be divided into data layer and mode layer from logical structure. The data layer contains a lot of factual information, mainly composed of many original data sets, which contain data from various modes such as text, image, video, radar,

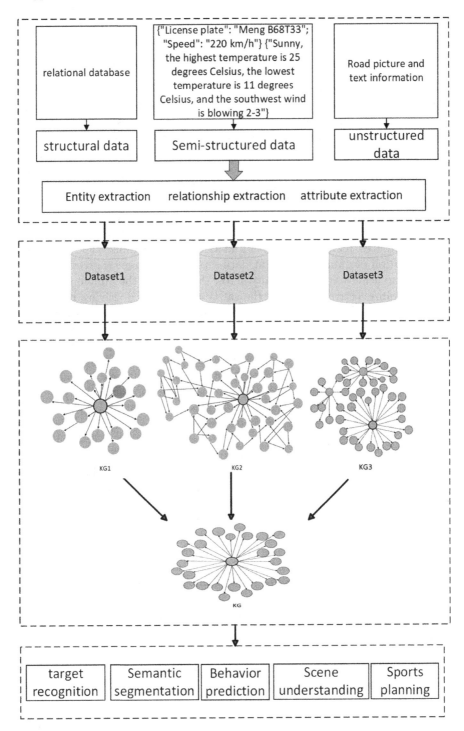

Fig. 1. Construction process of automatic driving knowledge graph

lidar and some high-quality annotations; The schema layer is based on the data layer, in which nodes represent ontology concepts and edges represent the relationships between concepts. Ontology is usually applied to the schema layer, and the rules and axioms of ontology library are used to manage and constrain the facts in the data layer. Automatic driving knowledge modeling is the basis of constructing automatic driving knowledge graph, and it is the definition of ontology concepts and relationships. It is mainly used to establish conceptual models. A high-quality model can greatly improve the efficiency of knowledge graph construction. There are two ways of knowledge modeling: top-down and bottom-up. Because the knowledge structure in the field of autonomous driving is complex, it is usually constructed by top-down method. Firstly, the ontology is defined, which can be compiled manually by domain experts or by computers, and then subdivided and adjusted step by step through algorithms, and finally a tightly structured hierarchical structure is formed. Ontology describes entities, classes, attributes, relationships, rules and axioms, and organizes and constructs the information of data sets, so that data sets from different sources can share resources and achieve interoperability across data sets. In the unmanned driving system, it can effectively improve the safety of unmanned driving by describing information such as graph, driving environment and driving road through ontology. In [2], a core ontology for developing advanced driver assistant system is introduced. These ontologies can fill the gap between perception-driven environment and knowledge processing, and can be used to build the knowledge base of intelligent vehicles, so that intelligent vehicles can understand the driving environment and realize different types of advanced driver assistance systems. A driving ontology focusing on automobile signals and sensors is proposed in [3], which focuses on narrowing the gap between the demand for data interoperability and the technical level in automobile modeling. In [4], a hybrid intelligent framework for vehicle driver classification based on context-aware ontology is proposed, which is a driver classification system for driver attributes modeled in driving ontology. In [5] An ontology of intelligent driver assistance system is proposed, which focuses on generating warning messages based on context-aware parameters, such as driving situation, vehicle dynamics, driver activities and environment. [6] describes the automobile ontology, and puts forward an ontology that focuses on managing the knowledge in the car and sharing the knowledge among cars. In order to facilitate the representation and extraction of knowledge, and the feasibility of using ontology to model and process graph data in automobiles, people put forward an ontology model for graph data in automobile systems [7]. Relevant ontology about driving can be found in [8–11]. Ontology organizes and constructs AD data sets through its classes, attributes and relationships to describe information such as graphs, driving environments and driving roads, which can effectively improve the safety of unmanned driving.

2.2 Automatic Driving Knowledge Extraction

Automatic driving knowledge extraction is the core link and key technology of constructing automatic driving knowledge graph, and it is the process of detecting, analyzing and understanding automatic driving information. Knowledge extraction is mainly divided into automatic driving entity identification, automatic driving entity relationship extraction and automatic driving entity attribute extraction according to extraction elements. Automatic driving entity extraction involves names, places, lanes, signs, dates and so on, and the relationships among these entities are also complicated. At present, the main research methods can be divided into rule-based methods, traditional machine learning methods and deep learning methods. The relationship extraction method [12] can be divided into supervised, unsupervised, semi-supervised, open-domain oriented, remote supervision method applied, and relationship extraction based on deep learning. At present, there is relatively little research on automatic driving attribute extraction. At present, the commonly used attribute extraction methods [13] can be divided into rule-based attribute extraction, attribute extraction-oriented machine learning mode, data-driven attribute extraction method and attribute extraction method combined with emotional words. The main problem of knowledge extraction in the field of autonomous driving is that there are too many unstructured data to improve the extraction accuracy.

2.3   Automatic Driving Knowledge Fusion

In order to solve the problems of ambiguity, redundancy and different structures in multiple knowledge graphs of autonomous driving, the automatic driving knowledge fusion technology has emerged, aiming at combining multi-source information into a knowledge base with unified structure and convenient use by disambiguating, eliminating differences and other ways. Automatic driving knowledge fusion involves the fusion of mode layer and data layer, which involves automatic driving entity alignment, automatic driving knowledge processing and automatic driving knowledge updating.

2.4   Automatic Driving Knowledge Reasoning

Automatic driving knowledge reasoning, that is, on the basis of the existing automatic driving knowledge base, according to the known factual information, carries out reasoning analysis in order to get new facts. Automatic driving knowledge reasoning can not only help the driver to operate the vehicle, but also help the driver to drive on the basis of driving, relieve the driver's fatigue and avoid some traffic accidents. The goal of automatic driving knowledge reasoning in the future is to improve the accuracy, completeness and interpretability of knowledge reasoning. Automatic driving knowledge reasoning, that is, on the basis of the existing automatic driving knowledge base, according to the known factual information, carries out reasoning analysis in order to get new facts. Automatic driving knowledge reasoning can not only help the driver to operate the vehicle, but also help the driver to drive on the basis of driving, relieve the driver's fatigue and avoid some traffic accidents. The goal of automatic driving knowledge reasoning in the future is to improve the accuracy, completeness and interpretability of knowledge reasoning.

3   Research Status

3.1   Application in Target Detection

In the automatic driving system, target detection is a common computer vision task, which mainly includes the detection of traffic participants, road signs, traffic signs and other related objects. At present, the target detection technology is relatively mature, but there are still some problems that are difficult to completely solve, such as low resolution, difficulty in distinguishing similar objects and large occlusion area. At present, automatic driving perception systems are all based on sensor information for target detection. In [14], it is proposed that the representation based on knowledge graph can provide important relationship information. This method makes use of the semantic relationship between entities in the knowledge graph, and when combined with target detection, it can improve its performance. Kim et al. [15] put forward a new target recognition method, which combines knowledge graph with machine learning to help human annotators distinguish road signs, which not only reduces the search space and human resources, but also simplifies the learning curve of road sign classification. Monka et al. [16] proposed a knowledge graph neural network (KG-NN) method to identify targets by using the invariant auxiliary knowledge of image data to supervise training. KG-NN, using a KG and an appropriate KG embedding algorithm, constructs a domain-invariant embedding space, and trains KG-NN to adapt its visual embedding to hKG given by KG, and associates between semantically similar categories, thus realizing the learning of image association features. In [17], a hierarchical semantic segmentation network for reliable scene feature extraction and a new lane feature representation graph structure are designed. The graph can flexibly describe the geometric shape and topological structure of lanes, and is not limited by strong geometric assumptions, so it is compact, efficient and flexible.

3.2   Application in Semantic Segmentation

Semantic segmentation refers to the process of classifying each pixel in an image into class labels, which can include a person, a vehicle, a road, a road sign and so on. Although the deep learning algorithm provides great help for accurate semantic segmentation, it can only achieve pixel-level output and does not support advanced reasoning and planning under complex road conditions. In [18], a hierarchical scene graph based on graphic representation is introduced, which uses the segmentation entities of road markers and curbs to reconstruct the semantic structure of road scenes, so as to combine the background knowledge for analysis. Bordes et al. [19] put forward a new framework called evidence grammar. This framework combines local information, prior information and semantic information, which can better express the prior uncertainty and improve the adaptability of the model.

3.3 The Application of Field Understanding

Scene understanding refers to the analysis of events in a given scene, such as vehicle position, vehicle driving conditions, road conditions and the positional relationship between vehicles, so as to understand the behavior of autonomous vehicles and avoid vehicle collisions. Scene understanding based on knowledge graph can represent the entities and their relationships in the scene through the form of knowledge graph. Wickramarachchi et al. [20] put forward knowledge-based entity prediction (KEP) and used it to predict the potential unidentified entities in the scene, given the current and background knowledge of the scene represented by knowledge graph, so as to improve the accuracy of scene understanding. Halilaj et al. [21] put forward a semantic and situational intelligence method based on knowledge graph, which is used to fuse and organize heterogeneous information. Then, an ontology is constructed to encapsulate the core concepts that are crucial to the driving context, which can better understand and expand the information of external ontologies and achieve better data management. Finally, the embedding methods such as graph neural network are introduced to use the knowledge in KG to realize typical classification and prediction tasks.

3.4 Application in Behavior Prediction

Behavior prediction of self-driving vehicles refers to analyzing the current position, speed, moving direction and behavior of traffic participants after understanding the scene of the self-driving vehicles, and then predicting their future trajectory. The most important point is his foresight of the unknown. In [22], an ontology-based reasoning method is proposed to predict the long-term behavior of road users, which considers the interaction between different objects. Ontology provides a conceptual description of all road users and their interactions, and then uses Markov logic network to infer possible behaviors and their related probabilities. Li et al. [23] put forward a graph-based target trajectory prediction scheme for self-driving cars, which is called GRIP. This method can predict the trajectories of all observed objects at the same time, and can be trained end to end. The results show that the model has better prediction results than the existing methods, and its running speed is five times faster than the most advanced scheme. Zipfl et al. [24] used the spatial semantic scene atlas to model the traffic scene, and used the graph neural network (GCN) to learn and reason, so as to realize the motion prediction in the traffic scene. This is a graphical operation based on GCN and modeling the interaction between vehicles.

3.5 Application in Sports Planning

The purpose of sports planning is to plan and implement driving actions, such as steering, acceleration and braking, and take all the information of previous steps into account. At present, the motion planning method is usually based on end-to-end learning, and imitative learning is used to learn how to decide a given decision-making motion. However, these purely data-driven methods often lack security guarantees. We believe that KG provides a method to combine data-driven and knowledge-based methods, which can incorporate knowledge such as safety requirements, traffic rules or scene context. Regele et al. [25] put forward an ontology-based traffic model, which uses an advanced abstract world model to realize the decision-making process of autonomous driving system. The new method is based on topological lane segmentation and introduces relationships to represent the semantic information of traffic scenes, which can better improve the effectiveness of decision-making for autonomous vehicles. In [26], a semantic modeling method based on ontology is proposed, which can express traffic semantics, improve knowledge sharing ability, improve sensor modeling ability and realize better reasoning decision. Huang et al. [27] introduced an ontology-based driving scene modeling, situation assessment and decision-making method for self-driving vehicles in urban environment. The test shows that this method can meet the requirements. [28] describes an ontology-based semantic reasoning method to identify and extract driving situations, so as to improve the context awareness and semantic understanding ability of autonomous vehicles. The outstanding advantage of this method is that it can correctly infer driving behavior in any field without extra training. Hovi et al. [29] described a method of generating ontology-based rules for reasoning systems through machine learning. In [30], an ontology-based method is proposed to relax the traffic control of autonomous vehicles in the face of different scenarios in actual situations. This ontology includes topological knowledge and reasoning rules, which can realize better reasoning decision and help drivers to assist driving.

4   Automatic Driving Decision-Making System Based on Knowledge Graph

The automatic driving decision-making system based on knowledge graph is realized by the cooperation of knowledge graph technology and reinforcement learning technology, including environment perception system, behavior decision-making system, control system and knowledge base management system. Environmental awareness is the use of various technologies (such as sensor technology, communication technology, etc.) to obtain environmental information, and then extract, fuse, process and store the obtained information in the knowledge base, which is the basis of behavior decision-making and provides a basis for decision-making. Behavior decision system mainly processes and fuses various knowledge base information to generate local knowledge graph and global knowledge graph, and then applies decision model to the knowledge base storing global information and realizes complex knowledge reasoning. Reasoning can be divided into rule-based reasoning and neural network-based reasoning. This part can realize the integration of knowledge, the construction of multi-level knowledge graph and the realization of upper application. Finally, the decision results of the system are transmitted to the control system, which controls the vehicle to change lanes or turn. The control system can be divided into turning, throttle, braking and other parts, and the priority needs to be set. The knowledge base management system mainly stores data information, prior information, rules and model information. Storage can store data in their own knowledge unit libraries according to their categories, such as rule knowledge base, traffic information knowledge base and model knowledge base. The knowledge management system can also realize the temporary storage of data, the storage of intermediate step processing results and the updating of knowledge. Automatic driving system based on knowledge graph can efficiently fuse knowledge and realize knowledge storage, re-acquisition and updating. The automatic driving decision system based on knowledge graph is shown in "Fig. 2". The automatic driving decision-making system based on knowledge graph can contain more knowledge capacity and is interpretable, which can make designers better understand the process of automatic decision-making and make autonomous vehicles make accurate and safe decisions in complex and changeable traffic environment.

5   Prospect

Unmanned vehicles can perform tasks independently, and many tasks that human beings can't do independently due to safety problems can be replaced by unmanned vehicles. Unmanned vehicle technology is the core element of intelligent transportation and military development in the future, which is of great significance to the development of intelligent vehicles and national defense construction. In order to meet the demand of the rapid development of unmanned vehicle technology in the future, it is necessary to realize the high efficiency, accuracy and safety of autonomous decision-making of unmanned vehicles. At present, reinforcement learning technology has provided a strong technical support for the intelligent decision-making of unmanned vehicles, but there are still problems such as weak generalization ability, poor interpretability and poor security. We hope that the decision-making technology of unmanned vehicles based on knowledge graph can break through these bottlenecks.

5.1   Improve the Generalization Ability of Unmanned Vehicle Decision-Making

The current intelligent driving technology is generally based on reinforcement learning, and one of the biggest problems facing reinforcement learning is that the exploration and output of strategies are difficult to predict and control. When the original data such as graphics, video and radar are used as input, there are too many parameters, which may lead to the over-fitting of the model and cannot be applied to new driving scenes. In addition, the expressive force of the trained model in actual combat is far less than the accuracy in training because of the difficulty in solving the detailed problems such as light problems, vague details and difficulty in distinguishing similar objects. In order to improve the generalization ability of unmanned vehicle decision-making in unfamiliar scenes, we hope to use knowledge graph to improve the generalization ability of

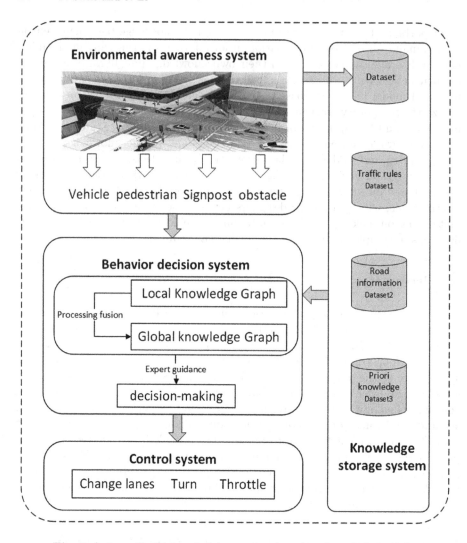

Fig. 2. Automatic driving decision system based on knowledge graph

unmanned vehicle decision-making. People usually think that when more data are available, the training results will perform better. However, it turns out that when multiple data sets are combined, the performance of the system does not necessarily improve. The aggregation of different data sets and fields leads to the problem of deep learning. The global knowledge graph mentioned in [14] can solve the above problems by using data sets from multiple fields. The global knowledge graph contains the information of all data sets, so it can guide the neural network to learn similar representations of all data sets independently of their implicit information. This method can alleviate the problems in training neural networks on aggregated data sets and achieve better overall performance.

In [31], a new metalearner Meta-iKG based on subgraph is proposed. The learner uses local subgraphs to convey specific information of subgraphs, and learns transferable patterns faster through metagradients. In this way, we find that the model can quickly adapt to the use of a few known facts and inductive settings, which greatly improves the generalization ability. A few shots relation reasoning method based on pre-training of connected subgraphs is proposed in [32], which can predict rare relations and effectively merge new relations into KG, which is the most difficult and important part in the process of KG completion. To sum up, I think that applying knowledge graphping technology to the unmanned field can effectively improve the generalization ability of unmanned driving and achieve new breakthroughs.

5.2   Improve the Interpretability of Unmanned Vehicle Decision-Making

At present, the research on unmanned vehicle decision-making is mostly based on the method of complex depth neural network. Because most neural network decision-making models are complex in structure, and the methods proposed at present have limitations, which can not provide a reasonable explanation for the deep reinforcement learning model, making it difficult for people to understand the decision-making process in the model from the network. Improving interpretability can not only enhance the understanding of model designers and users, but also better optimize the model performance. Improving the interpretability of unmanned vehicle decision is the focus of future research. Deep reinforcement learning is mostly data-driven, and data-driven modeling is mainly based on observed data. This method cannot naturally integrate external knowledge. However, the logical reasoning based on knowledge and axiomatic principle has strong interpretability, but this method often lacks the ability to estimate the statistical significance of reasoning. [33] expounds the related work of interpretable knowledge reasoning for knowledge graph and introduces the necessity of interpretability in reasoning tasks. I think this method can be applied to unmanned driving to improve the interpretability of unmanned driving. The method of hybrid artificial intelligence proposed in [34] can organically combine the advantages of both. This method inherits the concept of neurosymbolism as a way to guide the learning progress of deep neural network by using knowledge base, which not only maintains interpretability but also realizes comparability. In [35], an interpretable multi-hop reasoning method of dynamic knowledge graph is proposed. On the one hand, multi-hop knowledge graph is more interpretable. On the other hand, this method is designed for dynamic knowledge graph, which can realize the processing of dynamic scenes. The knowledge of autonomous driving is also dynamic, and multi-hop reasoning based on dynamic knowledge graph can be better applied to autonomous driving decision-making, solving the problem of dynamic change of knowledge and improving interpretability.

5.3   Improve the Decision-Making Safety of Unmanned Vehicles

The strategy obtained by reinforcement learning often takes the state graphping as the optimal solution. Due to the sparsity of extreme events, it is difficult to balance the security and stability of the algorithm, and the strategy with security risks can also meet the requirements of the optimal solution. The existing security guarantee method mainly improves the performance of the algorithm by punishing extreme events, but this punishment method will increase the variation of the parameters to be estimated and the sample complexity of the training problem, thus reducing the stability of the algorithm. Therefore, unmanned vehicles need to constantly improve their autonomy and intelligence to cope with various complex task types and emergencies. How to improve the safety of unmanned vehicle decision-making has become a key issue in this field. In order to improve the security of decision-making in complex and unknown environment, I hope to use knowledge graph to combine some common sense knowledge with knowledge from sensors and make efficient and safe decisions. For example, when the vehicle walks to a crowded area, it can be considered in advance that someone may appear at the corner or fork in the road, and if the above situation occurs, the correct decision can be made. We hope that all kinds of emergencies can be taken into account in the intelligent decision-making of UAV using knowledge graph, so that UAV can achieve high intelligence and improve the safety of unmanned driving.

6   Conclusion

This paper introduces the related technology of unmanned vehicles based on knowledge graph and its research status in the field of unmanned vehicles, and emphatically points out the application of unmanned vehicles based on knowledge graph in typical application scenarios such as target recognition, semantic segmentation, scene understanding, behavior prediction and motion planning, and puts forward the automatic driving decision-making technology based on knowledge graph. Finally, the application prospect of knowledge graph in the field of unmanned vehicles is prospected. The unmanned vehicle technology based on knowledge graph will greatly improve the decision-making ability of unmanned vehicles and can better adapt to various complex scenarios.

Acknowledgements. This paper was funded by the National Natural Science Foundation of China (72101033 and 71831001), the Beijing Key Laboratory of Intelligent Logistics Systems (BZ0211), the Canal Plan-Youth Top-Notch Talent Project of Beijing Tongzhou District (YHQN2017014), the Scheduling Model and Method for Large-scale Logistics Robot E-commerce Picking System based on Deep Reinforcement Learning (KZ202210037046), the Fundamental Research Funds for the Central Universities No.2015JBM125 and the Beijing Intelligent Logistics System Collaborative Innovation Center (BILSCIC-2018KF-01).

References

1. Qi, G.L., Gao, H., Wu, T.X.: Research progress of knowledge graph. Inf. Eng. **3**(1), 4–25 (2017)
2. Zhao, L., Ichise, R., Mita, S., Sasaki, Y.: Core ontologies for safe autonomous driving. In: The 14th International Semantic Web Conference (2015)
3. Klotz, B., Troncy, R., Wilms, D., Bonnet, C.: VSSo: a vehicle signal and attribute ontology. In: 9th International Semantic Sensor Networks Workshop (2018)
4. Sarwar, S., et al.: Context aware ontology-based hybrid intelligent framework for vehicle driver categorization. Trans. Emerg. Telecommun. Technol. **33**(8), e3729 (2022)
5. Kannan, S., Thangavelu, A., Kalivaradhan, R.B.: An intelligent driver assistance system (I-DAS) for vehicle safety modelling using ontology approach. Int. J. Ubi-Comp **1**(3), 15–29 (2010)
6. Feld, M., Müller, C.: The automotive ontology: managing knowledge inside the vehicle and sharing it between cars. In: Proceedings of the 3rd International Conference on Automotive User Interfaces and Interactive Vehicular Applications, pp. 79–86 (2011)
7. Suryawanshi, Y., Qiu, H., Ayara, A., Glimm, B.: An ontological model for map data in automotive systems. In: 2019 IEEE Second International Conference on Artificial Intelligence and Knowledge Engineering (AIKE), pp. 140–147. IEEE (2019)
8. Henson, C., Schmid, S., Tran, A.T., Karatzoglou, A.: Using a knowledge graph of scenes to enable search of autonomous driving data. In: ISWC, pp. 313–314 (2019)
9. Pollard, E., Morignot, P., Nashashibi, F.: An ontology-based model to determine the automation level of an automated vehicle for co-driving. In: Proceedings of the 16th International Conference on Information Fusion, pp. 596–603. IEEE (2013)
10. Westhofen, L., Neurohr, C., Butz, M., Scholtes, M., Schuldes, M.: Using ontologies for the formalization and recognition of criticality for automated driving. IEEE Open J. Intell. Transp. Syst. **3**, 519–538 (2022)
11. Ulbrich, S., Nothdurft, T., Maurer, M., Hecker, P.: Graph-based context representation, environment modeling and information aggregation for automated driving. In: IEEE Intelligent Vehicles Symposium Proceedings, pp. 541–547. IEEE (2014)
12. Xie, D. P., Chang, Q.: Summary of relation extraction. Appl. Res. Comput. **37**(07), 1921–1924+1930 (2020)
13. Xu, Q.T., Hong, Y., Pan, Y.C., et al.: Summary of attribute extraction research. J. Softw. **34**(02), 690–711 (2023)
14. Halilaj, L., Luettin, J., Henson, C., Monka, S.: Knowledge graphs for automated driving. In: 2022 IEEE Fifth International Conference on Artificial Intelligence and Knowledge Engineering (AIKE), pp. 98–105. IEEE (2022)
15. Kim, J.E., Henson, C., Huang, K., Tran, T.A., Lin, W.-Y.: Accelerating road sign ground truth construction with knowledge graph and machine learning. In: Arai, K. (ed.) Intelligent Computing. LNNS, vol. 284, pp. 325–340. Springer, Cham (2021). https://doi.org/10.1007/978-3-030-80126-7_25
16. Monka, S., Halilaj, L., Schmid, S., Rettinger, A.: Learning visual models using a knowledge graph as a trainer. In: Hotho, A., et al. (eds.) ISWC 2021. LNCS, vol. 12922, pp. 357–373. Springer, Cham (2021). https://doi.org/10.1007/978-3-030-88361-4_21
17. Lu, P., Xu, S., Peng, H.: Graph-embedded lane detection. IEEE Trans. Image Process. **30**, 2977–2988 (2021)

18. Kunze, L., Bruls, T., Suleymanov, T., Newman, P.: Reading between the lanes: road layout reconstruction from partially segmented scenes. In: 2018 21st International Conference on Intelligent Transportation Systems (ITSC), pp. 401–408. IEEE (2018)

19. Bordes, J.B., Davoine, F., Xu, P., Denoeux, T.: Evidential grammars: a compositional approach for scene understanding. Application to multimodal street data. Appl. Soft Comput. **61**, 1173–1185 (2017)

20. Wickramarachchi, R., Henson, C., Sheth, A.: Knowledge-infused learning for entity prediction in driving scenes. Front. Big Data **4**, 759110 (2021)

21. Halilaj, L., Dindorkar, I., Lüttin, J., Rothermel, S.: A knowledge graph-based approach for situation comprehension in driving scenarios. In: Verborgh, R., et al. (eds.) ESWC 2021. LNCS, vol. 12731, pp. 699–716. Springer, Cham (2021). https://doi.org/10.1007/978-3-030-77385-4_42

22. Fang, F., Yamaguchi, S., Khiat, A.: Ontology-based reasoning approach for long-term behavior prediction of road users. In: 2019 IEEE Intelligent Transportation Systems Conference (ITSC), pp. 2068–2073. IEEE (2019)

23. Li, X., Ying, X., Chuah, M.C.: GRIP: graph-based interaction-aware trajectory prediction. In: 2019 IEEE Intelligent Transportation Systems Conference (ITSC), pp. 3960–3966. IEEE (2019)

24. Zipfl, M., et al.: Relation-based motion prediction using traffic scene graphs. In: 2022 IEEE Intelligent Transportation Systems Conference (ITSC), pp. 825–831. IEEE (2022)

25. Regele, R.: Using ontology-based traffic models for more efficient decision making of autonomous vehicles. In: International Conference on Autonomic & Autonomous Systems, pp. 94–99. IEEE (2008)

26. Xiong, Z., Dixit, V. V., Waller, S. T.: The development of an ontology for driving context modelling and reasoning. In: 2016 IEEE 19th International Conference on Intelligent Transportation Systems (ITSC), pp. 13–18. IEEE (2016)

27. Huang, L., Liang, H., Yu, B., Li, B., Zhu, H.: Ontology-based driving scene modeling, situation assessment and decision making for autonomous vehicles. In: In 2019 4th Asia-Pacific Conference on Intelligent Robot Systems (ACIRS), pp. 57–62. IEEE (2019)

28. Dianov, I., Ramirez-Amaro, K., Cheng, G.: Generating compact models for traffic scenarios to estimate driver behavior using semantic reasoning. In: IROS, International Conference on Intelligent Robots and Systems, pp. 69–74 (2015)

29. Hovi, J., Ichise, R.: Feasibility study: rule generation for ontology-based decision-making systems. In: Wang, X., Lisi, F.A., Xiao, G., Botoeva, E. (eds.) JIST 2019. CCIS, vol. 1157, pp. 88–99. Springer, Singapore (2020). https://doi.org/10.1007/978-981-15-3412-6_9

30. Morignot, P., Nashashibi, F.: An ontology-based approach to relax traffic regulation for autonomous vehicle assistance. arXiv preprint arXiv:1212.0768 (2012)

31. Zheng, S., Mai, S., Sun, Y., Hu, H., Yang, Y.: Subgraph-aware few-shot inductive link prediction via meta-learning. IEEE Trans. Knowl. Data Eng. (2022)

32. Huang, Q., Ren, H., Leskovec, J.: Few-shot relational reasoning via connection subgraph pretraining. Adv. Neural. Inf. Process. Syst. **35**, 6397–6409 (2022)

33. Hong, Z.N., Jin, X.L., Chen, J.Y., et al.: A summary of the research on interpretable reasoning of knowledge graph. J. Softw. **33**(12), 4644–4667 (2022)

34. Oltramari, A., Francis, J., Henson, C., Ma, K., Wickramarachchi, R.: Neuro-symbolic architectures for context understanding. arXiv preprint arXiv:2003.04707 (2020)

35. Yan, C., Zhao, F., Jin, H.: ExKGR: explainable multi-hop reasoning for evolving knowledge graph. In: Bhattacharya, A., et al. (eds.) DASFAA 2022. LNCS, vol. 13245, pp. 153–161. Springer, Cham (2022). https://doi.org/10.1007/978-3-031-00123-9_11

Historical Location Information Based Improved Sparrow Search Algorithm for Microgrid Optimal Dispatching

Ting Zhou[1,2], Bo Shen[1,2(✉)], Anqi Pan[1,2], and Jiankai Xue[1,2]

[1] College of Information Science and Technology, Donghua University, Shanghai 201620, China
[2] Engineering Research Center of Digitalized Textile and Fashion Technology, Ministry of Education, Shanghai 201620, China
bo.shen@dhu.edu.cn

Abstract. The sparrow search algorithm (SSA), as an efficient meta-heuristic algorithm, has been widely used on practical problems in various fields. Nevertheless, the basic SSA is prone to fall into local optimum, which weakens the optimization ability. In order to address this problem, a novel improved SSA, called the historical location information based sparrow search algorithm (HLI-SSA), is presented. In order to solve the problem that the original sparrow search algorithm will miss part of the information during the iteration process, the historical useful information is fully utilized by creating a memory bank, which can make more population information available to individual sparrows. In addition, the Lévy stable distribution strategy is applied to improve the ability of jumping out of the local optimum. The adaptive quadratic interpolation mechanism and the use of randomness are introduced to enhance the algorithm diversity. The proposed HLI-SSA is then validated on the CEC benchmark functions. The experimental results indicate that the HLI-SSA can improve the optimization performance of the basic SSA. Finally, the method is successfully employed to the microgrid optimal dispatching problem under extreme conditions.

Keywords: Sparrow search algorithm · memory base · Lévy stabilized distribution · adaptive quadratic interpolation mechanism · microgrid optimal dispatching

1 Introduction

With the rapid progress of the times, there are more and more real-life problems to be optimized. A metaheuristic algorithm is an algorithm that is designed based on empirical knowledge to provide a practical solution to a problem within reasonable computational time and space constraints. Researchers have employed a range of metaheuristic algorithms in recent years to solve optimization problems, such as chaotic sparrow search algorithm (CSSA), Harris Hawks Optimizer (HHO), Differential Evolution algorithm (DE) and so on [5,6,12].

L. Pan et al. (Eds.): BIC-TA 2023, CCIS 2062, pp. 242–255, 2024.
https://doi.org/10.1007/978-981-97-2275-4_19

The sparrow search algorithm is a metaheuristic approach developed by drawing inspiration from the predatory and anti-predatory behavior of sparrow populations [16]. In contrast to other metaheuristic algorithms, SSA offers a unique search framework [15]. The sparrow population is categorized into producers, scroungers and vigilantes using pre-established rules. At the beginning of predation, the position of each individual sparrow is randomly generated, and then it starts to explore the problem space and needs to get as much food as possible [7]. At the same time, the vigilantes make to fly away from the original position according to their situation after detecting danger. According to [16], SSA is recognized for its benefits of high stability, powerful global search capability, and minimal parameter requirements.

Yet, the original SSA has limitations such as low diversity and a tendency to get trapped in local optima, for which many other strategies have been introduced to enhance it. For example, chaotic sparrow search algorithm (CSSA) designed by He et al. improves SSA's global optimization ability by using a sine chaotic map to initialize the sparrow population and introducing an adaptive T distribution [5]. Yang et al. developed the Tent Lévy Flying Sparrow Search Algorithm (TFSSA) to improve the exploration behavior of the sparrow population by utilizing an enhanced Tent chaos strategy to initialize the population and integrating the Lévy flight mechanism and adaptive hyperparameters [17]. To enhance the global search capability, Sun et al. used an inverse learning scheme, and in addition, nonlinear adaptive inertia weights and improved step control parameters were used to avoid falling into local optimum [13]. Chen et al. improved the algorithm performance by introducing a dyadic-based learning strategy to initialize the population and utilizing a nonlinear exponential decreasing strategy and a vertical and horizontal crossover strategy to enhance the population diversity [4].

However, upon further analysis of the original SSA's update strategy, it is found that further improvements could still be made through a number of strategies. The first is that SSA underutilizes the information during the update process, which can leave out useful information and thus cause the algorithm to find the optimal solution more slowly. Following, the producer update formula intentionally converges to zero, which can increase the likelihood of the algorithm getting trapped in a local optimum and experiencing premature convergence. To address these issues, the algorithm update strategy of SSA is improved in this study. First, the original SSA's judgment conditions for determining whether a location is updated may result in the omission of valid historical information, a memory base mechanism is used to further update the population based on historical information. Second, to avoid the behavior that the producer update formula in the original SSA intentionally converges to zero, the Lévy stable distribution is introduced. In addition, only the best and worst individual information of the sparrow population is used for follower update in the original SSA. In order to avoid the waste of group information during the updating process, the follower update formula is improved using an adaptive quadratic interpolation mechanism (AQIM) [18]. Finally, to enhance the algorithm's randomness, random variables are introduced when the worst individual information is used for vigilante position updating, which in turn improves the algorithm diversity.

The paper is structured as follows. The algorithm update strategy of the original SSA is described in detail in Sect. 2. The improved sparrow search algorithm is described in Sect. 3. Section 4 describes the parameter settings, benchmark functions, experimental results and analysis of the algorithm. In Sect. 5, the proposed improved SSA is successfully applied to the optimal scheduling of microgrids. Finally, conclusions are presented in Sect. 6.

2   Sparrow Search Algorithm

The SSA is a population-based metaheuristic algorithm that simulates the foraging mechanism in sparrows. The sparrow population is categorized into producers and scroungers based on the numerical magnitude of the fitness value. At the same time, some individuals are randomly selected as vigilantes. The position update formulas for individual sparrows are as follows [16]:

$$X_{pro,j}^{t+1} = \begin{cases} X_{i,j}^t \cdot \exp(-\frac{i}{\eta \cdot t_{max}}), & \text{if } R_2 < ST \\ X_{i,j}^t + R \cdot L, & \text{otherwise.} \end{cases} \tag{1}$$

$$X_{scr,j}^{t+1} = \begin{cases} R \cdot \exp(\frac{X_{worst} - X_{i,j}^t}{i^2}), & \text{if } i > n/2 \\ X_p^{t+1} + |X_{i,j}^t - X_p^{t+1}| \cdot E^+ \cdot L, & \text{otherwise.} \end{cases} \tag{2}$$

$$X_{vig,j}^{t+1} = \begin{cases} X_{best}^t + \beta \cdot |X_{i,j}^t - X_{best}^t|, & \text{if } f_i > f_g \\ X_{i,j}^t + K \cdot (\frac{|X_{i,j}^t - X_{worst}^t|}{(f_i - f_w) + \varepsilon}), & \text{if } f_i = f_g. \end{cases} \tag{3}$$

where $X_{i,j}^t$ denotes the position information of the ith sparrow in the jth dimension at tth iteration, $\eta \in (0,1]$ is a random number (RN), $R_2 \in [0,1]$ is a RN indicating the warning value, $ST \in [0.5,1]$ is a RN indicating the safety value, R is a RN, and L denotes a $1 \times d$ matrix with all elements of 1, n denotes the number of individuals in the sparrow population, X_p denotes the optimal position currently occupied by the discoverer, X_{worst} denotes the current global worst position, E denotes a $1 \times d$ matrix where each element is randomly assigned to 1 or -1, and $E^+ = E^T(EE^T)^{-1}$, X_{best} is the global optimal position, β is a RN which denotes the step control parameter, $K \in [-1,1]$ is a RN, f_i is the current fitness value of the individual sparrow, f_g and f_w denote the global best and global worst fitness values, respectively, and ε indicates the smallest constant to avoid the case of zero denominator.

3   Proposed Improved Sparrow Search Algorithm

The SSA demonstrates superior performance compared to many classical algorithms. Despite its advantages, the SSA algorithm still faces challenges such as premature convergence and a susceptibility to getting trapped in local optima. In addition, its ability to solve larger scale problems is poor. Therefore, four modifications will be made to the SSA to improve its algorithmic performance, which are described in detail in the following subsections.

3.1 Application of Historical Memory

In SSA, the update of an individual sparrow's position is determined by whether its current fitness value is superior to its previous value. If an individual's current fitness value is superior to its previous value, the previous fitness value is discarded. However, the location information of those globally optimal individuals with worse fitness values than the next generation is still of great importance.

To fully utilize the location information of the optimal individual in each generation during the iterative process, a memory base is established. The number of individuals stored in the memory base is N times the population size. The initial positions of the individuals in the memory base are generated randomly. During each iteration of the sparrow population, the current global optimum individual is compared with the individual in the memory base that has the worst fitness value (f_{MBw}). When the current global best individual has a superior fitness value, the memory base replaces the location information of the worst individual with the current global best individual's information. About the process of applying the memory base to update the sparrow population is shown in Fig. 1. In this way, the position information of all globally optimal individuals during the iteration is retained, and the position information in the memory base is continuously updated as the iteration progresses.

After completing the creation of the memory base, the information contained in the memory base is used when the sparrow population is updated. Every m iterations, v individuals are randomly chosen from the memory base and sequentially compared with the worst individual in the current population. If the fitness value of the individual randomly selected from the memory base is superior, it will replace the worst individual in the current population. Using the above methods, it is possible to renew and optimize the global worst individual which has little influence on the whole population. Further optimizing the position of the global worst individual can lead to improved results in the subsequent iteration of the population.

3.2 Improvement Based on Lévy Stable Distribution

Lévy stable distribution was proposed in 1937. The probability density of the symmetric Lévy stabilization process can be derived by [3]:

$$L_{\mu,\xi}(z) = \frac{1}{\pi} \int_0^\infty \exp(-\xi q^\mu) cos(qz) dq \tag{4}$$

where $\mu = 1.5$ and $\xi = 1$ are the two parameters characterizing the distribution. More specifically, μ governs the scaling properties of the stochastic process $\{z\}$, while ξ determines the scale unit of the stochastic process. It is challenging to manage Lévy stabilization processes both theoretically and numerically. Generating a RN that adheres to the Lévy stable distribution is also a challenging task. Many implementations for generating RNs obeying this distribution have been proposed by related scholars. One of the more accurate and straightforward

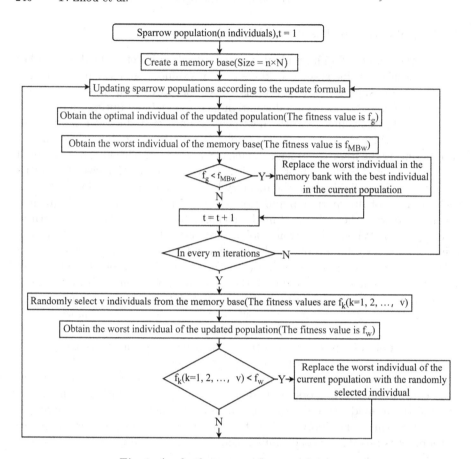

Fig. 1. Apply the memory base mechanism.

methods is known as the Mantegna method, which uses the normal distribution to generate RNs [10]. The RN s is generated according to Eq. (5).

$$s = \frac{x}{|y|^{\frac{1}{\mu}}} \tag{5}$$

where $x \sim N(0, \sigma^2)$, $y \sim N(0, 1)$, σ is obtained according to Eq. (6).

$$\sigma = \left[\frac{\Gamma(1 + \mu)sin(\pi\mu/2)}{\Gamma((1 + \mu)/2)\mu 2^{(\mu-1)/2}} \right]^{\frac{1}{\mu}} \tag{6}$$

From Eq. (1), we can see that in traditional SSA, when the warning value is less than the safety value, the position update of the discoverer intentionally converges to the origin. However, when the optimal point is non-origin, the population may oscillate between the origin and the optimal point, leading to a decrease in the algorithm's convergence. Referring to the aforementioned issue, the Lévy stable distribution can be applied to the discoverer update formula. The improved discoverer update formula is as follows:

$$X_{pro,j}^{t+1} = \begin{cases} X_{i,j}^t \cdot (1+s), & \text{if } R_2 < ST \\ X_{i,j}^t + R \cdot L, & \text{otherwise.} \end{cases} \quad (7)$$

When enough RNs obeying Lévy distribution are generated, "unusually large values" will appear from time to time. The emergence of these "unusually large values" can aid the sparrow search algorithm in escaping from local optima.

3.3 Improvement Based on Adaptive Quadratic Interpolation Mechanism (AQIM)

The quadratic interpolation method is used to search for the minimum value point of the objective function. Proposed as an extension of the quadratic interpolation method, the adaptive quadratic interpolation mechanism has been demonstrated to enhance the algorithm's local search capability [18]. Applying AQIM to SSA, the follower update formula for the improved SSA is as follows:

$$X_{scr} = \begin{cases} Y_i, & \text{if } rand \le p_t \\ Z_i, & \text{otherwise.} \end{cases} \quad (8)$$

$$Y_{i,j} = \frac{(X_{i,j}^{t\,2} - X_{m,j}^{t\,2}) \times f_p + (X_{m,j}^{t\,2} - X_{p,j}^{t\,2}) \times f_i + (X_{p,j}^{t\,2} - X_{i,j}^{t\,2}) \times f_m}{2((X_{i,j}^t - X_{m,j}^t) \times f_p + (X_{m,j}^t - X_{p,j}^t) \times f_i + (X_{p,j}^t - X_{i,j}^t) \times f_m)} \quad (9)$$

$$Z_{i,j} = \frac{(X_{i,j}^{t\,2} - X_{k1,j}^{t\,2}) \times f_{k2} + (X_{k1,j}^{t\,2} - X_{k2,j}^{t\,2}) \times f_i + (X_{k2,j}^{t\,2} - X_{i,j}^{t\,2}) \times f_{k1}}{2((X_{i,j}^t - X_{k1,j}^t) \times f_{k2} + (X_{k1,j}^t - X_{k2,j}^t) \times f_i + (X_{k2,j}^t - X_{i,j}^t) \times f_{k1})} \quad (10)$$

where p_t is set to 0.4, X_m represents the mean position information of all individuals, f_p is the current optimal individual fitness value, X_p denotes the position information of the current best individual, and f_m is the average of the fitness values of all individuals in the current population. $k1$ and $k2$ are the adaptive random integers generated in the overall of $[1, N_{ada}]$, where N_{ada} is derived according to the following equation:

$$N_{ada} = N \times \left(\omega_a - \frac{(\omega_a - \omega_b) \times t}{iter_{max}} \right) \quad (11)$$

where $\omega_a = 0.7$, $\omega_b = 0.2$.

From Eqs. (8–10), it is evident that when the random number generated in accordance with the uniform distribution is greater than p_t, the follower will be updated according to the current information of itself and the information of two individuals in the current population randomly selected according to Eq. (11). As the iteration proceeds, N_{ada} will then become smaller, and the range of the generated RNs $k1$ and $k2$ will be continuously reduced. This means that at the beginning of the iteration, the algorithm focuses more on diversity, while at the end of the iteration, the algorithm will focus more on convergence.

3.4　Utilization of Randomness

In the previous subsection, the algorithm design of traditional SSA was introduced, in which the vigilantes in the sparrow population fly elsewhere after realizing the danger. From Eq. (3), we know that when the vigilante is the current global best individual, it needs to fly toward other sparrows to avoid predation, and the position is updated in the traditional SSA based on the current global worst individual position information and its fitness value.

However, to enhance the algorithm's level of randomness and prevent premature convergence, the direction in which the current global optimal individual approaches, acting as a vigilant, can be determined by considering multiple individuals. First, a certain individual is randomly selected from the worst $(0.2 \times n)$, and then the position of the current globally optimal individual, acting as a vigilant, is updated by utilizing its position information and fitness value. The improved formula for the vigilante position update is as follows:

$$
X_{i,j}^{t+1} = \begin{cases} X_{best}^t + \beta \cdot |X_{i,j}^t - X_{best}^t|, & \text{if } f_i > f_g \\ X_{i,j}^t + K \cdot \left(\frac{|X_{i,j}^t - X_{rw}^t|}{(f_i - f_{rw}) + \varepsilon}\right), & \text{if } f_i = f_g. \end{cases} \tag{12}
$$

where $rw \in [0.8n + 1, n]$ is a random integer, X_{rw}^t denotes the position information of an individual randomly selected from the worst $(0.2 \times n)$ individuals at moment t, and f_{rw} represents the fitness value of the sampled sparrow.

4　Experimental Results and Discussions

To assess the effectiveness of the HLI-SSA, it was evaluated by IEEE CEC2017 benchmark functions [14]. Four other advanced algorithms were chosen for comparison.

4.1　Experimental Setting

The performance of the HLI-SSA is compared with the SSA, the CSSA, the HHO, the DE. To be fair, four algorithms' parameters are set the same as in their respective original papers. In addition, due to the existence of the vigilante update mechanism in SSA, the number of population individuals is set to 25 for SSA and the two improved SSAs, and 30 for the rest of the algorithms, in order to ensure that each algorithm calculates the fitness value the same number of times. The experiments are implemented with MATLAB R2021a running on a PC with Intel(R) Core(TM) $i7$-$7500U$ CPU @ 2.70GHz 2.90 GHz and 8 GB RAM in the windows 10. All algorithms run independently 30 times and the resulting data is recorded.

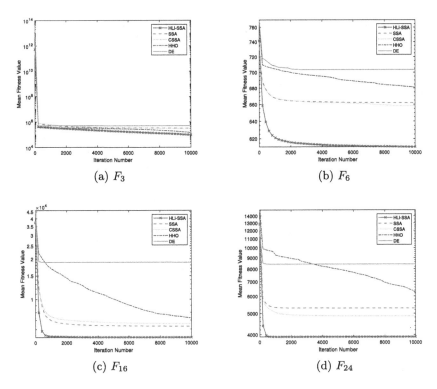

Fig. 2. The convergence curve of mean fitness value on some selected functions with 100D.

4.2 Benchmark Functions

IEEE CEC2017 benchmark functions include four types of functions, which are unimodal functions, simple multimodal functions, hybrid functions and composition functions. Different kinds of benchmark functions can examine and test the algorithms in terms of global search ability, local exploitation ability, convergence speed and ability to jump out of local optimum.

4.3 Comparison of Simulation Results

The simulation result of benchmark functions with dimension D = 100 is shown in Tables 1. The values in the table are the average of the results of 30 experiments. For simple multimodal functions 4–10, the algorithmic performance of HLI-SSA outperforms the other comparative algorithms, which verifies that HLI-SSA more adequately performs a global search while avoiding falling into a local optimum. In addition, for the hybrid functions 11–20, HLI-SSA's algorithmic accuracy is better than the other algorithms, and this integrated empirical orthogonal function, as an optimization problem with hybrid characteristics, requires the algorithms to overcome the competition between multiple locally optimal and globally optimal solutions, which suggests that HLI-SSA is more prone to jumping out of the local optimum. The composite functions 21–30 are designed to simulate real-world optimization problems, for composite functions, HLI-SSA's algorithmic accuracy is higher, which indicates HLI-SSA has better convergence and higher search ability at the same time.

4.4   Convergence Rate Comparison

The mean convergence curves for 30 times are shown in Fig. 2. The selected functions F3, F6, F16 and F24 are unimodal function, simple multimodal function, hybrid function and composition function respectively. As shown in Fig. 2, the HLI-SSA guarantees fast convergence with higher accuracy.

5   Application to Microgrid Optimal Dispatching

In this section, the HLI-SSA is applied to the optimal scheduling of microgrids in the face of extreme environments or man-made attacks. The distributed generation of microgrids in this paper includes wind power, photovoltaic power, fuel cells (FC), diesel power generation (DG), micro gas turbines (MT) and energy storage cells (ES).

5.1   Objective Function and Constraints for Microgrid Scheduling Problem

In order to fully utilize the renewable energy, the output power generated by wind and photovoltaic sources is directed towards the load side. The output power of wind generation is dependent on the wind speed, while the output power of photovoltaic generation is determined by factors such as light intensity, temperature, and wind speed [2,9]. A large amount of data shows that the wind speed in a region approximately obeys the bolometric parameter Weibull distribution, and this property is utilized to generate the wind speed for each hour of the day [8]. In addition, light intensity and temperature data are obtained from the software Meteonorm.

Moreover, under extreme conditions, the input power at the load side is determined based on the results of microgrid scheduling optimization. The microgrid optimization objectives in this paper include the operating cost and the input power at the load side. The objective function is shown as follow:

$$F = w_1 \cdot C + w_2 \cdot P_{load} \tag{13}$$

where w_1 and w_2 denote the weights and their sum is one. C is the running cost, which is calculated by the following equation:

$$C = C_{MT} + C_{FC} + C_{DG} + C_M \tag{14}$$

where C_{MT} denotes the total fuel cost of the MT, C_{FC} denotes the total fuel cost of the FC, C_{DG} denotes the total fuel cost of the DG, the fuel cost of MT,

Table 1. Optimization results and comparison for CEC2017 test functions with 100D.

Functions	HLI-SSA	SSA	CSSA	HHO	DE
Func1	1.0003E+4	9.2215E+03	**5.5322E+03**	3.1867E+08	2.8591E+11
Func3	**9.4321E+04**	3.1804E+05	3.0366E+05	1.3336E+05	5.1344E+05
Func4	**6.3925E+02**	6.8041E+02	6.8235E+02	9.4958E+02	9.7467E+04
Func5	**9.9443E+02**	1.3741E+03	1.3578E+03	1.4784E+03	2.0307E+03
Func6	**6.1061E+02**	6.6179E+02	6.5810E+02	6.8055E+02	7.0345E+02
Func7	**1.5641E+03**	3.1698E+03	3.1021E+03	3.6704E+03	4.8036E+03
Func8	**1.2404E+03**	1.8331E+03	1.7678E+03	1.9274E+03	2.4768E+03
Func9	**1.3549E+04**	2.3186E+04	2.4488E+04	4.1608E+04	7.5801E+04
Func10	**1.4631E+04**	1.6379E+04	1.5637E+04	1.9371E+04	3.0548E+04
Func11	**1.4018E+03**	3.6357E+03	8.6009E+03	3.4120E+03	2.0531E+05
Func12	**1.9857E+06**	1.3297E+07	1.3671E+08	3.6816E+08	1.8350E+11
Func13	**5.3132E+03**	1.7616E+04	3.2533E+04	5.5388E+06	3.9388E+10
Func14	**2.7413E+05**	3.9818E+05	6.2730E+05	1.5130E+06	2.9926E+07
Func15	**2.7715E+03**	5.8495E+03	9.3423E+03	1.5446E+06	1.2364E+10
Func16	**5.2206E+03**	6.3043E+03	6.5893E+03	7.1894E+03	1.8855E+04
Func17	**4.7298E+03**	5.9600E+03	6.0984E+03	6.5123E+03	1.0066E+06
Func18	**3.3364E+05**	6.8781E+05	1.6349E+06	2.9106E+06	5.8463E+07
Func19	**4.5024E+03**	6.0241E+03	8.0666E+03	6.5157E+06	1.3146E+10
Func20	**4.9504E+03**	6.2404E+03	5.7403E+03	5.8712E+03	6.4124E+03
Func21	**2.7424E+03**	3.9041E+03	3.8166E+03	4.0301E+03	4.5748E+03
Func22	**1.6630E+04**	2.0244E+04	1.9348E+04	2.3318E+04	3.4443E+04
Func23	**3.2588E+03**	4.4630E+03	3.8723E+03	4.8795E+03	5.7909E+03
Func24	**3.9097E+03**	5.2884E+03	4.8638E+03	6.1341E+03	8.4221E+03
Func25	**3.2770E+03**	3.3061E+03	3.3307E+03	3.5932E+03	3.1499E+04
Func26	**1.4680E+04**	2.0031E+04	2.3231E+04	2.4496E+04	5.7682E+04
Func27	**3.6703E+03**	3.9735E+03	3.9391E+03	4.2645E+03	3.8935E+03
Func28	**3.3598E+03**	3.3947E+03	3.4353E+03	3.6197E+03	1.2376E+04
Func29	**6.5069E+03**	7.4086E+03	7.7128E+03	9.1549E+03	2.0283E+05
Func30	**1.0594E+04**	2.0772E+04	1.8836E+05	3.4324E+07	2.9742E+10

FC and DG is determined by their respective output power and C_M denotes the maintenance cost of the distributed generation source, the mathematical model is shown in the following equation [1, 19, 20]:

$$C_M = \sum_{i=1}^{N} K_{M,i} P_i \tag{15}$$

where $K_{M,i}$ denotes the maintenance factor of the ith distributed generation source and P_i denotes the output power of the ith distributed generation source [11].

In addition, P_{load} indicates the input power at the load side, and its operation formula is shown in the following equation:

$$P_{load} = w_{l1} \sum_{i=1}^{H} P_{Hload_i} + w_{l2} \sum_{j=1}^{L} P_{Lload_j} \tag{16}$$

where w_{l1} and w_{l2} denote the load weights, the loads in this paper are divided into primary and secondary loads, and P_{Hload_i} and P_{Lload_j} are the input power of the primary load and the input power of the secondary load, respectively.

In this section, the constraints are processed using a penalty function to embed the constraints into the objective function, and the equation is described as shown below:

$$\hat{F} = F + G \tag{17}$$

where $\hat{F}$ is the penalty objective function, F is the original fitness function, G denotes the weighted sum of constraints.

In order to ensure that the microgrid system can operate properly, necessary constraints need to be imposed on the distributed generation sources in the microgrid, which is as follow:

$$\sum_{o=1}^{O} P_o(t) = \sum_{l=1}^{L} P_l(t) \tag{18}$$

where $P_o(t)$ denotes the output power of the oth distributed generation source in time period t and $P_l(t)$ denotes the input power of the lth load in time period t. In addition, it should be noted that the output power and input power should be within the specified range.

For the energy storage cells, the constraint to be satisfied is as follow:

$$SOC_{min} \leq SOC \leq SOC_{max} \tag{19}$$

where SOC indicates the state of charge in the energy storage cells, which is the ratio of the current capacity C_r in the battery to the rated capacity C_n of the battery, and its calculation formula is shown in the following equation:

$$SOC = (C_r/C_n) \times 100 \tag{20}$$

The SOC needs to be guaranteed to be within the set interval.

5.2   Parameter Setting and Simulation Results

The model parameters of distributed generation units and energy storage devices in the microgrid system are shown in Table 2.

Table 2. Model parameters

Power supply type	WT	PV	MT	FC	DG	ES
Maximum output power (kW)	40	20	65	50	50	100
Minimum output power (kW)	0	0	0	0	0	−100
Operation and maintenance costs (\$/kW)	0.0296	0.0096	0.088	0.0293	0.1	0.0012

Apply HLI-SSA, SSA, CSSA, HHO and DE to the microgrid scheduling optimization problem in this section. To ensure fairness, due to the presence of vigilantes in the sparrow population, the number of population individuals is set to 25 for HLI-SSA, SSA, and CSSA, and 30 for HHO and DE, in order to ensure that each algorithm calculates the fitness value the same number of times. To ensure the accuracy of the simulation results, all algorithms were run independently for 30 times. The final simulation results obtained are shown in Table 3.

Table 3. Simulation results of microgrid optimal dispatching

	HLI-SSA	SSA	CSSA	HHO	DE
$\hat{F}$	4.6000E−2	3.8411E+01	1.3450E−1	4.2297E+11	1.0429E+13

From the experimental results, it can be concluded that the HLI-SSA proposed in this paper can be successfully applied to the microgrid optimal scheduling problem in extreme environments, and compared with other algorithms, HLI-SSA can find lower fitness values for the objective function. In 30 independently run experiments, both HLI-SSA and CSSA are guaranteed to satisfy the constraints, SSA is guaranteed to satisfy the constraints in 60% of the cases, while neither HHO nor DE can satisfy the constraints.

6 Conclusion

In this paper, the HLI-SSA is proposed to compensate for the lack of performance of the SSA. The algorithmic performance of HLI-SSA is improved by designing a memory bank method to update the population position based on historical data, introducing Lévy stable distribution, adopting adaptive quadratic interpolation mechanism and introducing random variables to update the position of vigilantes. Applying the existing excellent algorithms and the HLI-SSA proposed in this paper to the CEC2017 benchmark function, the experimental results show that HLI-SSA has great advantages in convergence speed and optimization accuracy.

In addition, the HLI-SSA is successfully applied to the microgrid optimal scheduling problem in extreme environments, and the HLI-SSA has better algorithmic performance compared to the other four existing excellent algorithms.

References

1. Azmy, A.M., Erlich, I.: Online optimal management of PEMFuel cells using neural networks. IEEE Trans. Power Deliv. **20**(2), 1051–1058 (2005)
2. Bouraiou, A., Hamouda, M., Chaker, A., Sadok, M., Mostefaoui, M., Lachtar, S.: Modeling and simulation of photovoltaic module and array based on one and two diode model using Matlab/Simulink. Energy Procedia **74**, 864–877 (2015)
3. Brookes, B.: Théorie de l'addition de variables aléatoires. By Paul Lévy pp. xx 385. 1954. 1200f. (Gauthier-villars, paris). Math. Gazette **39**(330), 344 (1955)
4. Chen, P., Wang, H., Yan, H., Du, J., Ning, Y., Wei, J.: sEMG-based upper limb motion recognition using improved sparrow search algorithm. Appl. Intell. **53**(7), 7677–7696 (2023)
5. He, D., Liu, C., Jin, Z., Ma, R., Chen, Y., Shan, S.: Fault diagnosis of flywheel bearing based on parameter optimization variational mode decomposition energy entropy and deep learning. Energy **239**, 122108 (2022)
6. Heidari, A.A., Mirjalili, S., Faris, H., Aljarah, I., Mafarja, M., Chen, H.: Harris hawks optimization: algorithm and applications. Futur. Gener. Comput. Syst. **97**, 849–872 (2019)
7. Hong, J., Shen, B., Xue, J., Pan, A.: A vector-encirclement-model-based sparrow search algorithm for engineering optimization and numerical optimization problems. Appl. Soft Comput. **131**, 109777 (2022)
8. Justus, C., Hargraves, W., Mikhail, A., Graber, D.: Methods for estimating wind speed frequency distributions. J. Appl. Meteorol. **17**(3), 350–353 (1978)
9. Kreishan, M.Z., Zobaa, A.F.: Scenario-based uncertainty modeling for power management in islanded microgrid using the mixed-integer distributed ant colony optimization. Energies **16**(10), 4257 (2023)
10. Mantegna, R.N.: Fast, accurate algorithm for numerical simulation of levy stable stochastic processes. Phys. Rev. E **49**(5), 4677 (1994)
11. Premkumar, M., et al.: An efficient and reliable scheduling algorithm for unit commitment scheme in microgrid systems using enhanced mixed integer particle swarm optimizer considering uncertainties. Energy Rep. **9**, 1029–1053 (2023)
12. Storn, R., Price, K.: Differential evolution-a simple and efficient heuristic for global optimization over continuous spaces. J. Glob. Optim. **11**, 341–359 (1997)
13. Sun, L., Si, S., Ding, W., Xu, J., Zhang, Y.: BSSFS: binary sparrow search algorithm for feature selection. Int. J. Mach. Learn. Cybern. **14**, 1–25 (2023)
14. Wu, G., Mallipeddi, R., Suganthan, P.N.: Problem definitions and evaluation criteria for the CEC 2017 competition on constrained real-parameter optimization. National University of Defense Technology, Changsha, Hunan, PR China and Kyungpook National University, Daegu, South Korea and Nanyang Technological University, Singapore, Technical report (2017)
15. Wu, R., et al.: An improved sparrow search algorithm based on quantum computations and multi-strategy enhancement. Expert Syst. Appl. **215**, 119421 (2023)
16. Xue, J., Shen, B.: A novel swarm intelligence optimization approach: sparrow search algorithm. Syst. Sci. Control Eng. **8**(1), 22–34 (2020)

17. Yang, Q., Gao, Y., Song, Y.: A tent Lévy flying sparrow search algorithm for wrapper-based feature selection: a COVID-19 case study. Symmetry **15**(2), 316 (2023)
18. Yang, X., et al.: An adaptive quadratic interpolation and rounding mechanism sine cosine algorithm with application to constrained engineering optimization problems. Expert Syst. Appl. **213**, 119041 (2023)
19. Yao, J.: Optimal scheduling of microgrid based on improved particle swarm optimization algorithm. Master's thesis, Liaoning University of Technology (2016)
20. Zhang, Z.: Economic dispatch of microgrid based on improved quantum particle swarm optimization. Master's thesis, North China Electric Power University (2021)

Optimization of Large-Scale Distribution Center Location Selection in Fresh Produce Transportation

Wenhao Jia[1], Yang Lin[2], Junyuan Ding[2], Guoan Qin[2], Shuai Shao[3(✉)], and Ye Tian[4]

[1] School of Artificial Intelligence, Anhui University, Hefei 230601, China
[2] School of Internet, Anhui University, Hefei 230601, China
[3] Institutes of Physical Science and Information Technology, Anhui University, Hefei 230601, China
freshshao@gmail.com
[4] School of Computer Science and Technology, Anhui University, Hefei 230601, China

Abstract. The transportation process of fresh food often experiences spoilage, leading to a negative impact on its freshness and sales. Although the existing logistics distribution centers can transport a large number of goods, they have poor timeliness, so they are not suitable for large-scale fresh food transportation. In order to ensure the freshness and sales of fresh food, it is necessary to establish multi-level logistics distribution centers, hence the location of distribution centers has become a key issue. Such facility location problems are challenging in both modeling and optimization, especially when facing thousands of communities in a city that is ubiquitous in China. In this paper, a large-scale multi-objective optimization model for distribution center location problem is formulated and solved by sparse multi-objective optimization evolutionary algorithms (sparse MOEAs). Experimental results on the formulated optimization model show that the center locations obtained by the state-of-the-art sparse MOEAs can effectively reduce the cost of logistics construction, optimize the loss in the transportation process, and improve the overall benefit.

Keywords: Evolutionary computation · Facility location · Sparse Pareto optimal solutions · Multi-objective optimization

1 Introduction

With the improvement of residents' living standards, the demand for fresh products is on the rise, both online and offline. As a result, people's quality expectations for fresh products have also been increasing. This has resulted in an expanding market demand for fresh products. To meet this growing demand, it has become crucial to continuously upgrade the logistics and distribution methods of fresh products, particularly in terms of achieving more precise delivery of customer needs. By doing so, we can ensure that the vast fresh market is capable of meeting the needs of consumers. In recent years, people have begun to try using cold chain transportation [1] to ensure the freshness of fresh food. As revealed in [2], the transportation cost of fresh product logistics is 40%–60% higher than that of ordinary commodity logistics, while the cost is rising. For this reason, scientifically and reasonably planning transportation methods can reduce distribution costs and total logistics costs, providing better service for consumers.

As the capital of Anhui Province in China, Hefei has developed rapidly since 2010. According to data from the Hefei Bureau of Statistics [3], by the end of 2021, the permanent population of Hefei had reached 9.465 million, an increase of 95,000 from the previous year. It is estimated that by 2030, the urbanization rate of Hefei will reach 90%–95%. With the increase in population and the expansion of the city scale, traffic congestion in Hefei is becoming more and more serious. Therefore, when carrying out fresh food distribution in Hefei, the importance of establishing a multi-level distribution center's fresh food fixed-point transportation optimization scheme is becoming increasingly prominent.

In recent years, the facility location problem has been extensively researched, and researchers have explored numerous approaches to address this problem in different scenarios [4, 5]. For example, Zafar et al. [6] integrated Geographic Information Systems with Multichannel Linear Prediction (MCLP) [7] to ultimately develop a platform for optimizing the location of electric vehicle charging station facilities. Wu et al. [8] proposed a multi-stage facility location problem based on clustering, and Tang et al. [9] proposed a method to solve the multi-objective facility location problem in green logistics based on the fast Non dominated Sorting Genetic Algorithm (NSGA-II) [10] and the greedy algorithm.

Transporting fresh products within cities is a complex decision-making process that involves multiple objectives and discrete search spaces. Consequently, solving this problem using MCLP can lead to high computational complexity. While MCLP can identify the most essential points, it does not find the optimal solution. Although cluster-based methods are effective for large-scale facility selection problems, they do not handle multi-objective problems well. Recently, heuristic algorithms such as NSGA-II have been explored to solve facility location problems with multiple objectives. However, these methods tend to converge slowly in large-scale search spaces [11].

In recent years, many sparse multi-objective evolutionary algorithms (MOEAs) have been studied to solve some large-scale multi-objective sparse problems [12, 13]. These sparse MOEAs excel at solving problems with sparse optimal solutions under large-scale binary optimization problems (i.e., the values of most decision variables in the optimal solution are zero), making them very suitable for handling various subset selection problems [14]. Therefore, this paper first establishes an optimization model for the location problem of fresh distribution centers, and then solves the problem through sparse MOEAs. Research experiments in Hefei show that this method reduces the cost of logistics construction and optimizes losses during transportation, achieving an overall benefit increase, which is characterized by reducing economic consumption, facilitating citizens, and improving the efficiency of fresh food transportation.

The rest of this paper is organized as follows. In Sect. 2, we detail the acquisition and processing of the dataset and the construction of the optimization model. In Sect. 3, we introduce the sparse MOEAs used. In Sect. 4, we describe the experimental setup and compare the results obtained by the MOEAs. Finally, in Sect. 5, we draw some analyses and conclusions.

2 Data Acquisition, Processing, and Model Construction

2.1 Dataset

This paper targets Hefei city in China, selecting the location of a large logistics distribution center with an area of 11445 km^2 and a population of 9.37 million as the research theme. We obtain the traffic routes and intersection data of the target districts from OpenStreetMap [15] using the osmnx library [16] in Python. The dataset includes 773 communities, 79, 591 traffic routes, and 34,481 intersection coordinates, distributed across four urban districts: Shushan, Yaohai, Luyang, and Baohe. In addition, the dataset also contains the population numbers of each community.

To reflect the actual situation, this paper considers the Dijkstra distance between nodes. However, directly using the osmnx library to calculate the Dijkstra distance between nodes would take several days for this dataset, which is inefficient. To improve efficiency, this paper uses the Shortest Path Faster Algorithm (SPFA) [17] to calculate the Dijkstra distance between nodes, which significantly improves efficiency compared to using the osmnx library.

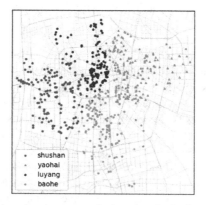

Fig. 1. Traffic network and main community locations in the four districts of Hefei city.

As shown in Fig. 1, an undirected graph is constructed from the intersection data in the dataset according to the traffic route information, and each community is associated with its nearest intersection node according to coordinates, thereby obtaining the Dijkstra distance from all intersections to each community. Distribution centers of various levels are established among all intersections, thereby transforming the selection of distribution center locations into the selection of nodes in the graph.

2.2 Encoding Scheme

Fresh produce in supermarkets has a limited shelf-life and depreciates in quality over time. As a result, daily restocking is necessary based on historical sales data and demands for each product. However, frequent restocking can lead to cost losses and transportation issues such as untimely supply and product loss during transportation. Thus, selecting optimal transportation methods and routes can ensure the timely restocking of fresh produce while minimizing losses, increasing sales, and ultimately boosting profits for the supermarket. In urban transportation, considering that the common point-to-point delivery method in logistics is not only inefficient but also incurs significant losses during the delivery process. This paper establishes a hierarchical logistics transportation method based on the coordination principle of logistics distribution centers, thereby improving the turnover speed of logistics to ensure the quantity and freshness of fresh food. The multi-level logistics distribution network is shown in Fig. 2, where each district has a primary distribution center and several secondary distribution centers. The primary distribution center in each district is responsible for receiving fresh food transported from the source, and the secondary distribution centers receive fresh food transported from the primary distribution center and are responsible for transporting it to the communities they serve.

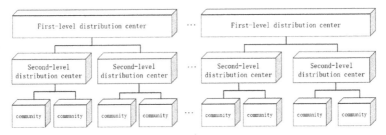

Fig. 2. The multi-level logistics transportation layout used in this paper.

To find the best location plan for each district, this paper divided each district, established four subgraphs, and processed each subgraph separately. First, an optimization algorithm is used to determine the best location for the secondary distribution center for all communities in this subgraph. Then, based on the location of the secondary distribution center, the location of the primary distribution center with the smallest total distance to all secondary distribution centers is determined. Given that the count of secondary distribution centers in hierarchical transportation is not fixed, and there exists only a single primary distribution center, the pivotal objective of this study is to ascertain the optimal location for the secondary distribution center. The optimal location for the primary distribution center is relatively simple. Once the location of the secondary distribution center is determined, the location of the primary distribution center can be determined through linear time complexity calculations.

Therefore, let x_j represent whether the j-th intersection is used as a secondary distribution center. If it is, then $x_j = 1$, otherwise $x_j = 0$. This forms the location scheme set $\mathbf{x} = \{x_1, x_2, \cdots, x_j, \cdots, x_r\}, j = 1, 2, \cdots, r$, and r represents the total number of intersections in the district.

2.3 Objective Functions

In existing location problem optimization models, objectives related to total distance [18], time [19], and the number of facilities [20] are usually considered. In most cases, the transportation of fresh produce is carried out once a day, so the time requirement is ignored. For other influencing factors, this paper comprehensively considered and customized three minimization objectives. The symbols involved are shown in Table 1.

Table 1. Formal notations used in the proposed optimization.

Notation	Description
C	Set of Community
D	A group of secondary distribution center site selection plans
$dis(i, j)$	Dijkstra distance between nodes i and j
P_i	Population of the ith community
$f_{load}(x)$	Effort defined in Eq. (1)
$f_{cost}(x)$	Costs during Transportation defined in Eq. (3)
$f_{sum}(x)$	The construction cost of distribution centers as defined in Eq. (4)
b	Number of distribution centers in the plan
d	Number of decision variables
x	A solution represented by a $1 \times d$ binary vector

The total amount of fresh food that needs to be transported in each community is related to the population of the community. The first objective is to evaluate the amount of work by considering the population size of the community and the distance between the secondary distribution center and the community it is responsible for. Defined as:

$$f_{load}(x) = \sum_{i \in C} \min_{j \in D} dis(i, j) \times P_i \tag{1}$$

where C represents the set composed of all communities, D is the set of secondary distribution centers determined by x, $dis(i, j)$ is the calculated Dijkstra distance between community i and secondary distribution center j, P_i represents the population of community i. This means that the amount of work in each community is positively correlated with its distance from the nearest secondary distribution center and the population of the community.

The second objective is to evaluate the transportation cost of fresh products. Fresh food usually has a short shelf life, so it often results in the loss of goods during transportation. At the same time, in order to ensure the freshness of fresh food, cold chain transportation is used throughout transportation, resulting in an expensive cost. Therefore, we define the formula for transportation cost per kilometer as:

$$c_t = c_1 + c_2 + c_3 \tag{2}$$

After investigation, the cost during transportation per kilometer involved in this paper mainly includes transportation cost $c_1 = 1144.6$, cargo damage cost $c_2 = 623.5$, power consumption cost $c_3 = 916.6$. Therefore, the total cost during transportation is obtained by multiplying the cost during transportation per kilometer by the total distance, which is:

$$f_{cost}(x) = \sum_{i \in C} \min_{j \in D} dis(i, j) \times c_t \tag{3}$$

where C represents the set composed of all communities, D is the set of secondary distribution centers determined by x. $dis(i, j)$ is the calculated Dijkstra distance between community i and secondary distribution center j, that is, total costs of transportation are related to the total distance from the community to the nearest distribution center.

The third objective is to consider the construction cost of the facility. For each distribution center, there are fixed infrastructure construction costs and later maintenance costs. Therefore, the evaluation function is defined as:

$$f_{sum}(x) = b \tag{4}$$

where, b denotes the aggregate count of secondary distribution centers incorporated in the scheme. Given that the expenditure associated with facility establishment is directly proportional to the number of facilities, the cumulative number of secondary distribution centers can be interpreted as a representation of the facility construction expenditure.

In summary, the following multi-objective optimization problem is constructed for the location of distribution centers in a district:

$$\text{Minimize} \quad \begin{aligned} f_1(x) &= f_{load}(x) \\ f_2(x) &= f_{cost}(x) \\ f_3(x) &= f_{sum}(x) \end{aligned} \tag{5}$$
$$\text{Subject to } x = (x_1, x_1, \cdots, x_d) \in \{0, 1\}^d$$

The first and second objectives mentioned above are to some extent conflicting with the third objective, because adding more distribution centers can reduce the distance between the distribution center and each community (means lesser f_1 and f_2), but it will increase the construction cost of facilities (means more f_3). Furthermore, Shushan, Yaohai, Luyang, and Baohe districts have 5443, 2763, 1287, and 4507 candidate locations respectively. This makes the optimization model a large-scale optimization problem.

Considering the difficulty of solving the proposed optimization model, we will use several more advanced sparse MOEAs to solve this problem, which will be introduced in the next section.

3 Optimization Algorithms

As a type of heuristic algorithm inspired by natural, MOEAs provide a new way for us to solve multi-objective optimization problems with conflicting objectives. As shown in Fig. 3, the randomly generated original population obtained a better population solution after multiple iterations. The general process of MOEA is shown in the Algorithm.

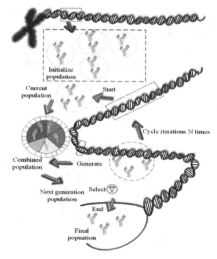

Fig. 3. General process of multi-objective evolutionary algorithms.

Algorithm: General Process of MOEA

Intput: N (population size)
Output: P (final population)
1    $P \leftarrow$ Randomly generate N solutions;
2   **while** *termination criterion not fulfilled* **do**
3        $P' \leftarrow$ Select $2N$ parents from solutions in P;
4        $P \leftarrow P \cup Variation(P')$
5   Delete duplicated solutions from P;
6        $[F_1, F_2, \ldots] \leftarrow$ Do non-dominated sorting on P;
7        $CrowdDis \leftarrow CrowdingDistance(F)$
8        $k \leftarrow argmin_i |F_1 \cup \ldots \cup F_i| \geq N$
9   Delete $|F_1 \cup \ldots \cup F_k| - N$ solutions from F_k with the smallest $CrowdDis$;
10       $P \leftarrow F_1 \cup \ldots \cup F_k$;
11   **return** P;

During the past twenty years, many MOEAs algorithms have been developed based on this idea in order to solve various types of optimization problems, such as large-scale variable problems [21], adaptive weight adjustment [22], multi-objective decomposition optimization [23], multiple selection [24], to name a few. For the optimization model constructed by Formula (5), there are many intersections can be chosen, but in some excellent solutions, most candidate points will not be selected as distribution centers. Therefore, although the dimensions of decision variables are relatively high, there are few meaningful dimensions, and they have multiple objectives. These problems are called a general designation as sparse multi-objective optimization problems (SMOPs) [12]. Large-scale sparse multi-objective optimization problems are prevalent in real-life scenarios. These problems are typically characterized by a large number of decision variables and sparse optimal solutions. Some common questions like neural network training [25], portfolio optimization [26], pattern mining [27], and some others. At the same time, the traditional MOEAs algorithm are difficult to find a better sparse solution on this kind of problems. Therefore, in order to solve such problems, many people began to upgrade MOEAs to solve various sparse multi-objective problems [28] and compare their advantages and disadvantages together. This paper selected three sparse MOEAs that make good performance (i.e., SparseEA [12], MOEA/PSL [25], and PM-MOEA [29]) to find the required optimal solutions.

SparseEA considers the sparse nature of the optimal solution and proposes a new population initialization strategy, the calculation method of crossover operator and mutation operator, which can ensure the sparsity of the solution. In SparseEA, each solution is composed of real coded dec and binary coded mask. The real coded dec of each solution records the best decision variable, and the binary coded mask records the decision variable that should be set to 0, and the final decision variable is the inner product of two vectors, which not only ensures the sparsity of the solution, but also makes it have better convergence and diversity.

MOEA/PSL proposes an evolutionary algorithm based on Pareto optimal subspace learning to generate the optimal solutions. It uses the restricted Boltzmann machine (RBM) [30] and denoising autoencoder (DAE) [31] to learn the sparse distribution and compact representation of decision variables, and the sparse distribution and compact representation are regarded as the approximation of the optimal subspace. The genetic operator is carried out in the obtained subspace, so that it greatly reduces the original search space. At the same time, it designs a parameter adaptive strategy to automatically determine the parameters in the optimal subspace learning.

PM-MOEA considers that the number of non-zero variables in each optimal solution is far less than their total number of decision variables, which uses an evolutionary pattern mining method to find the maximum and minimum candidate sets of non-zero variables. The variables in the maximum candidate set should have a value of zero or a non-zero value in the optimal solution, and the variables in the minimum candidate set should be zero. Therefore, it just needs to search variables within the maximum candidate set and variables outside the minimum candidate set, which reduces the number of decision variables. The crossover operator and mutation operator proposed can effectively search the best value of the variables in the maximum candidate set, and change each decision variable with different probabilities, which ensures the sparsity of the offspring solution and keeps the total crossover and mutation probabilities unchanged at the same time.

Therefore, the above sparse MOEAs use different strategies to generate better sparse solutions and use novel mutation and crossover operator calculation methods or use neural network learning and evolutionary pattern mining methods to reduce the search space. These sparse MOEAs are suitable for finding the best location for the establishment of distribution centers in a large number of candidate intersections. In the next section, we will introduce and analyze the location of the distribution center found by the above MOEAs.

4 Result and Analysis

4.1 Experimental Design

To highlight the performance difference between sparse MOEAs and conventional MOEAs in solving sparse multi-objective problems, this paper employs NSGA-II [13] for comparison. This comparison underscores the effectiveness of sparse MOEAs in identifying sparse solutions for large-scale sparse multi-objective problems.

In order to compare MOEAs, this paper sets the population size to 100 and the evaluation times of the function to 100000 to handle the region of each city. The experiment was carried out on PlatEMO [32] and a personal computer equipped with AMD Ryzen 7 5800H with Radeon Graphics 3.20 GHz 16 GB and 64-bit Windows 10 OS.

4.2 Experimental Results

Table 2 and Table 3 compare the performance of the four algorithms in four districts and list the average and standard deviation of hypervolume (HV) [33] and runtime obtained in 20 independent runs. It can be found that the three sparse MOEAs have better performance than NSGA-II, because they are specifically used to solve the problems with sparse optimal solutions. At the same time, it can be found that MOEA/PSL can get better HV value than the other algorithms, and MOEA/PSL and PM-MOEA are better than SparseEA. However, from the perspective of runtime, SparseEA has much better running speed than MOEA/PSL and PM-MOEA. Therefore, SparseEA is inferior to the latter in solving quality, but it has higher running efficiency and may be better in some cases.

Table 2. The HV value obtained by NSGA-II, SparseEA, MOEA/PSL, and PM-MOEA

District	NSGA-II	SparseEA	MOEA/PSL	PM-MOEA
Shushan	7.0414e−1 (8.51e−3)	9.9874e−1 (6.04e−5)	**9.9923e−1 (9.65e−4)**	9.9877e−1 (1.22e−4)
Yaohai	7.4935e−1 (6.36e−3)	9.9692e−1 (1.40e−4)	**9.9784e−1 (3.96e−3)**	9.9731e−1 (2.85e−4)
Luyang	8.1648e−1 (1.02e−2)	9.9806e−1 (1.48e−4)	**9.9930e−1 (4.95e−5)**	9.9851e−1 (1.18e−4)
Baohe	7.1925e−1 (7.01e−3)	9.9742e−1 (1.05e−4)	**9.9861e−1 (1.99e−3)**	9.9760e−1 (4.90e−4)

Table 3. The runtime consumed by NSGA-II, SparseEA, MOEA/PSL, and PM-MOEA

District	NSGA-II	SparseEA	MOEA/PSL	PM-MOEA
Shushan	**13.644** **(1.25)**	58.844 (5.44)	392.105 (36.14)	1597.306 (94.39)
Yaohai	**6.726** **(0.63)**	9.161 (2.32)	192.999 (18.32)	344.217 (23.32)
Luyang	**2.325** **(0.46)**	2.497 (0.068)	31.680 (1.49)	70.071 (5.03)
Baohe	**10.593** **(0.73)**	61.213 (2.57)	400.321 (75.85)	1021.435 (60.99)

For a more intuitive comparison, Fig. 4 shows the performance of the non-dominated solutions of the four algorithms in the Yaohai district in the three-dimensional axis. It is evident that the solutions obtained by sparse MOEAs have a wider propagation range than those obtained by NSGA-II, and NSGA-II cannot converge well in the large-scale search space. In conclusion, sparse MOEAs can better solve the proposed optimization model than conventional MOEAs.

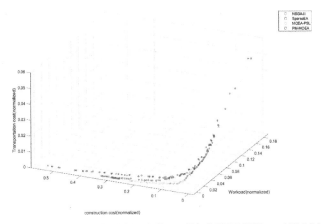

Fig. 4. Non-dominated solutions of NSGA-II, SparseEA, MOEA/PSL, and PM-MOEA obtained on the Yaohai district.

In order to further investigate the advantages of sparse MOEAs, Table 4, Table 5, and Table 6 list the total workload of all MOEAs (i.e., the first objective), the loss of road transportation (i.e., the second objective), and the number of secondary distribution centers (i.e., the third objective).

Table 4. The total workload obtained by NSGA-II, SparseEA, MOEA/PSL, and PM-MOEA.

District	NSGA-II	SparseEA	MOEA/PSL	PM-MOEA
Shushan	**1.4310e+8** **(6.91e+5)**	1.4203e+9 (1.60e+8)	1.0293e+9 (4.99e+7)	1.7504e+9 (1.21e+8)
Yaohai	**5.1514e+7** **(1.13e+6)**	1.2739e+9 (1.27e+8)	9.3793e+8 (9.88e+7)	1.5202e+9 (1.59e+8)
Luyang	**5.5598e+7** **(1.10e+6)**	3.2795e+8 (3.08e+7)	2.8216e+8 (1.87e+7)	4.1917e+8 (5.40e+7)
Baohe	**2.3134e+7** **(6.87e+5)**	8.7726e+8 (8.56e+7)	5.3891e+8 (1.45e+8)	1.0477e+9 (1.46e+8)

Table 5. The loss of road transportation obtained by NSGA-II, SparseEA, MOEA/PSL, and PM-MOEA.

District	NSGA-II	SparseEA	MOEA/PSL	PM-MOEA
Shushan	**2.9185e+4** **(6.30e+2)**	3.8091e+5 (4.01e+4)	3.1348e+5 (1.30e+4)	4.3918e+5 (3.46e+4)
Yaohai	**1.3786e+4(7.59e+2)**	2.9069e+5 (2.77e+4)	2.6676e+5 (2.71e+4)	3.4225e+5 (2.62e+4)
Luyang	**2.3251e+4(6.43e+2)**	1.9925e+5 (1.32e+4)	1.9255e+5 (1.46e+4)	2.5463e+5 (3.37e+4)
Baohe	**1.2631e+4** **(9.66e+2)**	2.7376e+5 (2.34e+4)	2.4103e+5 (2.82e+4)	3.3195e+5 (3.90e+4)

Table 6. The number of secondary distribution centers obtained by NSGA-II, SparseEA, MOEA/PSL, and PM-MOEA.

District	NSGA-II	SparseEA	MOEA/PSL	PM-MOEA
Shushan	1774.2 (48.6)	**26.8** **(11.0)**	32.9 (6.62)	61.9 (13.1)
Yaohai	762.5 (19.3)	**37.1** **(7.00)**	38.7 (9.14)	54.6 (9.28)
Luyang	260.6 (14.5)	**14.9** **(1.48)**	15.7 (1.49)	18.6 (4.70)
Baohe	1392.5 (34.7)	49.75 (20.5)	**47.8** **(10.5)**	81.0 (14.2)

It can be seen that compared with sparse MOEAs, NSGA-II gets smaller values on the first target and the third target, while it gets larger values on the second target, indicating that NSGA-II cannot deal with the problem of mutual restriction of multiple targets properly. Moreover, from Table 3, it can be seen that the number of intersection nodes selected by NSGA-II is much larger than other algorithms, where more distribution centers correspond to more construction costs. However, it is contrary to the objective of reducing construction costs expected in this paper. That is to say, although conventional MOEAs can well reduce the workload of facility location and the loss on the transportation path, they cannot meet the core requirement of the proposed optimization model.

Finally, this paper draws the optimal locations of distribution centers among all solutions obtained from all sparse MOEAs in four districts, and the optimal solution shows that they are closest to the origin in the target space. Figure 5 illustrates that each district is centered around a primary distribution center, with a number of evenly distributed secondary distribution centers surrounding it. Therefore, the optimization model in this paper can well solve the location problem of large-scale distribution centers for fresh food transportation in cities.

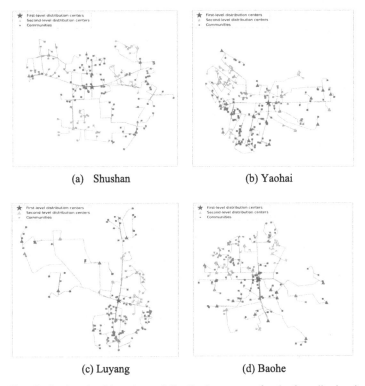

(a)   Shushan   (b) Yaohai

(c) Luyang   (d) Baohe

Fig. 5. The obtained optimal locations of distribution centers for the four districts in Hefei.

5 Conclusion

As people's living standards continue to improve, the demand for fresh produce is also increasing. Ensuring the quantity and freshness of these products depends on the strategic placement of distribution centers. This paper proposes a multi-objective optimization model to determine the optimal locations for distribution centers in large cities across multiple communities. The model aims to minimize the workload of fresh produce transportation, losses incurred during transportation, and construction costs. In order to solve the proposed optimization problem, this paper uses sparse MOEAs to find the optimal location of distribution centers in a large number of nodes. The experimental results demonstrate that sparse MOEAs outperform conventional algorithms in solving the optimization problem. They are particularly effective for determining the optimal locations of distribution centers for fresh produce. In the future, it is urgent to develop more delicate models and methods to solve the problem of optimizing distribution routes [34].

Acknowledgements. This work was supported in part by the innovation and Entrepreneurship Training Program for College Students of Anhui University (No. S202310357235), in part by the National Natural Science Foundation of China (No. 62276001, No. 62136008, No. U21A20512), in part by the Anhui Provincial Natural Science Foundation (No. 2308085J03), and in part by the Excellent Youth Foundation of Anhui Provincial Colleges (No. 2022AH030013).

References

1. Li, X., Huo, L.: Study on logistics information packaging technology of fresh food in cold chain. In: Ouyang, Y., Xu, M., Yang, Li., Ouyang, Y. (eds.) Advanced Graphic Communications, Packaging Technology and Materials. LNEE, vol. 369, pp. 579–584. Springer, Singapore (2016). https://doi.org/10.1007/978-981-10-0072-0_72
2. Yu, H.: Research on fresh product logistics transportation scheduling based on deep reinforcement learning. Sci. Program. **2022**, 8750580 (2022)
3. CSY: China Statistical Yearbook. China Statistical Publishing House, Hefei (2022)
4. Zhen, L., Wang, W., Zhuge, D.: Optimizing locations and scales of distribution centers under uncertainty. IEEE Trans. Syst. Man Cybern. Syst. **47**(11), 2908–2919 (2016)
5. Yu, H., Gao, L., Lei, Y.: Model and solution for capacitated facility location problem. In: 2012 24th Chinese Control and Decision Conference (CCDC), pp. 1773–1776. IEEE (2012)
6. Zafar, U., Bayram, I.S., Bayhan, S.: A GIS-based optimal facility location framework for fast electric vehicle charging stations. In: 2021 IEEE 30th International Symposium on Industrial Electronics (ISIE), pp. 1–5. IEEE (2021)
7. Church, R., ReVelle, C.: The maximal covering location problem. Pap. Reg. Sci. Assoc. **32**(1), 101–118 (1974)
8. Wu, K., Guan, Y., Li, J., Lu, F., Hu, Y.: A clustering-based approach to the multi-stage facility location problem. In: 2019 IEEE 6th International Conference on Industrial Engineering and Applications (ICIEA), pp. 536–540. IEEE (2019)
9. Tang, X., Zhang, J.: The multi-objective capacitated facility location problem for green logistics. In: 2015 4th International Conference on Advanced Logistics and Transport (ICALT), pp. 163–168. IEEE (2015)

10. Deb, K., Pratap, A., Agarwal, S., Meyarivan, T.A.M.T.: A fast and elitist multiobjective genetic algorithm: NSGA-II. IEEE Trans. Evol. Comput. **6**(2), 182–197 (2002)
11. Tian, Y., et al.: Evolutionary large-scale multi-objective optimization: a survey. ACM Comput. Surv. (CSUR) **54**(8), 1–34 (2021)
12. Tian, Y., Zhang, X., Wang, C., Jin, Y.: An evolutionary algorithm for large-scale sparse multiobjective optimization problems. IEEE Trans. Evol. Comput. **24**(2), 380–393 (2019)
13. Tian, Y., Feng, Y., Zhang, X., Sun, C.: A fast clustering based evolutionary algorithm for super-large-scale sparse multi-objective optimization. IEEE/CAA J. Automatica Sinica **10**(4), 1048–1063 (2022)
14. Qian, C., Yu, Y., Zhou, Z.H.: Subset selection by Pareto optimization. In: Advances in Neural Information Processing Systems, vol. 28, 1774–1782 (2015)
15. Haklay, M., Weber, P.: OpenStreetMap: user-generated street maps. IEEE Pervasive Comput. **7**(4), 12–18 (2008)
16. Boeing, G.: OSMnx: new methods for acquiring, constructing, analyzing, and visualizing complex street networks. Comput. Environ. Urban Syst. **65**, 126–139 (2017)
17. Zhang, H., Liu, X., Xiang, L.: Improved SPFA algorithm based on Cell-like P system. In: 2019 10th International Conference on Information Technology in Medicine and Education (ITME), pp. 679–683. IEEE (2019)
18. Daskin, M.S.: What you should know about location modeling. Naval Res. Logist. (NRL) **55**(4), 283–294 (2008)
19. Zhang, L., Rushton, G.: Optimizing the size and locations of facilities in competitive multi-site service systems. Comput. Oper. Res. **35**(2), 327–338 (2008)
20. Plastria, F.: Static competitive facility location: an overview of optimisation approaches. Eur. J. Oper. Res. **129**(3), 461–470 (2001)
21. Hong, W., Tang, K., Zhou, A., Ishibuchi, H., Yao, X.: A scalable indicator-based evolutionary algorithm for large-scale multiobjective optimization. IEEE Trans. Evol. Comput. **23**(3), 525–537 (2018)
22. Qi, Y., Ma, X., Liu, F., Jiao, L., Sun, J., Wu, J.: MOEA/D with adaptive weight adjustment. Evol. Comput. **22**(2), 231–264 (2014)
23. Zhang, Q., Li, H.: MOEA/D: a multiobjective evolutionary algorithm based on decomposition. IEEE Trans. Evol. Comput. **11**(6), 712–731 (2007)
24. Cui, Z., Zhao, L., Zeng, Y., Ren, Y., Zhang, W., Gao, X.Z.: Novel PIO algorithm with multiple selection strategies for many-objective optimization problems. Complex Syst. Model. Simul. **1**(4), 291–307 (2021)
25. Tian, Y., Lu, C., Zhang, X., Tan, K.C., Jin, Y.: Solving large-scale multiobjective optimization problems with sparse optimal solutions via unsupervised neural networks. IEEE Trans. Cybern. **51**(6), 3115–3128 (2020)
26. Streichert, F., Ulmer, H., Zell, A.: Evaluating a hybrid encoding and three crossover operators on the constrained portfolio selection problem. In: Proceedings of the 2004 Congress on Evolutionary Computation, pp. 932–939. IEEE (2004)
27. Zhang, X., Duan, F., Zhang, L., Cheng, F., Jin, Y., Tang, K.: Pattern recommendation in task-oriented applications: a multi-objective perspective [application notes]. IEEE Comput. Intell. Mag. **12**(3), 43–53 (2017)
28. Su, Y., Jin, Z., Tian, Y., Zhang, X., Tan, K.C.: Comparing the performance of evolutionary algorithms for sparse multi-objective optimization via a comprehensive indicator [research frontier]. IEEE Comput. Intell. Mag. **17**(3), 34–53 (2022)
29. Tian, Y., Lu, C., Zhang, X., Cheng, F., Jin, Y.: A pattern mining-based evolutionary algorithm for large-scale sparse multiobjective optimization problems. IEEE Trans. Cybern. **52**(7), 6784–6797 (2020)

30. Fischer, A., Igel, C.: An introduction to restricted Boltzmann machines. In: Alvarez, L., Mejail, M., Gomez, L., Jacobo, J. (eds.) CIARP 2012. LNCS, vol. 7441, pp. 14–36. Springer, Heidelberg (2012). https://doi.org/10.1007/978-3-642-33275-3_2

31. Vincent, P., Larochelle, H., Bengio, Y., Manzagol, P.A.: Extracting and composing robust features with denoising autoencoders. In: Proceedings of the 25th International Conference on Machine Learning, pp. 1096–1103 (2008)

32. Tian, Y., Cheng, R., Zhang, X., Jin, Y.: PlatEMO: a MATLAB platform for evolutionary multi-objective optimization [educational forum]. IEEE Comput. Intell. Mag. 12(4), 73–87 (2017)

33. While, L., Hingston, P., Barone, L., Huband, S.: A faster algorithm for calculating hypervolume. IEEE Trans. Evol. Comput. 10(1), 29–38 (2006)

34. Zhou, A., Jin, Y., Zhang, Q., Sendhoff, B., Tsang, E.: Combining model-based and genetics-based offspring generation for multi-objective optimization using a convergence criterion. In: 2006 IEEE International Conference on Evolutionary Computation, pp. 892–899. IEEE (2006)

Research on the Influence of Twin-Bow Appendage with Drum on Ship Speed and Resistance

Guangwu Liu[1] , Chao Lu[2(✉)], Si Chen[2], Lixiang Guo[2], Jin Wang[2], and Zhangrui Wu[1]

[1] Green & Smart River-Sea-Going Ship, Cruise and Yacht Research Center, Wuhan University of Technology, Wuhan 430063, China
[2] China Ship Development and Design Center, Wuhan 430064, China
609511886@qq.com

Abstract. Speed and Resistance are very critical general parameters for ship hull forms design. Reducing resistance is one of the main methods for obtaining higher speed. Optimal hull form contributes a lot to the resistance and wave-making of a vessel. In recent research, the effectiveness of resistance reducing by twin-bow appendage for monohull ship has been paid high attention. The twin-bow appendage has been developed for high-speed monohull vessels to reduce the resistance. However, for some monohull ships such as monohull with bulbous bow it is impossible to deploy twin-bow appendage since it is necessary to consider the load capacity of bulbous bow. For exploring the applicability of twin-bow appendage, a twin-bow appendage with drum is proposed for replacement of bulbous bow in this article. The hydrostatic resistance of normal vessel with bulbous bow and that of twin-bow appendage with drum will be analyzed and compared by numerical simulation. RANS equation is applied for numerical computation and the CFD tool Star CCM+ is used for simulation.

Keywords: Ship Speed and Resistance · Bulbous Bow · Numerical Simulation

1 Introduction

For numerical simulation, the following main terminologies will be used in this paper.

L_{OA} Length overall (m)

L_{wl} Length of water line (m)

B Breadth (m)

T Draft (m)

This work was supported by Key Research and Development Plan of Hubei Province (Project Research and Application of Green and Smart River-Sea-Going Ship Technologies, No.20211g0104).

L. Pan et al. (Eds.): BIC-TA 2023, CCIS 2062, pp. 271–280, 2024.
https://doi.org/10.1007/978-981-97-2275-4_21

F_{nl} Length Froude number

U Current speed in x direction

R Resistance (N)

ω Percentage of resistance difference between prototype and modified vessel

$Y+$ Dimensionless wall distance

Twin-bow appendage is an appendage with narrow section arranged close to water line in the bow part of the vessel. It is declined with same angle as the bow. The profile of this appendage makes the vessel have two bows for such reason comes the name twin-bow appendage. Its main role is to reduce resistance. In recent studies, it is found that the twin-bow appendage showed good performance for reducing resistance in line sailing and it can be expected to be widely used for high-medium vessels. Nevertheless, the twin-bow appendage cannot be implemented directly in monohull ships with bulbous bow because the cargo load function of bulbous bow. In order to investigate the further applicability in practice of twin-bow appendage we propose a twin-bow appendage with drum. With this type of appendage, the hull forms in longitudinal direction can be maintained as twin-bow appendage and a drum with the same capacity as the bulbous bow will be used. Thus, the forms characteristics and the bulbous bow's capacity are taken into consideration. Numerical calculation with scaled model will be deployed in the article for fully using the calculation resources. The same calculation field will be used for both prototype and the modified ship model but only replacing with the modified model during calculation in which way the consistency of numerical calculation conditions can be kept. After comparison of hydrostatic resistance under different speeds, the results show that swirl will appear under medium speed for vessels of twin-bow appendage with drum and resistance will slightly increase.

2 Theory and Methodology

The main methods for ship resistance study include model test and numerical simulation. Chunyu Guo etc. [1] studied the resistance forecast by using Ayre method and Lap-Keller method and made some amendment. Model test is method of testing the small, scaled ship model or rudder model made by similarity theory in the laboratory of tanks to obtain problem solutions [2–4]. Model test is being carried out for almost all the new innovation vessels with good performance and important task vessels. In the history of research on ship speed model test is always the most important method [5–7]. But model test has its limitations as well due to incomplete real condition simulation and its high cost.

In the phase of new vessel innovation, numerical calculation and simulation is widely used since its easiness for modelling [5], fast calculation and low cost etc. 3-D viscous flow method based on Reynolds-Averaged Navier-Stokes equations (RANS) [6, 7] becomes more popular for ship resistance calculation. We have used the state-of-the-art CFD tools as Fluent and Star CCM + for investigation on ship resistance. Results of calculation models, mesh generation and other simulation experience have been obtained.

In this article, a well-designed monohull prototype ship with bulbous bow is chosen as research object and Star CCM + as simulation platform [8, 9]. In the simulation, only the bow part mesh model is replaced for getting the new ship model without modification of other appendages. The same calculation field, turbulence model [10] and boundary conditions will be used. Calculations under nine different velocities and result comparison will be carried out.

3 Numerical Computation Model

3.1 Control Equation

In this article, RANS equation [6, 11, 12] is applied for numerical computation. The RANS equations representing continuous parameter and incompressible fluid are shown by Eq. (1) and (2) respectively.

$$\frac{\partial U_i}{\partial x_i} = 0 \tag{1}$$

$$\frac{\partial U_i}{\partial t} + U_j \frac{\partial U_i}{\partial x_j} = -\frac{1}{\rho}\frac{\partial \overline{P}}{\partial x_i} + \frac{1}{V}\frac{\partial^2 U_i}{\partial x_j^2} - \frac{\partial}{\partial x_j}\overline{u_i' u_j'} \tag{2}$$

where:
  ρ: the fluid density;
  x_i : coordination value, $i = 1, 2, 3$;
  U_i : Reynolds velocity, $i = 1, 2, 3$;
  $-\rho \overline{u_i' u_j'}$: Reynolds potential energy;
  $\overline{P} = \left(\frac{P-P_\infty}{\rho U_0^2} + \frac{Z}{F_r^2}\right)$: osmotic pressure parameter;

3.2 Turbulence Model

For fully representing the influence of appendage to the turbulence behavior [13], turbulence model of second-order eddy viscosity is chose as depicted by formulas (3) and (4).

$$\frac{D\rho k}{Dt} = \tau_{ij}\frac{\partial \overline{u_i}}{\partial x_j} - \beta^* \rho \omega k + \frac{\partial}{\partial x_j}\left[(\mu + \sigma_k \mu_t)\frac{\partial k}{\partial x_j}\right] \tag{3}$$

$$\frac{D\rho\omega}{Dt} = \frac{\gamma}{V_t}\tau_{ij}\frac{\partial \overline{u_i}}{\partial x_j} - \beta\rho\omega^2 + \frac{\partial}{\partial x_j}\left[(\mu + \sigma_k \mu_t)\frac{\partial \omega}{\partial x_j}\right] \\ +2\left[1 - F_t \rho \sigma_{\omega 2}\frac{1}{\omega}\frac{\partial k}{\partial x_j}\frac{\partial \omega}{\partial x_j}\right] \tag{4}$$

3.3   3-D Model

For monohull vessels with big ratio of length/breadth, the comparison of resistance is very representative for evaluation of the vessel performance. In this article, a standard ship model of monohull with bulbous bow and round bilge is selected and research object. Meanwhile, small scale model is used for saving computation resources, improving calculation speed and to get enough comparison date for further research. The main objective in his phase is to investigate the influence of appendage change on the resistance without considering the vessel sizes. The following table shows the sizes of the two 3-D models (Table 1).

Table 1. Size Comparison of the 3-D Models of Prototype Vessel and Modified Vessel

Item/Model Name	Prototype with Bulbous Bow	Twin-bow Appendage with Drum Vessel
L_{OA}(m)	7.38	7.38
L_{wl}(m)	6.84	7.34
B(m)	0.78	0.78
T(m)	0.29	0.24

Since the twin-bow appendage shall be similarly declined as the stem, its upmost point shall be the same height as the bow's highest point which makes the water line length increase of 7.3% comparing to the prototype vessel. On the other side, a drum is added at the location of 1.4 m above the appendage for capacity compensation due to the cancellation of bulbous bow. The diameter and height of the drum shall maintain the same as the bulbous bow. The bottom of the drum shall be in the line with the bottom line. It makes 17.2% loss of draft for the modified vessel with drum appendage comparing to prototype vessel.

The 3-D model of the vessel with appendage with drum is built with Catia system based on the prototype vessel by deletion of the bow part. The 3-D models of the prototype and vessel with drum appendage are shown in Fig. 1. And Fig. 2. Respectively. The yellow part in Fig. 2 is the part of appendage with drum replacing the bulbous bow.

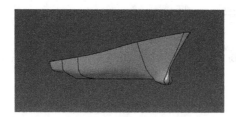

Fig. 1. 3-D Model of Prototype with Bulbous Bow

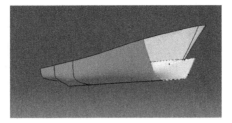

Fig. 2. 3-D Model of Twin-bow Appendage with Drum Vessel

3.4 Calculation Fields

The main objective of the simulation computation in this paper is to compare the resistance performance of the two models for wave-making study. So, it is necessary set up hydrostatic calculation field condition with two phases of sea water and free air. Reasonable flow ranges shall be configured to eliminate influence of boundary condition on the calculation result. In the simulation, the pressure-inlet position is set at 1.5 L_{OA} in front of the bow and the pressure-outlet position is set at 4.5 L_{OA} after the aft part. Pressure side boundary is located at the position of L_{OA} from vessel profile as shown in below Fig. 3.

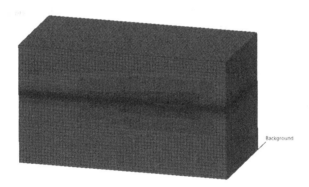

Fig. 3. Calculation Fields

3.5 Boundary Conditions

Boundary conditions are important for the simulation process. In this paper, the inlet point and outlet point are set at the pressure-inlet position and pressure-outlet position defined above respectively. The shell and flow field is selected as calculation wall and the flow velocity at wall is settled as 0. The turbulent kinetic energy k = 0 at this place because of the molecular viscosity. Nine velocities shown in Table 2.

Table 2. Velocity Configuration

Velocity	v1	v2	v3	v4	v5	v6	v7	v8	v9
v(m/s)	3.1	2.8	2.6	2.4	2.1	1.9	1.7	1.5	1.3

At the free flow surface, the x direction mesh size shall be less than $0.15F_{nl^2} \cdot L$ and the z direction mesh size shall be less than $0.03F_{nl^2} \cdot L$. Boundary layers shall be set for hull shell surface. The value of Y+ shall be from 30 to 300.

3.6 Simulation Calculation

After meshing, calculations under nine different conditions are carried out for both prototype and modified vessel with appendage replacement and 18 conditions are in total. The hydrostatic resistances for these 18 conditions are obtained.

4 Results Analysis

After simulation calculation made above, the calculation results are studies and 18 groups of resistance data have been obtained. In this section, the length Froude Number and hydrostatic resistance are calculated and analyzed as shown in Table 3. The hydrostatic curves of the two vessels are shown in below Fig. 4.

Table 3. Hydrostatic Resistance Comparison between Prototype and Modified Vessel

Velocity	v1	v2	v3	v4	v5	v6	v7	v8	v9
Modified Vessel Resistance (N)	65.88	−54.92	−44.90	−36.57	−29.64	−24.31	−20.12	−14.26	−10.14
Prototype Resistance (N)	−58.76	−50.51	−41.68	−34.18	−28.24	−22.79	−18.33	−13.15	−7.30
F_{nl}	0.37	0.34	0.32	0.29	0.26	0.24	0.21	0.18	0.16
Resistance Difference (%)	12.11	8.72	7.72	6.99	4.97	6.69	9.75	8.44	38.92

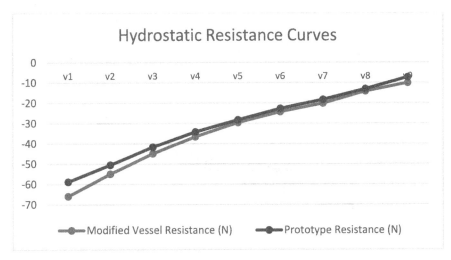

Fig. 4. Comparison of the Hydrostatic Resistance Curve

From the above table and figure it can be found that the hydrostatic resistance of vessel twin-bow appendage vessel is bigger than the prototype due to replacement of the appendage. The resistance difference under condition v5 and Froude number 0.266714 is the smallest, and the resistance difference under lowest velocity v9 and Froude number 0.160028 is the biggest 38.92%. Furthermore, the free surface patterns of the prototype and modified vessel are drawn and depicted in Fig. 5. It can be seen that the free flow surface patterns are very similar but the pattern for prototype is smoother. Turbulent may exist in the fore part of the twin-bow appendage with drum vessel.

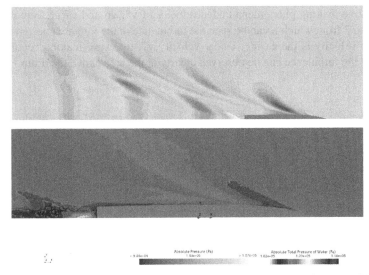

Fig. 5. Free Surface Pattern of Prototype (above) and Twin-bow Appendage Vessel (below)

The Fig. 6 below shows the pressure distribution on the shell surface of the two vessels. It shows the pressure stress appears in the peak area of the bulbous bow for the first one, and the drum fore area for the second one.

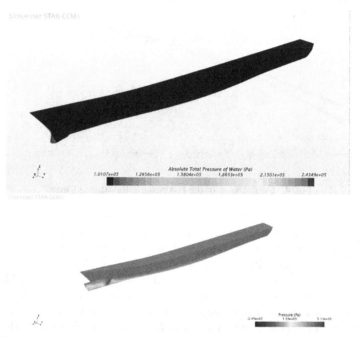

Fig. 6. Pressure Distribution of Prototype (above) and modified vessel (below)

The wave-making phenomena under velocity of v5, v6 and v7 are represented in picture Fig. 7 from which it can be seen the turbulence in the fore area becomes clearer which the velocity is increasing. Under velocity v7, the wave-making becomes very weak, and the turbulence can be observed clearly in the area above the drum.

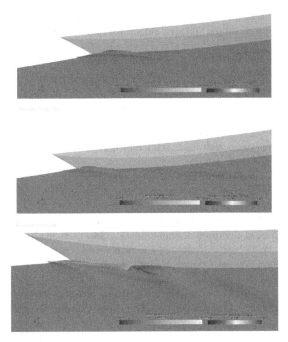

Fig. 7. Wave-making under velocity v5 (above), v6 (middle) and v7 (below)

5 Summary

The simulation research results show that the twin-bow appendage with drum has negative influence on the resistance. The resistance increase is big when the velocity is low but the increase is very small when the velocity is rather high. The study shows the main reason is because of the continuity and regularity of the bow part forms line by introducing of the drum. The whirl will occur when the vessel sails in medium speed which will bring negative influence on the sailing performance and equipment. Thus, the twin-bow appendage with drum is not expected to have good performance for resistance reducing for those vessels with bulbous bow or other big appendage under water. This means the twin-bow appendage with drum is not recommended to be implemented in the stem of the ship. In further research, streamline appendage with the same capacity will be considered to replace bulbous bow instead of drum.

References

1. Guo, C., Liu, G., Zhou, G., Huang C.: Ship resistance forecast based on weighting factor correction with Ayre method and Lap-Keller method. Ship Ocean Eng. 34–37 (2014)
2. Gao, F., Wang, L.: Numerical study on synthetic flow field of adjacent microjet actuators. Acta Aerodynamica Sinica **21**(3), 267–274 (2003)
3. Wang, W., Tang, T., Lu, S.: Numerical simulation and analysis of active jet control of hydrofoil cavitation. Chin. J. Theoret. Appl. Mech. **51**(6), 1752–1760 (2019)

4. Lu, C., Liu, G., Zhang, W., Wang, J.: Simulation of Airflow Characteristics of a Seabird Following a Ship Based on Steady State. In: Pan, L., Zhao, D., Li, L. (eds.) BIC-TA 2022. CCIS, vol. 1801, pp. 591–604. Springer, Singapore (2023). https://doi.org/10.1007/978-981-99-1549-1_47

5. Ma, Z., Pei, Z., Liu, G., Wu, W.: Dynamic stability research on stiffened plate of ship structures under harmonic excitation. Ships Offshore Struct. https://doi.org/10.1080/17445302.2023.2225946

6. Zhang, Z., Ye, S., Yue, J., Wang, Y.: A combined neural network and multiple modification strategy for Reynolds-averaged Navier-Stokes turbulence modeling. Chin. J. Theoret. Appl. Mech. 53(6), 1533–1542 (2021)

7. Menter, F.: Two-equation eddy-viscosity turbulence models for engineering applications. AIAA J. 32(8), 1598–1605 (1994)

8. Pei, Z., Yang, B., Liu, G., Wu, W.: Experimental research on the stiffness step between the main hull and superstructure of cruise ships. J. Mar. Sci. Eng. 11(7), 1264 (2023). https://doi.org/10.3390/jmse11071264

9. Liu, J., Zhang, B., Xu, N.: Research on design optimization of high-speed ship bulbous bow based on Nelder-mead algorithm. J. Ship Prod. Des. 38(1), 28–38 (2022)

10. Lin, J., Pan, L.: Multiobjective trajectory optimization with a cutting and padding encoding strategy for single-UAV-assisted mobile edge computing system. Swarm Evol. Comput. 75, 101163 (2022)

11. Rajabi, E., Kavianpour, M.: Intelligent prediction of turbulent flow over backward-facing step using direct numerical simulation data. Eng. Appl. Comput. Fluid Mech. 6(4), 490–503 (2012)

12. Weatheritt, J., Sandberg, R.: A novel evolutionary algorithm applied to algebraic modifications of the RANS stress–strain relationship. J. Comput. Phys. 325, 22–37 (2016)

13. Mahmood, S., Huang, D.: Computational fluid dynamics based bulbous bow optimization using a genetic algorithm. J. Mar. Sci. Appl. 11, 286–294 (2012)

Application of Multi-fidelity Surrogate Models to the Noisy Optimization Problems of Carbon Fiber Polymerization Process Parameters

Yilin Fang[1]($\boxtimes$) (ID), Xinwei Lu[1], and Liang Jin[2]

[1] School of Information Engineering, Wuhan University of Technology, Wuhan, China
`fangspirit@whut.edu.cn`
[2] School of Mechanical and Electronic Engineering, Wuhan University of Technology, Wuhan, China

Abstract. Carbon fiber is an innovative and strategic material for aeros-pace and other critical areas, and the polymerization process is one of the most critical processes for the production of carbon fiber. To improve the quality of precursor fibers and save costs, the optimization of process parameters is one of the important aspects. Considering the complex production plant environment and other circumstances of the carbon fiber polymerization process, a noisy multi-objective optimization problem model on the molecular weight of polyacrylonitrile (PAN) and the monomer conversion rate is developed. In order to solve the noisy optimization problems of carbon fiber polymerization process parameters (NOPCFPPP), this paper proposes an improved multi-fidelity optimization (IMFO) algorithm, aiming to reduce the impact of the noise on the evolution of a smaller evaluation budget, so as to further solve and optimize the process parameters, and hope to be able to provide some theoretical guidance in the actual polymerization process of carbon fibers. Experimental results show that the algorithm proposed in this paper is highly competitive with existing algorithms for noisy multi-objective optimization problems.

Keywords: Polymerization process · Noisy multi-objective optimization · Multi-fidelity optimization · Surrogate model

1 Introduction

PAN-based carbon fiber has excellent mechanical and functional properties and is an irreplaceable material in aerospace, transportation, civil construction, sports equipment, and other fields [8]. Based on the theory and technology of polymer materials, PAN-based carbon fiber starts from the basic organic monomer small molecules, and strives to synthesize a perfect graphite ring structure through the main steps such as polymerization, spinning, pre-oxidation, and carbonization. Homogeneous, high-quality PAN fibers are the prerequisite for the preparation of high-performance carbon fibers. Because the polymerization

L. Pan et al. (Eds.): BIC-TA 2023, CCIS 2062, pp. 281–295, 2024.
https://doi.org/10.1007/978-981-97-2275-4_22

method and process determine the average molecular weight and distribution of precursors, the composition and distribution of comonomers, and thus determine the physical and chemical properties of precursors. Therefore, it is of great significance to study the optimization of carbon fiber polymerization process parameters.

However, the carbon fiber polymerization process is complex, the production environment is noisy, incomplete data, inaccurate mathematical modeling, sensor measurement errors and changes in environmental conditions may cause noise, so we consider the optimization problem of carbon fiber polymerization process parameters as a noisy multi-objective optimization problem (NMOP).

Since the fitness evaluation of noisy multi-objective optimization problems (NMOPs) is both noisy and expensive, the presence of noise interferes with the search process of the multi-objective evolutionary algorithm and the performance of the algorithm is significantly degraded. In order to discover Pareto solutions with superior performance, it becomes crucial to find a balance between minimizing the number of actual evaluations and avoiding misleading optimization processes caused by noise. Addressing this challenge requires robust optimization techniques to handle noisy fitness evaluations. In this paper, we propose IMFO algorithm to allocate the computational budget to high-quality Pareto solutions. As a result, the performance of the algorithm can be improved and high-performance Pareto solutions can be found in as few real evaluations as possible.

2 Preliminaries

2.1 Noisy Multi-objective Optimization Problems

A mathematical formalism for minimizing a multi-objective optimization problem can be formulated as

$$\min_{x \in \Omega} \quad F(x) = (f_1(x), f_2(x), \ldots, f_m(x))^T, \tag{1}$$

where F is an objective function composed of m targets, and at least two targets are contradictory. $x = (x_1, x_2, \ldots, x_n)^T \in \Omega$ is a n-dimensional decision variable, and Ω is the search space. For two decision variables x and y, if $f_i(x) \leq f_i(y)$, $\forall\, i = 1, \ldots, m$, and with at least one index $j \in \{1, \ldots, m\}$ such that $f_j(x) \prec f_j(y)$, x is said to dominate y. The x^* is called the Pareto optimal solution, if there is no other $x \in \Omega$ dominating x^*, and all Pareto optimal solutions form the Pareto optimal set, whose corresponding set of target values constitutes the Pareto front (PF).

Concerning NMOPs [15], we consider additive noise in this paper. A minimized noisy multi-objective optimization problem is described as Eq. 2.

$$\min_{x \in \Omega} \quad F(x) = (f_1(x), f_2(x), \ldots, f_m(x))^T + z, \tag{2}$$

where $z = (z_1, z_2, \ldots, z_m)^T$, z_i, $i = 1, 2, \ldots, m$, is often assumed to be normally distributed with zero mean and variance σ^2.

2.2   Kriging Model

Kriging models, also known as Gaussian process regression models, are used to approximate each objective function [16]. A Gaussian process is a stochastic process subject to a joint normal distribution of random variables. Due to its high accuracy of model fitting and the ability to output variance as uncertainty information can show good results in model management, the Kriging model has become a very attractive surrogate model for many evolutionary algorithms to find optimal solutions. It is based on the assumption that the distribution of the objective function $y(x)$ obeys a joint normal distribution with constant mean μ and constant variance σ^2. For any solution x, the Kriging model approximates the individual objective function values as:

$$y(x) = \mu(\mathbf{x}) + \epsilon(\mathbf{x}), \epsilon(x) \sim N(0, \sigma^2) \tag{3}$$

where μ is the prediction of the regression model $F(\beta, x)$; $\epsilon(x)$ is a Gaussian distribution with zero mean and standard deviation σ^2. The regression model $F(\beta, x) = \beta_1 g_1(x) + \cdots + \beta_l g_l(x)$ is a linear combination of l selected functions with coefficients β.

2.3   Existing Noisy Multi-objective Evolutionary Algorithms

In recent years, several noisy multi-objective evolutionary algorithms (NMOEAs) have been proposed to solve noise. These techniques can be broadly categorized as follows.

The explicit averaging method is the first strategy, which is a crucial tool in noisy optimization. Its main idea involves computing the fitness value multiple times for the same variable and subsequently taking the average as the final fitness value for the solution. However, the cost of resampling is expensive, which requires a lot of evaluation to achieve the desired result, the main key challenge of this method is to identify the optimal trade-off between the resampling size and the computation time. In [13], the effect of resampling on noise and the effectiveness of sampling were analysed based on rigorous runtimes. The researchers devised methods to reduce the total number of evaluations to achieve the same performance. In each generation of the rolling tide evolutionary algorithm (RTEA) [5], only the number of non-dominated solutions sampled is increased to improve the accuracy of their active elite profiles by continuously resampling elite solutions, thus ensuring that high-quality individuals enter the evolutionary process. Afterwards, [10] adjusts the resampling rule based on the number of iterations and problem dimensionality dynamics.

Another averaging method is implicit averaging. The greater the number of sampled individuals, the more likely the algorithm is to locate near the optimal solution, thus implicit averaging reduces the effect of noise on the optimization process by increasing the number of individuals searched. For a sufficiently large population, [4] demonstrated that it can maintain more diversity, and thus larger differences in ability can be identified with a small number of evaluations. [14]

observed evolution on a smooth fitness landscape and nine deceptive landscapes and analysed the relationship between population size, noise evaluations, and match size, finding experimentally that large populations in narrow landscapes lead to rapid evolution. However, increasing population size increases computational costs, similar to the explicit averaging technique.

The third strategy is to increase the diversity of the population, which facilitates the provision of more traits for individuals during selection. In [6], an experiential learning directed perturbation (ELDP) operator was introduced, where the ELDP adapts the magnitude and direction of the changes based on past experience and introduces a momentum term based on personal history information to create new individuals for fast convergence.

Some scholars have attempted to modify deterministic selection strategies to preserve more promising solutions. [2] used NSGA-II with the α-dominant operator to classify the sampled fitness values of two individuals at the α-confidence level, and then determines the dominant relationship between these two solutions. An indicator-based multiobjective optimization [1]proposed a fitness evaluation method for selecting the best solution in the environmental selection process based on ε-index.

Currently, model-based approaches are popular in NMOPs. Based on NSGA-II, [11] combined a regularity model using a probabilistic ordering method and a Gaussian model using a non-dominated ordering to deal with NMOPs, thus improving the robustness of NMOEAs. Inspired by the generalised approximation theorem, [9] used a radial basis network (RBN) to estimate the noise-free fitness and embedded it into the classical multi-objective algorithm NSGA-II to solve NMOPs, called RBN-NSGA-II. [19] proposed a two-stage evolutionary algorithm (TSEA), which designs an adaptive switching strategy with two switches to adaptively switch between different noise treatments based on the effect of noise in order to solve a bi-objective optimization problem with noisy optimization. In addition, a data selection strategy and a model performance estimation method are proposed to enhance the model-based denoising approach.

3   Problem Statement

The polymerization process is the initial and crucial step in the manufacturing of carbon fibers, directly influencing their properties. Two important parameters for evaluating the polymerization process and its products are the monomer conversion rate and the average molecular weight of PAN. Increasing the average molecular weight of PAN enhances the performance of carbon fibers, as higher molecular weights lead to stronger PAN precursor products [12]. Achieving a higher monomer conversion rate improves product utilization; however, it also affects the molecular weight and distribution of PAN. Thus, there exists a trade-off between increasing the conversion rate and maintaining a desired molecular weight. To address this, a multi-objective model is developed in this study, utilizing the molecular weight of PAN and monomer conversion during polymerization as optimization objectives. The decision variables in the model are the polymerization temperature (T), initiator concentration (I), and polymerization time (t). In real production environments, uncertainties such as changes in environmental conditions can introduce noise. To simulate the approximation error of the model, additive noise is introduced, following a Gaussian distribution with zero mean and variance σ^2. Consequently, the optimization objectives for monomer conversion and average molecular weight of PAN during polymerization can be approximated as

$$\max \ C = 1 - e^{k_p(\frac{2fA_de^{-\frac{E}{RT}}I}{k_t})^{\frac{1}{2}} \frac{2(e^{(-\frac{A_de^{-\frac{E}{RT}}t}{2})}-1)}{A_de^{-\frac{E}{RT}}}} + N(0,\sigma^2), \qquad (4)$$

$$\max \ M_n = 2k_pM_rM_0 \times \frac{1-C}{(2fk_tIA_de^{-\frac{E}{RT}})^{\frac{1}{2}}} + N(0,\sigma^2), \qquad (5)$$

$$\text{s.t.} \begin{cases} T_{\min} \leq T \leq T_{\max}, \\ I_{\min} \leq I \leq I_{\max}, \\ t_{\min} \leq t \leq t_{\max}, \end{cases} \qquad (6)$$

where $T_{\min}$, $I_{\min}$, $t_{\min}$ represent the lower boundaries of T, I, t, respectively, while $T_{\max}$, $I_{\max}$, $t_{\max}$ represent the upper boundaries of T, I, t.

The notation and implications of the models used in our formulation are summarized in Table 1. Additionally, Table 2 presents the fixed parameters and their corresponding values. In industrial production, it is crucial to ensure the safety and reliability of the aggregation process. Therefore, each decision variable must adhere to specific ranges. The allowable range of variation for each decision variable is outlined in Table 3.

Table 1. Symbols and meanings.

Symbol	Meaning	Symbol	Meaning
C	Monomer conversion	k_t	Chain termination rate constant
M_n	Average molecular weight of PAN	M_r	Molecular weight of acrylonitrile
T	Polymerization temperature	A_d	Decomposition frequency factor
I	Initiator concentration	R	Molar gas constant
t	Reaction time of polymerization process	E	Activation energy
k_p	Chain growth rate constant	M_0	Initial monomer concentration
f	Initiator efficiency		

4   Proposed Algorithm

In this paper, we adopt an averaging approach where the average of several evaluations can be regarded as an approximation of the exact fitness, and then

Table 2. Fixed parameters and their values.

Parameter	Value	Parameter	Value
k_p	1960 L/(mol · s)	M_0	10 mol/L
f	0.9	A_d	6.98×10^{14}
k_t	7.82×10^8 L/(mol · s)	R	8.314 J/(mol · K)
M_r	53.06	E	1.21×10^5 J/mol

Table 3. Boundaries of decision variables.

Name	T	I	t
lower bound	323	0.1	3600
upper bound	343	2	36000
unit	K	mol/L	s

embed the averaging method of noise reduction into our proposed multi-fidelity optimization framework to tackle the challenge that the fitness evaluation for NOPCFPPP is both noisy and costly. In this section, we give the implementation of IMFO and discuss ordinal transformation strategy and optimal sampling strategy in detail.

4.1   Framework of the Proposed Algorithm

The pseudocode of the proposed framework is represented in Algorithm 1, which can be divided into the following phases:

Phase I: Initialization. In this phase, the initial training data is obtained by sampling the decision space. Each individual in the sampled data is evaluated once using the expensive true objective function. To generate the initial training data, we employ the Latin hypercube sampling method, which ensures a more representative and evenly distributed set of samples. Moreover, the initial data obtained from the sampling process is also utilized as the initial population, denoted as P_{FE}, for the subsequent evolutionary optimization process.

Phase II: Building Surrogate Models. For expensive optimization tasks, evaluating m objectives, $f_1(x), f_2(x), ..., f_m(x)$, is expensive, so we need to build a surrogate model for each objective function. When initially building the surrogate model, the kriging surrogate model is trained using sampled individuals and their fitness values that are evaluated once.

Phase III: Solution Space Searching. In this phase, the kriging model is considered as an approximate objective function in NSGA-II, and multiple iterations are performed on the progeny, i.e., the kriging model is used instead of the expensive objective function. The Pareto solutions obtained from each iteration

---

Algorithm 1. Pseudocode of the proposed framework

Input: The maximum number of expensive function evaluations FE_{max}; Size of population N; Total sampling budget for each iteration T; Initial local sampling budget for each iteration C; The amount of local sampling budget growth per iteration k; Number of re-evaluations per individual m.

Output: The set of nondominant solutions.

1: Generate an initial training data set Arch using Latin hypercube sampling, initialize the number of function evaluations: FE = N; P_{FE} = Arch;

2: Train a surrogate model for each objective function by Arch;

3: **while** $FE < FE_{max}$ **do**

4:   Obtain the solution space P_g and the local solution set $P_{t_{max}}$ through the solution space searching phase;

5:   Fast non-dominated sorting and crowding distance computation for set $P_{t_{max}}$ with convergence and diversity as objectives;

6:   Select $C+k$ individuals P_c from $P_{t_{max}}$ to conduct m real evaluations respectively, calculate the mean of the m evaluation results, and use the mean as the fitness value of P_c, recorded as Offspring;

7:   Solution set P_{OS} obtained through the OS stage;

8:   Offspring= Offspring $\cup$ P_{OS}, use Arch to update surrogate models;

9:   Combine parents and offspring populations, $C_{FE}= P_{FE} \cup$ Offspring ;

10: Fast non-dominated sorting and crowding distance computation for C_{FE}, $P_{FE}=$ elite selection (C_{FE});

11: $FE = FE + |$Offspring$| * m$, $C = C + k$;

12: **end while**

---

are stored in the constructed solution space to form the final feasible solution space P_g. The population of the last iteration is considered as the local solution set $P_{t_{max}}$. The pseudocode for solution space searching is given in Algorithm 2.

Phase IV: Ordinal Transformation (OT). Considering convergence and diversity as objective functions, non-dominated sorting and crowding distance computation are performed for $P_{t_{max}}$. The first $C + k$ solutions s_i of the sorted $P_{t_{max}}$ are chosen to be evaluated m times realistically, respectively, and the mean value of the solutions of the m evaluations is used as the fitness value of s_i. Non-dominated sorting and crowding distance computation on P_g to form a one-dimensional ordered solution space.

Phase V: Optimal Sampling (OS). The P_g is divided into k groups, n_0 solutions are taken from each group and m evaluations are performed taking the mean of the m evaluated solutions as the fitness value. Then follow the optimal computing budget allocation (OCBA) rules for assigning sampling budget to each group until the iterative sampling budget is exhausted.

Phase VI: Updating. When new expensive evaluation solutions are obtained, the kriging model is updated and the accuracy of the model will be further improved. If only the lowest accuracy surrogate model is used throughout the evolutionary process, the optimization process can easily be misled by noise.

If only the highest precision surrogate model is used, the process of collecting data is time-consuming and costly, so building a multi-precision surrogate model strikes a dynamic balance between these two conflicting needs. Moreover, when the number of solutions evaluated by expensive functions increases, then the environmental selection is performed to select N solutions from the current population P_{FE} and new evaluated solutions.

Phase III, IV, V, and VI will be repeated until the maximum number of expensive functional evaluations is reached, and then the Pareto solution will be output.

Algorithm 2. Pseudocode of solution space searching

Input: Initial population P_{FE}; The number of iterations in the solution space searching stage t_{max}.

Output: Solution space P_g; The local solution set $P_{t_{max}}$.

1: $P_t = P_{FE}$, t = 1;
2: **while** $t < t_{max}$ **do**
3:     Generate offspring Q_t by crossover and mutation operations with Kriging models;
4:     Combine parents and offspring populations, $C_t = P_t \cup Q_t$;
5:     Fast non-dominated sorting and crowding distance computation for C_t;
6:     $P_{t+1} =$ elite selection(C_t) ;
7:     $P_g = P_g \cup P_{t+1}$;
8: **end while**

4.2   Ordinal Transformation Strategy

The conventional multi-fidelity optimization algorithm based on OT and OS uses the idea of the NSGA-II algorithm to order the solutions to form a one-dimensional ordered space. However, due to the low accuracy of the surrogate model and the large error with the real model, the non-dominated sorting according to the approximation obtained from the surrogate model may not give good results. Inspired by the Pareto-based bi-indicator infill sampling criterion (PBISC) [17], convergence and diversity are considered as two objectives that need to be optimized, and then non-dominated ordering and crowding distance computation of solutions are performed based on these two objectives. In PBISC, a convergence indicator (CI) is introduced, ideal points in the plane are introduced for calculating the Euclidean distances to the experimental solutions, and the distance between the computed experimental solutions and the true Pareto front is computed in an approximate ideal method. The ideal point is formed from the experimental solution and the best available objective function value in the current aggregate. The CI is a minimization objective and the smaller the CI, the closer the solution is to the true PF. PBISC uses the distance between the experimental solution and its nearest parent solution as a diversity indicator (DI) to measure the increase in maintaining diversity. In addition, since the diversity measure is expected to be maximized, this paper uses negative values to convert the problem uniformly to a minimization indicator. It can be formulated as follow:

$$\min_{x^*} \quad G(x^*) = (\text{CI}(\text{x}^*), -\text{DI}(\text{x}^*)). \tag{7}$$

In this paper, a dynamic sampling strategy is used for sampling. The sampling phase is divided into global sampling in the OS phase and local sampling in the OT phase. In the global sampling process, the OS process selects solutions from the entire feasible solution space according to the OCBA rule. In the OT phase, sampling can be performed from the set of local solutions $P_{t_{max}}$. The more global solutions are sampled, the easier it is for the alternative model to fit the overall landscape of the objective function. In the early stages of optimization, the surrogate model is more likely to capture the overall trend of the objective function. Therefore, when the level of fidelity is small, the number of global solutions sampled is greater than the number of local solutions. In the later stages of optimization, the surrogate model is more likely to capture the trend of changes around the optimum. Therefore, when the fidelity level of the surrogate model is large, more solutions are sampled from the set of local solutions. To implement this dynamic sampling strategy, the number of samples in the OT phase is gradually increased during the iteration process. Specifically, the number of samples in the OT stage during each iteration is denoted as $C + k$, while the number of samples in the OS stage is calculated as $T - (C + k)$.

4.3   Optimal Sampling Strategy

Since solving the noisy optimization problem using the explicit averaging method takes a lot of computational resources, we use the OCBA rule for optimal sampling based on the idea of MO^2TOS [18]. A finite number of solutions are extracted from the feasible solution space for multiple evaluations, and the average value is used as the fitness approximation.

Chen [3] proposed the OCBA method, the core idea of OCBA is to improve the efficiency of simulation experiments by efficiently allocating computational resources in order to find the optimal solution faster or to obtain a more accurate estimation accelerating the optimization process under given resource constraints while minimizing the computational cost. Let b be the best design, for any two non-best designs $i \neq j \neq b$, $\bar{J}_b$, $\bar{J}_i$ and $\bar{J}_j$ are the mean of design b, i, j, respectively, while s_b, s_i, s_j are the standard deviation of design b, i, j. N_i, N_j, and N_b represent the budget allocated to design i, j, and b respectively. The budget allocation rules for OCBA are given by

$$\frac{N_i}{N_j} = \left(\frac{s_i(\bar{J}_b - \bar{J}_j)}{s_j(\bar{J}_b - \bar{J}_i)} \right)^2, \tag{8}$$

$$N_b = s_b \sqrt{\sum_{i=1,i\neq b}^{k} \left(\frac{N_i}{s_i} \right)^2}. \tag{9}$$

According to Eqs. 8 and 9, under a fixed budget, most of the budget will be allocated to the good performance design, and a small amount of the budget will be allocated to the poor performance design. The steps of the optimal sampling process are given in Algorithm 3.

Algorithm 3. The steps of the optimal sampling process

Input: The feasible solution space Γ_g; Total sampling budget for each iteration T; Initial local sampling budget for each iteration C; The amount of local sampling budget growth per iteration k; The number of groups h; The initial sampling budget M_0; The iterative sampling budget ΔM; Number of reevaluations per individual m.

Output: The set of non-dominated solutions P_{OS}.

1: $M = T - (C + k)$;
2: Divide the set P_g into h groups;
3: $r = 0$, and randomly extract M_0/h solutions from each group;
4: **while** $\sum\limits_{i=1}^{r} \Delta M_i \leq M - M_0$ **do**
5:     Conduct m real evaluations respectively, use the mean as the fitness value of each individual and store m individuals into the set P_{OS};
6:     Calculate the group mean μ_i and standard deviation σ_i for each group i;
7:     Set $r = r + 1$, select m_i^r solutions from each group i according to the OCBA rule;
8: **end while**

---

5   Experimental Results and Analysis

In order to verify the effectiveness of the IMFO algorithm in solving NOPCFPPP, the convergence and diversity of the obtained final solution set were evaluated employing hypervolume (HV) metrics. HV represents the volume of the hyper-cube enclosed by the individual in the solution set and the reference point in the target space [7]. The larger HV values indicate better diversity and convergence, which is calculated as

$$\text{HV}(\text{S}, \text{r}) = volume \left(\bigcup_{i=1}^{|S|} v_i \right), \tag{10}$$

where S represents the optimal solution set, r represents the reference point, and HV(S, r) gives the volume (in the objective space) that is dominated by the optimal solution set S. Besides, for each solution $\vec{s}_i \in S$, a hypercube $\vec{v}_i$ is constructed with the reference point r and the solution $\vec{s}_i$ as the diagonal corners of the hypercube.

5.1   Parameter Settings

In the following experiments, the population size is $N = 100$ and the number of iterations in the solution space searching phase is $t_{max} = 20$. The range of hyperparameter of the design and analysis of computer experiments (DACE) for building Kriging models is set to $[10^{-5}, 10^5]$. The total sampling budget per iteration is $T = 250$, the local sampling budget per iteration is $C = 40$, the iterative growth local sampling budget is $k = 10$, and the number of reevaluations per individual is $m = 4$. In the OS phase, the number of subgroups is $h = 10$, and the initial sampling budget per subgroup is $M_0 = 5$, with the iterative sampling budget $\Delta M = 10$.

And other three comparative algorithms are used, including RBN-NSGA-II [9], TSEA [19], RTEA [5] to solve NOPCFPPP. Furthermore, we adopt the recommended setting in the original literature for specific parameters of the compared algorithms. Then, we used HV as the evaluation metric. The accuracy of calculating the HV index depends on the choice of reference points. First, we normalize the Pareto solutions obtained in each experiment and multiply the maximum value of the Pareto solutions boundary by 1.1 as the reference point for each experiment. Next, we run each algorithm independently 20 times under different budgets, calculate the average value of HV and compare it.

5.2   Comparison with Other Algorithms

Table 4 shows the mean and standard deviation values of the HV value of all algorithms after 20 independent runs. To analyze the obtained results, Wilcoxon's rank sum test is used to statistically analyze the experimental results, and the significance level is set to be 0.05, where the symbol "+", "−", and "≈" indicate that the result of the corresponding algorithm is significantly better, significantly worse, and comparable to IMFO, respectively. We highlight the best result of each condition in bold.

When the noise level is 0.05, 0.10, and 0.20, respectively, the number of fitness evaluations is set to 1000, 2000, and 3000, and the results obtained by these four algorithms are shown in Table 4. It can be seen that IMFO outperforms RBN-NSGA-II, TSEA, RTEA. In this paper, we combine the surrogate model and adaptive sampling technique to construct the surrogate model to approximate the fitness assessment and guide the search process. By iteratively searching the surrogate model in the solution space search stage, we can construct the solution space with finite solutions and valuable solutions, and provide the sampling process with good Directions. In the OT and OS process, the dynamic sampling technique is used to gradually increase the number of local samples in the iterative process, thus sampling more Pareto solutions. As the budget increases, IMFO gets better results, indicating that as the algorithm iterates, the algorithm gradually overloads from global search to local search to better find the location of the optimal solution. Therefore, it is the reason why IMFO performs better with a smaller evaluation budget.

Table 4. The Mean and Standard Deviation of Four Algorithms to HV.

σ	FE_{max}	RBN-NSGA-II		TSEA		RTEA		IMFO
	1000	5.6404E-01 (5.9115E-02)	−	6.0162E-01 (3.4838E-02)	−	6.3041E-01 (2.8334E-02)	≈	**6.4730E-01 (2.1754E-02)**
0.05	2000	6.0725E-01 (3.5431E-02)	−	6.4125E-01 (2.1535E-02)	≈	6.5161E-01 (3.0851E-02)	≈	**6.5493E-01 (2.3497E-02)**
	3000	6.2280E-01 (2.7259E-02)	−	6.4621E-01 (1.7653E-02)	−	6.5382E-01 (1.9070E-02)	≈	**6.6308E-01 (1.7750E-02)**
	1000	5.6648E-01 (6.5922E-02)	−	6.2962E-01 (4.7434E-02)	−	6.5759E-01 (3.5618E-02)	≈	**6.7524E-01 (5.4025E-02)**
0.10	2000	6.1656E-01 (3.0998E-02)	−	6.7284E-01 (5.1337E-02)	≈	6.6900E-01 (3.6503E-02)	≈	**6.8420E-01 (3.2778E-02)**
	3000	6.1642E-01 (2.4370E-02)	−	6.8043E-01 (4.1322E-02)	≈	6.6275E-01 (2.5929E-02)	−	**6.8523E-01 (2.7460E-02)**
	1000	5.4795E-01 (5.9235E-02)	−	6.3767E-01 (5.5314E-02)	−	6.7731E-01 (2.6879E-02)	−	**7.0262E-01 (4.4679E-02)**
0.20	2000	6.3247E-01 (6.4041E-02)	−	6.3314E-01 (3.4865E-02)	−	6.8247E-01 (2.7250E-02)	−	**7.0852E-01 (3.0900E-02)**
	3000	6.2161E-01 (5.2822E-02)	−	6.7388E-01 (2.6506E-02)	−	6.7040E-01 (2.8609E-02)	−	**7.2545E-01 (5.0704E-02)**

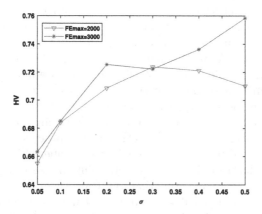

Fig. 1. The curve of change of the mean of HV corresponding the noisy level σ by IMFO.

To figure out whether IMFO is effective at higher noise levels, we plotted Fig. 1 and observed the change in the mean value of HV relative to σ from 0.05 to 0.5. For each noise case, the number of function evaluations was set to 2000 or 3000, and the algorithm was run independently 20 times to compute the average value of HV and compare it. As shown in Fig. 1, the HV of IMFO scores well when the noise level increases to 0.3, 0.4, or 0.5, indicating that IMFO is still effective at higher noise levels.

6   Conclusions

In this paper, we investigated the effects of polymerization process conditions, such as temperature, reaction time, and initiator concentration, on monomer conversion and product molecular weight, as well as modeled the NOPCFPPP considering the complex carbon fiber production environment. To address the challenge that the fitness evaluation of the NOPCFPPP is both noisy and expensive, we propose an improved multi-fidelity framework to achieve solving the noisy optimization problem with a small evaluation budget. This paper combines surrogate models to approximate the fitness landscape and guide the search process, and an adaptive sampling strategy to dynamically adjust the sampling density to allocate the computational budget to high-quality Pareto solutions. Numerical simulation results show that the IMFO algorithm proposed in this paper is effective for solving real-world problems compared to the other three algorithms. It provides a theoretical and practical basis for the industrialized large-scale production of PAN and process improvement with a view to improving the quality of carbon fiber raw silk with some reference inspiration.

In future work on the carbon fiber polymerization process, the dynamic constraint optimization problem may be one of the more interesting research directions for us. For IMFO, although good results have been achieved at different noise levels, we believe that improvements can still be made. How to further eliminate the effect of noise on the training process of surrogate model, and how to improve the accuracy of the training data while minimizing the time cost of obtaining the training data. Meanwhile, how to improve the search performance under unknown noise conditions is also an important direction for future research.

Acknowledgements. This research has been supported by the National Major Technology Equipment Research Program of China (No. 2021-1635-01) and the National Natural Science Foundation of China (Grant No. 52075402).

References

1. Basseur, M., Zitzler, E.: A preliminary study on handling uncertainty in indicator-based multiobjective optimization. In: Rothlauf, F., et al. (eds.) EvoWorkshops 2006. LNCS, vol. 3907, pp. 727–739. Springer, Heidelberg (2006). https://doi.org/10.1007/11732242_71

2. Boonma, P., Suzuki, J.: A confidence-based dominance operator in evolutionary algorithms for noisy multiobjective optimization problems. In: 2009 21st IEEE International Conference on Tools with Artificial Intelligence, pp. 387–394. IEEE (2009)

3. Chen, C.H., Lin, J., Yücesan, E., Chick, S.E.: Simulation budget allocation for further enhancing the efficiency of ordinal optimization. Discret. Event Dyn. Syst. **10**(3), 251–270 (2000)

4. Darwen, P.J., Pollack, J.B.: Co-evolutionary learning on noisy tasks. In: Proceedings of the 1999 Congress on Evolutionary Computation-CEC99 (Cat. No. 99TH8406), vol. 3, pp. 1724–1731. IEEE (1999)

5. Fieldsend, J.E., Everson, R.M.: The rolling tide evolutionary algorithm: a multiobjective optimizer for noisy optimization problems. IEEE Trans. Evol. Comput. **19**(1), 103–117 (2014)

6. Goh, C.K., Tan, K.C.: Noise handling in evolutionary multi-objective optimization. In: 2006 IEEE International Conference on Evolutionary Computation, pp. 1354–1361. IEEE (2006)

7. Jiang, S., Ong, Y.S., Zhang, J., Feng, L.: Consistencies and contradictions of performance metrics in multiobjective optimization. IEEE Trans. Cybern. **44**(12), 2391–2404 (2014). https://doi.org/10.1109/TCYB.2014.2307319

8. Khayyam, H., et al.: Pan precursor fabrication, applications and thermal stabilization process in carbon fiber production: experimental and mathematical modelling. Prog. Mater Sci. **107**, 100575 (2020). https://doi.org/10.1016/j.pmatsci.2019.100575

9. Li, Y., Liu, R., Chen, W., Liu, J.: Radial basis network simulation for noisy multiobjective optimization considering evolution control. Inf. Sci. **609**, 1489–1505 (2022)

10. Liu, J., Teytaud, O.: A simple yet effective resampling rule in noisy evolutionary optimization. In: 2019 IEEE Symposium Series on Computational Intelligence (SSCI), pp. 689–696. IEEE (2019)

11. Liu, R., Li, N., Wang, F.: Noisy multi-objective optimization algorithm based on gaussian model and regularity model. Swarm Evol. Comput. **69**, 101027 (2022)

12. Moskowitz, J.D., Abel, B.A., McCormick, C.L., Wiggins, J.S.: High molecular weight and low dispersity polyacrylonitrile by low temperature raft polymerization. J. Polym. Sci., Part A: Polym. Chem. **54**(4), 553–562 (2016). https://doi.org/10.1002/pola.27806

13. Qian, C., Yu, Y., Jin, Y., Zhou, Z.-H.: On the effectiveness of sampling for evolutionary optimization in noisy environments. In: Bartz-Beielstein, T., Branke, J., Filipič, B., Smith, J. (eds.) PPSN 2014. LNCS, vol. 8672, pp. 302–311. Springer, Cham (2014). https://doi.org/10.1007/978-3-319-10762-2_30

14. Ragusa, V.R., Bohm, C.: Connections between noisy fitness and selection strength. In: ALIFE 2021: The 2021 Conference on Artificial Life. MIT Press (2021)

15. Rakshit, P., Konar, A., Das, S.: Noisy evolutionary optimization algorithms-a comprehensive survey. Swarm Evol. Comput. **33**, 18–45 (2017)

16. Seeger, M.: Gaussian processes for machine learning. Int. J. Neural Syst. **14**(02), 69–106 (2004)

17. Song, Z., Wang, H., Xu, H.: Pareto-based bi-indicator infill sampling criterion for expensive multiobjective optimization. In: Ishibuchi, H., et al. (eds.) EMO 2021. LNCS, vol. 12654, pp. 531–542. Springer, Cham (2021). https://doi.org/10.1007/978-3-030-72062-9_42

18. Xu, J., Zhang, S., Huang, E., Chen, C.H., Lee, L.H., Celik, N.: Mo2tos: multi-fidelity optimization with ordinal transformation and optimal sampling. Asia-Pac. J. Oper. Res. **33**(03), 1650017 (2016). https://doi.org/10.1142/S0217595916500172
19. Zheng, N., Wang, H.: A two-stage evolutionary algorithm for noisy bi-objective optimization. Swarm Evol. Comput. **78**, 101259 (2023)

Multi-strategy Improved Kepler Optimization Algorithm

Haohao Ma[ID] and Yuxin Liao[(✉)] [ID]

School of Automation, Central South University, Changsha 410083, China
{224611099,liaoyuxin}@csu.edu.cn

Abstract. To solve the problem of falling into local optima and slow convergence speed in the standard Kepler optimization algorithm, the improved Kepler optimization algorithm is proposed. Firstly, the random number generator is replaced by the circle mapping and reverse learning strategy in the initial population stage. Secondly, the introduction of adaptive mutation rate and arctangent decay factor in the stage that planets update positions, which adjust the search space of the algorithm in the early and late stages, improves the convergence speed of the algorithm. Thirdly, Gauss mutation is performed on the position of the optimal solution to improve the algorithm's tendency to fall into local optima. Finally, improved Kepler optimization algorithm, and four other optimization algorithms, are comprehensively tested on 8 benchmark functions. The experimental results show that the proposed algorithm has higher convergence speed and accuracy.

Keywords: Kepler optimization algorithm · Circle mapping · Reverse learning · Adaptive mutation · Arctangent de-cay factor · Gauss mutation

1 Introduction

Intelligent optimization algorithm, which is characterized by its strong flexibility and reliability, is widely used in engineering optimization problems [1]. Common optimization algorithms include evolutionary algorithms such as genetic algorithm (GA) and differential evolution algorithm; Algorithm based on swarm intelligence, such as ant colony algorithm (ACA) and bee colony algorithm; Algorithms based on physical rules, such as gravity search algorithm, black hole algorithm, and Kepler optimization algorithm (KOA) [2]. These algorithms have different optimization abilities due to their unique structure, among which the KOA has the simple structure and strong optimization ability.

The basic principle of the KOA, which abstracts planets and stars into the solution of the problem, is derived from the Kepler's law of planetary motion. Combining with Kepler's three laws, KOA utilizes the motion of planets around stars, and adjusts parameters such as gravity to control the search direction, ultimately achieves the overall evolution of the solution and the goal of finding the optimal solution. Similar to some current intelligent optimization algorithms, KOA is also prone to problems such as falling into local optima and slow convergence speed. Many efforts have been made by domestic and foreign researchers to improve the above two problems in optimization algorithms.

In terms of improving the convergence speed of the algorithm, Lin et al. [3] introduce the concept of cultural algorithms into particle swarm optimization, which accelerated the convergence speed of the algorithm; Zhang et al. [4] introduce the nearest neighbor strategy of the seeker optimization algorithm into GA and proposed a seeker genetic algorithm; Liu et al. [5] use the lion optimization algorithm to endow different wolf packs with active search capabilities, improving the problem of slow convergence speed of the algorithm; Yue et al. [6] change the initial pheromone concentration of the ACA to avoid blind search in the early stage and improve the efficiency of the algorithm.

In terms of improving the algorithm's ability to jump out of local optima, Li et al. [7] add nonlinear cloud-transfer to the emperor butterfly algorithm to improve the accuracy of the solution; Sun et al. [8] introduce Lévy flight into the migration phase of the crow tern algorithm, enhancing its global search ability; Tang et al. [9] introduce the elite reverse learning strategy into the firefly algorithm to improve the diversity of solutions; Jiang et al. [10] introduce an acceleration factor of an increasing exponential function in particle swarm optimization(PSO) to control the even inclusion between particle swarms and effectively controlled the algorithm's precocity.

In response to the above two problems, this article analyzes the basic principle of the KOA and proposes the improved Kepler optimization algorithm (IKOA). The main work of this article are as follows:

(1) The IKOA is proposed. In the initial population stage, the random number generator is replaced by the circle mapping [11–13] and the reverse learning strategy [14–16]; In the stage that planets update positions, adaptive mutation strategy and arctangent decay factor are added to improve the convergence speed of the algorithm; Gauss mutation strategy [17, 18] is introduced to improve the position of the sun and enhance the algorithm's ability to jump out of local optima.
(2) Comparative tests are conducted on 8 test functions with 5 algorithms including IKOA, KOA, PSO, adaptive genetic algorithm (AGA), and coati optimization algorithm (COA), proving that the proposed algorithm has higher convergence efficiency.

The remaining parts of this article are as follows: The basic principle of the KOA is introduced in Sect. 2; Specific improvement strategies is introduced in Sect. 3; the experimental setup and results are implemented in Sect. 4; In Sect. 5, the analysis and summary are showed.

2   The Basic Principle of KOA

The KOA, which is based on the behavior of the planets and stars, was firstly proposed by Abdel-Basset et al. in 2023 [2]. As the work of this paper is based on the standard KOA, this section mainly introduces the main principle of this algorithm.

2.1 Initialize Population

As shown in Eq. (1), the KOA generates the initial solution through random numbers.

$$X_i^j = X_{i,low}^j + rand \bullet (X_{i,up}^j - X_{i,low}^j), \quad \begin{cases} i = 1, 2, \ldots, N \\ j = 1, 2, \ldots, d \end{cases} \tag{1}$$

where X_i^j is the j-th dimension of the i-th planet in the search space, N is the number of planets, d is the dimension of the problem. Moreover, $X_{i,low}^j$ represents the lower limit of the j-th dimension whereas $X_{i,up}^j$ is the upper limit of this dimension.

2.2 Definition of Important Parameters

An important parameter $F_{g_i}(t)$ in the KOA is defined as follows:

$$F_{g_i}(t) = e_i \times \mu(t) \times \frac{\overline{M} \times \overline{m}_i}{\overline{R}_i^2 + \varepsilon} + r_1 \tag{2}$$

where $\overline{M}$ and $\overline{m}_i$ denote the normalized values of M_s and m_i, which are defined as follows:

$$M_s = r_2 \times \frac{fit_s(t) - worst(t)}{\sum\limits_{k=1}^{N} (fit_k(t) - worst(t))} \tag{3}$$

$$m_i = \frac{fit_i(t) - worst(t)}{\sum\limits_{k=1}^{N} (fit_k(t) - worst(t))} \tag{4}$$

where $fit_s(t) = best(t) = \min\limits_{k \in 1,2,\ldots,N} fit_k(t)$, $worst(t) = \max\limits_{k \in 1,2,\ldots,N} fit_k(t)$. $\mu(t)$ is the constant of universal gravitation, as shown in Eq. (5).

$$\mu(t) = \mu_0 \times \exp\left(-\gamma \frac{t}{T_{max}}\right) \tag{5}$$

where e_i is the eccentricity of the i-th orbit, and r_1 is a random number of [0, 1], μ_0 is the initial constant of universal gravitation, γ is a constant, t is the number of iterations, and T_{max} is the maximum number of iterations.

2.3 Planets Update Positions

Each planet represents the position of a solution, and the velocity of a planet decreases with increasing distance from the sun. The formula for updating the velocity of a planet is shown in Eq. (6).

$$V_i(t) = \begin{cases} \varsigma \bullet (2r_4 \bullet X_i - X_b) + \ddot{\varsigma} \bullet (X_a - X_b) + (1 - R_{i-norm}(t)) \\ \bullet \xi \bullet U_1 \bullet r_5 \bullet (X_{i,up}^j - X_{i,low}^j), \ \text{if } R_{i-norm}(t) \le 0.5 \\ r_4 \bullet \zeta \bullet (X_a - X_b) + (1 - R_{i-norm}(t)) \\ \bullet \xi \bullet U_2 \bullet r_5 \bullet (r_3 X_{i,up}^j - X_{i,low}^j), \ \text{else} \end{cases} \tag{6}$$

where

$$\varsigma = U \times M \times \zeta \tag{7}$$

$$\ddot{\varsigma}_i = (1 - U_i) \times M_{1,i} \times \zeta \tag{8}$$

$$\zeta = \left[\mu(t) \times (M_s + m_i) \left| \frac{2}{R_i(t) + \varepsilon} - \frac{1}{a_i(t) + \varepsilon} \right| \right]^{\frac{1}{2}} \tag{9}$$

where

$$M = (r_3 \times (1 - r_4) + r_4), \ U = \begin{cases} 0, \ r_5 \le r_6 \\ 1, \ \text{else} \end{cases}, \ \xi = \begin{cases} 1, \ \text{if } r_4 \le 0.5 \\ -1, \ \text{else} \end{cases}, \tag{10}$$

$$M_{1,i} = (r_3 \times (1 - r_{5,i}) + r_{5,i}), \ U_1 = \begin{cases} 0, \ r_5 \le r_4 \\ 1, \ \text{else} \end{cases}, \ U_2 = \begin{cases} 0, \ r_3 \le r_4 \\ 1, \ \text{else} \end{cases} \tag{11}$$

where $r_3 \sim r_6$ is the random number of $[0, 1]$, R_i represents the Euclidian distance between X_s and X_i, and a_i represents the semimajor axis of the elliptical orbit of object i at time t. The definitions of R_i and a_i are as follows:

$$R_i(t) = \|X_s(t) - X_i(t)\|_2 = \sqrt{\sum_{j=1}^{d} (X_{Sj}(t) - X_{ij}(t))^2} \tag{12}$$

$$a_i(t) = r_3 \times \left[T_i^2 \times \frac{\mu(t) \times (M_s + m_i)}{4\pi^2} \right]^{\frac{1}{3}} \tag{13}$$

The position of planets is updated by Eq. (14):

$$X_i(t + 1) = X_i(t) + \xi \times V_i(t) + (F_{g_i}(t) + |r|) \times U \bullet (X_s(t) - X_i(t)) \tag{14}$$

where $X_i(t + 1)$ is the position of the i-th solution in $t + 1$ generation, $V_i(t)$ is the velocity of the i-th solution, $X_s(t)$ is the position of the optimal solution, and r is a random number of $[0, 1]$.

2.4 Planets Update Distance with the Sun

Planets will change their distance from the sun. When planets are close to the sun, KOA focuses on optimizing the exploitation; When planets are far from the sun, KOA will optimize the exploration operator. The mathematical model of this step is shown in Eq. (15):

$$Y_i(t+1) = X_i(t) \bullet U_1 + (1 - U_1) \bullet (\frac{X_i(t) + X_s + X_a(t)}{3} + h \times (\frac{X_i(t) + X_s + X_a(t)}{3} - X_b(t))) \quad (15)$$

where $h = \frac{1}{e^{\eta r}}$ is the adaptive factor, $\eta = (a_2 - 1) \times r_4 + 1$ is the linear attenuation factor from 1 to -2, the expression of a_2 is showed in Eq. (16), and $\overline{T}$ is a loop control parameter which gradually decreases from -1 to -2 in the entire $\overline{T}$ loop.

$$a_2 = -1 - 1 \times (\frac{\text{mod}(t, T_{max}/\overline{T})}{T_{max}/\overline{T}}) \quad (16)$$

Finally, as shown in Eq. (17), the new position of the planet is replaced by the greedy principle to ensure the optimal position of the planet and the sun.

$$X_i(t+1) = \begin{cases} X_i(t+1), & \text{if } f(X_i(t+1)) \leq f(X_i(t)) \\ X_i(t), & \text{else} \end{cases} \quad (17)$$

3 Multi-strategy Improved KOA

3.1 Circle Mapping and Reverse Learning Strategy

The difference in the distribution of initial solutions may lead to the algorithm falling into different local optima. Circle mapping to replace random numbers in generating initial solutions is introduced in this article. As shown in Fig. 1, compared to random number, the initial solutions generated by Circle mapping can be better distributed in the solution space. In addition, the reverse learning strategy is introduced in this article in the initialization process of the population. The specific operations are as follows: N individuals are generated by Circle mapping, and N individuals are generated by reverse learning strategy, then the first N individuals with fitness are selected from these $2N$ individuals to participate in the iterative process. The mathematical formulas for Circle mapping and reverse learning strategy are shown in Eq. (18) and (19).

$$x_{i+1} = \text{mod}(x_i + 0.2 - (\frac{0.5}{2\pi}) \sin(2\pi x_i), 1) \quad (18)$$

$$X_i = 1 - x_i \quad (19)$$

After generating the initial solution through circle mapping and reverse learning strategy, it is necessary to map it into the solution space according to Eq. (20).

$$x_{i,j}^* = lb_j + x_{i,j} \times (ub_j - lb_j) \quad (20)$$

where $x_{i,j}^*$ is the solution after mapping, $x_{i,j}$ is the solution before mapping, ub_j and lb_j is the upper and lower limits of the j-th dimension.

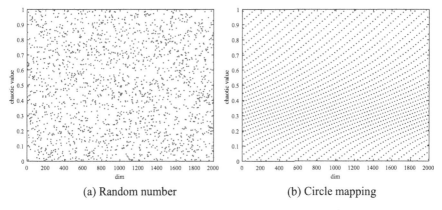

(a) Random number   (b) Circle mapping

Fig. 1. Comparison of chaotic value for random numbers and circle mapping

3.2 Adaptive Mutation Strategy Combines with Arctangent Decay Factor

The essence that planets move around the sun is to add a random disturbance to the current position of the planet, and the disturbance is uncertain in the standard KOA and exists in every dimension, which cannot ensure sufficient search of the solution space. Therefore, during the stage that planets move around the sun, adaptive mutation rate [19] and a decay factor of the arctangent function are added in this article. The formula for updating the new position of the planet is shown in Eq. (21).

$$X_i(t+1) = \begin{cases} X_i(t) + \omega \times F, \text{if } rand < pm \\ X_i(t), \text{else} \end{cases} \tag{21}$$

where

$$\omega = \omega_0 \times (\arctan)(1.56 \times (1 - (\frac{t}{T_{max}})^3)) \tag{22}$$

$$pm(i) = \begin{cases} pm_0 \times \frac{fit(i)-fit_{min}}{f_{avg}-fit_{min}}, & fit(i) \leq f_{avg} \\ pm_0, & \text{else} \end{cases} \tag{23}$$

In Eq. (22) ω_0 is a constant is the initial attenuation factor, and ω varies with the number of iterations as shown in Fig. 2. In Eq. (23), pm_0 is the initial mutation rate, $pm(i)$ and $fit(i)$ is the mutation rate and fitness of individual i, f_{avg} and fit_{min} is the average fitness and minimum fitness of population.

As the number of iterations increases, ω gradually approaches 0, which can result in a smaller change in each position of the planet, and the search space is larger in the early stage of population iteration; the search space is smaller the later stage of population iteration, therefor, the algorithm's fine search ability is enhanced.

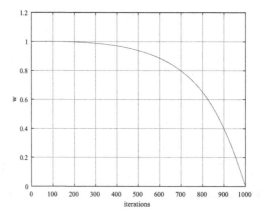

Fig. 2. The figure of ω varies with the number of iterations

3.3 Gauss Mutation Strategy

During the operation of the algorithm, the position of the sun may stagnate, which causes the algorithm to fall into the local optima. In order to improve the position of the sun, the Gauss mutation strategy is introduced in this article. The mathematical formula of Gauss mutation is as follows:

$$X_{i,j}^{new}(t) = \begin{cases} X_{i,j}(t) \times [1 + \sigma N(0, 1)], & \text{if } rand < rand \\ X_{i,j}(t), & \text{else} \end{cases} \tag{24}$$

where $X_{i,j}(t)$ is the original position of the sun, $\sigma N(0, 1)$ is a natural number that follows Gauss distribution with a mean of 0 and a variance of 1. The new individual generated by mutation is accepted according to the greedy principle. The pseudo code of the IKOA is shown in Algorithm 1.

Algorithm 1: Improved Keple r optimization algorithm

1. Input: Maximum number of iterations, number of individuals.
2. Set the parameters of the algorithm: e_i、 μ_0、 γ、 ω_0、 $\bar{T}$
3. Generate initial solution through Eqs. (13)-(15)
4. Update attenuation factor and mutation rate change according to Eq. (17) and (18)
5. Update the position of the planet according to Eq. (16)
6. Update the position of the planet according to Eq. (10) and (11)
7. Update the position of the sun according to Eq. (19)
8. Accept the new solution according to Eq. (12)
9. Repeat 4 to 8 until the maximum number of iterations
10. Output: the optimal solution and the position of the sun

---

4   Comparative Experiments and Analysis

4.1   Test Functions and Simulation Settings

In order to test the performance of the proposed algorithm, 8 benchmark test functions are set up in this article. The information of the test functions, which include two-dimensional and multi-dimensional and are all minimum values, are shown in the Table 1. This article introduces PSO, AGA, and COA, which is an emerging algorithm in 2023[20], for comparative testing. The parameter information of different algorithms is shown in Table 2. The maximum number of iterations for all algorithms is 1000, and the population size is 40.

The simulation codes are all written and executed in MATLAB R2016a using a PC with 2.70 GHz, 8 GB of RAM, Inter Core i7 of CPU and Windows 10.

Table 1. Test function information

Function	Formula	Dim	Best
F1	$f(x, y) = \sum\limits_{k=1}^{5} k\cos[(k+1)x + k] * \sum\limits_{k=1}^{5} k\cos[(k+1)y + k]$	2	-186.73
F2	$f(x, y) = x\sin(\sqrt{\lvert x \rvert}) + y\sin(\sqrt{\lvert y \rvert})$	2	-837.96
F3	$f(x) = -\sum\limits_{i=1}^{n} \sin(x_i)\sin^{20}(\frac{ix_i^2}{\pi})$	10	-9.66
F4	$f(x, y) = -(y+47)\sin\left(\sqrt{\lvert y + \frac{x}{2} + 47 \rvert}\right) - x\sin\left(\sqrt{\lvert x - (y+47) \rvert}\right)$	2	-959.64
F5	$f(x) = -20 \bullet \exp(-0.2 \bullet \sqrt{\frac{1}{10}\sum\limits_{i=1}^{10} x_i^2}) - \exp(\frac{1}{10}\sum\limits_{i=1}^{10}\cos 2\pi x_i) + 20 + e$	10	0
F6	$f(x, y) = 4x^2 - 2.1x^4 + \frac{1}{3}x^6 + xy - 4y^2 + 4y^4$	2	-1.03
F7	$f(x) = \sum\limits_{i=1}^{5}\left(\sum\limits_{j=1}^{5}(j+10)\left(x_j^i - \frac{1}{j^i}\right)\right)^2$	10	0
F8	$f(x, y) = [1 + (x + y + 1)^2(19 - 14x + 3x^2 - 14y + 6xy + 3y^2)] \bullet$ $[30 + (2x - 3y)^2(18 - 32x + 12x^2 + 48y - 36xy + 27y^2)]$	2	3

Table 2. Algorithm parameter settings

Algorithm	Parameter settings
IKOA	$\mu_0 = 0.1, \omega_0 = 0.2, Pm_0 = 0.5$
KOA	$\mu_0 = 0.1$
PSO	$\omega = 0.8, c_1 = 1.5, c_2 = 1.5$
AGA	$Pc = 0.8, Pm_0 = 0.1$
COA	–

4.2 Results Analysis

The simulation experiments are independently run 50 times, and get the best value, worst value, average value, and standard deviation obtained by each algorithm. The best values obtained by five algorithms are indicated in bold font, and the results are shown in Table 3. From Table 3, for the two-dimensional test function, IKOA obtains the best of all four test indicators except for the worst and standard deviation of f1, which are slightly lower than PSO.

For multi-dimensional functions, the four evaluation indicators obtained by IKOA have obvious advantages on f3. For f5, IKOA and COA obtained the same optimal value, but the other three indicators obtained by COA are better than IKOA, but the difference in magnitude is below 10^{-15}. For f7, the best value obtained by IKOA is slightly smaller than PSO, but it is indicated that IKOA is more stable by the results that the standard deviation obtained by IKOA is smaller than PSO.

Table 3. Benchmark function test results

Fun	best		IKOA	KOA	PSO	AGA	COA
F1	−186.73	best	**−186.73**	**−186.73**	**−186.73**	**−186.73**	**−186.73**
		worst	−186.72	−186.71	**−186.73**	−60.23	−186.62
		avg	**−186.73**	−186.72	−186.73	−169.54	−186.59
		std	1.2e−4	3.3e−3	**0**	28.69	0.38
F2	−837.96	best	**−837.96**	**−837.96**	**−837.96**	**−837.96**	**−837.96**
		worst	**−837.96**	**−837.96**	−719.52	−635.62	**−837.96**
		avg	**−837.96**	**−837.96**	−828.49	−809.27	**−837.96**
		std	**0**	**0**	32.45	41.88	2.0e−4
F3	−9.66	best	**−9.64**	−9.59	−9.42	−9.39	−6.64
		worst	**−9.36**	−8.94	−7.61	−7.29	−4.60
		avg	**−9.51**	−9.38	−8.77	−8.59	−5.44
		std	**0.05**	0.14	0.46	0.51	0.48
F4	−959.64	best	**−959.64**	**−959.64**	−959.58	−956.44	−959.07
		worst	**−959.64**	**−959.64**	−956.91	−531.64	−888.89
		avg	**−959.64**	**−959.64**	−957.81	−833.88	−956.07
		std	**0**	**0**	1.18	104.36	11.87
F5	0	best	**8.8e−16**	1.6e−12	4.5e−6	3.61	**8.8e−16**
		worst	4.4e−15	2.9e−11	5.09	12.52	**8.8e−16**
		avg	4.2e−15	8.0e−12	2.78	7.18	**8.8e−16**
		std	8.5e−16	5.8e−12	1.14	2.31	**0**
F6	−1.03	best	**−1.03**	**−1.03**	**−1.03**	**−1.03**	**−1.03**
		worst	**−1.03**	**−1.03**	**−1.03**	0.51	**−1.03**

(continued)

Table 3. (*continued*)

Fun	best		IKOA	KOA	PSO	AGA	COA
		avg	**−1.03**	**−1.03**	**−1.03**	−0.90	−1.03
		std	**0**	**0**	**0**	0.28	5.09e−5
F7	0	best	1.5e−3	4.19e−5	**4.28e−7**	309.93	1.39
		worst	**0.56**	3.55	27.42	5.65e16	288.21
		avg	**0.10**	0.45	4.51	1.53e15	123.16
		std	**0.12**	0.77	7.18	8.21e15	77.68
F8	3	best	**3**	**3**	**3**	**3**	**3**
		worst	**3**	**3**	**3**	94.90	5.32
		avg	**3**	**3**	**3**	12.40	3.20
		std	**0**	**0**	**0**	14.81	0.43

In order to compare and analyze the convergence process of algorithms more intuitively, the above five algorithms are used for experiments on the test function, and the convergence curve obtained are shown below.

From Fig. 3(a), IKOA converges to the optimal solution at a faster speed compared to PSO, KOA, and GA. IKOA has a slower convergence speed than COA, but its convergence accuracy is better than that of relativistic COA.

For Figs. 3(d), (h), the convergence speed and accuracy of IKOA are all best compared to the other four optimization algorithms.

For Fig. 3(e), the convergence accuracy of IKOA is higher than PSO, GA, and KOA, but not as good as COA. However, the magnitude of fitness is below 10^{-5}, and IKOA can converge to the same accuracy compared to COA during the iteration process.

For Figs. 3(b), (c), (g), and IKOA, the initial convergence speed is slower the other four optimization algorithms. However, during the iteration process, IKOA can jump out of the local optimum multiple times and finally converges to a best value relative to the other four algorithms at the fastest speed.

In summary, IKOA outperforms the other four optimization algorithms in terms of overall performance. The effectiveness of the improvements made in this article has been proven.

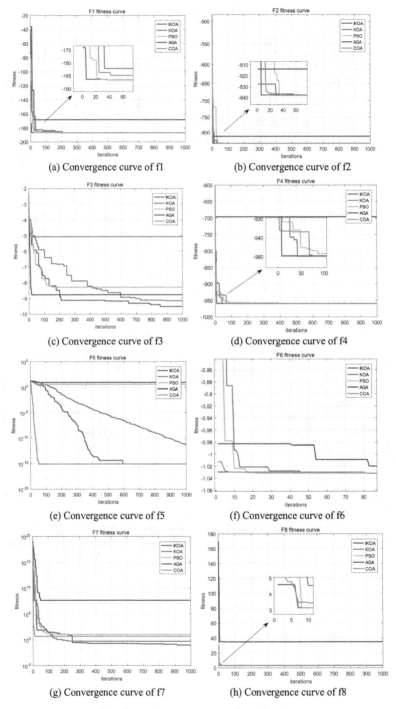

(a) Convergence curve of f1

(b) Convergence curve of f2

(c) Convergence curve of f3

(d) Convergence curve of f4

(e) Convergence curve of f5

(f) Convergence curve of f6

(g) Convergence curve of f7

(h) Convergence curve of f8

Fig. 3. Convergence curve of test function

5 Summary

The improved Kepler optimization algorithm is proposed in this article. Firstly, the quality of the initial solution is improved by replacing the random number generator with the circle mapping and reverse learning strategy in the initialization stage. Secondly, adaptive mutation rate and arctangent decay factor are introduced in the stage that planets update positions, which improves the performance of the algorithm. Thirdly, Gauss mutation is performed on the position of the optimal solution (i.e. the sun) to improve the algorithm's tendency to fall into local optima. Finally, IKOA, and four other optimization algorithms, are comprehensively tested on 8 benchmark test functions. The results verify the effectiveness and feasibility of the proposed algorithm.

References

1. Li, W., Wang, G., Gandomi, A.H.: A survey of learning-based intelligent optimization algorithms. Arch. Comput. Methods Eng. **28**(5), 3781–3799 (2021)
2. Abdel-Basset, M., Mohamed, R., Azeem, S.A.A., Jameel, M., Abouhawwash, M.: Kepler optimization algorithm: a new metaheuristic algorithm inspired by Kepler's laws of planetary motion. Knowl.-Based Syst. **268**, 110454 (2023)
3. Lin, S.W., Liu, A., Wang, J.G., Kong, X.Y.: An improved fault-tolerant cultural-PSO with probability for multi-AGV path planning. Expert Syst. Appl. **237**, 121510 (2024)
4. Zhang, L.Y., Gao, Y., Sun, Y.S., Fei, T., Wang, Y.J.: Application on cold chain logistics routing optimization based on improved genetic algorithm. Autom. Control. Comput. Sci. **53**(2), 169–180 (2019)
5. Liu, J., Wei, X., Huang, H.: An improved grey wolf optimization algorithm and its application in path planning. IEEE Access **9**, 121944–121956 (2021)
6. Yue, C., Huang, J., Deng, L.L.: Research on improved ant colony algorithm in AGV path planning. Comput. Eng. Des. **43**(9), 2533–2541 (2022)
7. Li, X.P., Du, B., Wang, X.W.: Self-adaptive monarch butterfly optimization based on nonlinear cloud-transfer. Control Decis. **38**(12), 3327–3335 (2023)
8. Sun, K.Q., Chen, Y.F.: A hybrid sine and cosine algorithm and Lévy flight adaptive tern algorithm and its application. Mach. Des. Manuf. **1**, 212–217 (2023)
9. Tang, W.L., Zhang, P., Tang, S.F.: Improved firefly k-means algorithm based on elite opposition-based learning. Comput. Eng. Des. **40**(11), 3164–3169 (2019)
10. Jiang, J.J., Wei, W.X., Sha, W.L., Liang, Y.F., Qu, Y.Y.: Research on large-scale bi-level particle swarm optimization algorithm. IEEE Access **9**, 56364–56375 (2021)
11. Hu, S.S., et al.: Tool wear prediction in glass fiber rein-forced polymer small-hole drilling based on an improved circle chaotic mapping grey wolf algorithm for BP neural network. Appl. Sci.-Basel **13**(5), 2811 (2023)
12. Li, Z., Wang, L., Zhang, X.: Improved salp swarm optimization K-means algorithm for image segmentation. Packaging Eng. **43**(9), 207–216 (2022)
13. Li, A.D., Liu, S.: Multi-strategy improved whale optimization algorithm. Appl. Res. Comput. **39**(5), 1415–1421 (2022)
14. Li, D., Zhang, C., Yang, X.: Improved slime mould algorithm fused with multi-strategy. Pattern Recogn. Artif. Intell. **36**(7), 647–660 (2023)
15. Ding, Y., Xia, Q., Zhang, R., Li, S.: Review of literature survey of butterfly optimization algorithm. Sci. Technol. Eng. **23**(7), 2705–2716 (2023)
16. Li, L., Huang, X., Qiang, S., Li, Z., Li, S., Mansour, R.: Fuzzy hybrid coyote optimization algorithm for image thresholding. CMC-Comput. Mater. Continua **72**(2), 3073–3090 (2022)

17. Shen, S., Du, Y., Xu, Z., Qin, X., Chen, J.: Temperature prediction based on STOA-SVR rolling adaptive optimization model. Sustainability **15**(14), 11068 (2023)
18. Jiang, Y., Xu, X., Xu, F., Gao, B.: Multi-strategy fusion improved adaptive mayfly algorithm. J. Beijing Univ. Aeronaut. Astronaut. 1–14 (2023)
19. Nie, W., Cai, L., Qiu, G., Li, C.: Adaptive genetic algorithm with density weighted. Comput. Syst. Appl. **27**(1), 137–142 (2018)
20. Dehghani, M., Montazeri, Z., Trojovska, E., Trojovsky, P.: Coati optimization algorithm: a new bio-inspired metaheuristic algorithm for solving optimization problems. Knowl.-Based Syst. **259**, 110011 (2023)

A Method of Pathing for Underwater Glider Cluster Based on Optimization Algorithm

Lihua Wu[✉], Gang Xie, Kaiyu Li, Yuncheng Lu, Chao Sui, and Shuang Huang

Wuhan Second Ship Design and Research Institute, Wuhan 430025, China
wlhcheers@126.com

Abstract. Underwater gliders have become one of the iconic technologies in ocean environmental monitoring, which relies on adjusting buoyancy to achieve buoyancy control and gliding through the water using hydrodynamics. They can be used for long-term and wide-range observation and detecting complex ocean environments. Meanwhile, underwater gliders have the advantages of low cost, long endurance, reusability, and certainly trajectory control capabilities. However, in the current cooperative detection, due to the complexity and large disturbances of ocean currents, there exists a problem of high energy consumption and a non-optimal path in the underwater glider cluster. To address this issue, this paper combines cluster size optimization and path planning, proposes a method that combines the classic path optimization problem in the plane with the special motion mode of the glider, and then extends it to the optimization method in three-dimensional motion, and conducts relevant simulation experiments for verification. The simulation results show that the proposed method can effectively obtain the path with the minimum energy consumption of the glider under a three-dimensional trajectory.

Keywords: Underwater gliders cluster · Path optimization · Cooperative planning

1 Introduction

As a long-term and long-distance autonomous underwater vehicle [1, 2], the formation, networking and cooperative observation of underwater gliders are among the most important applications. The formation networking of underwater gliders is increasingly applied to actual ocean exploration [3]. Underwater gliders have the advantages of low cost, long endurance, reusable, and specific flight path control [4, 5]. In order to determine the optimal path planning scheme, it is necessary to model and analyze various complex factors in the ocean [6], including energy management [7] and collaborative planning [8].

In 1989, plans for the development and application of underwater gliders were proposed [9, 10]. Since then, Marine science has increasingly used underwater gliders for extensive research [11–14]. Due to the relatively low speed of glider, complex ocean current environment and large disturbance, glider cluster is more sensitive to water flow,

L. Pan et al. (Eds.): BIC-TA 2023, CCIS 2062, pp. 309–321, 2024.
https://doi.org/10.1007/978-981-97-2275-4_24

and there is a problem of suboptimal coordinated control path. At present, underwater glider path planning is a complex optimization problem involving multiple objectives and constraints [15, 16]. Existing methods may have limitations in dealing with these complexities and cannot find the global optimal solution. Recent research on underwater gliders mainly includes depth deterministic strategy gradient [17, 18], ant colony optimization [19, 20], multi-formation coordination [21, 22], model combination method [23], mixed integer linear programming method [24], differential evolution algorithm, and frame optimization [25] and factor optimization method [26].

In these optimization processes, only trajectory motion in a two-dimensional plane is usually considered, which is applicable to most underwater vehicle path planning problems. However, due to the unique motion mode of gliders, trajectory planning in a two-dimensional plane cannot accurately describe the operational characteristics of gliders. The paper combines cluster scale optimization and path planning to propose an optimization method that combines the classic path optimization problem in the plane with the special motion mode of gliders, and then extends it to three-dimensional motion. This provides a new research direction for the optimization of cluster scale path planning in three-dimensional motion.

2 Underwater Glider Cluster Path Planning Model

2.1 Global Path Planning for Underwater Glider Clusters

Traditional grid-based methods based on the inertial coordinate system can only select adjacent eight directions of the grid, which limits the optimality of path selection. In this paper, three-dimensional underwater space is discretized, and a discrete space model based on environmental information is established. The line connecting the starting and the end points of the path is used as the abscissa, and the inertial coordinate system $O_E - X_E Y_E Z_E$ is transformed to represent the discrete space, which is used for the path planning of underwater gliders.

The horizontal coordinate transformation is shown in Fig. 1. In the $O_E - X_E Y_E$ plane, the transformed horizontal coordinate system $P_S - X'_E Y'_E$ is established with P_S as the coordinate origin and the line connecting points P_S and P_D as the X'_E axis. M segmentation points are set between line segment $P_S P_D$ to divide it into $M + 1$ segments. Typically, $M = 4[S_{P_S P_D} \tan \Theta / 2H] - 1$ and $S_{P_S P_D}$ are taken as the distance between line segment P_S and P_D. H is the diving depth, and $[\cdot]$ is rounded up.

The set of lines perpendicular to the X'_E axis and passing through the segmentation points is represented as $\{L_{y1}, L_{y2}, \ldots, L_{yM}\}$, which is referred to as the M_y dimension. Thus, a discrete spatial model of the horizontal plane is established. With the shortest underwater glider travel time as the optimization objective, an adaptive particle cluster optimization method is used to find an optimal path point set with the shortest travel time from the starting point to the target point among all potential paths. This set is denoted as $P_{xy} = \{P_S, P_{xy1}, P_{xy2}, \ldots, P_{xyM}, P_D\}$. Similarly, the M_z dimension and the optimal path set $P_{xz} = \{P_S, P_{xz1}, P_{xz2}, \ldots, P_{xzM}, P_D\}$ in the vertical plane can be obtained.

The optimal path point sets in the horizontal plane and vertical plane obtained using the path decoupling method have the same coordinates on the X'_E axis. Therefore, the path points in P_{xy} and P_{xz} can be combined into a three-dimensional optimal path point

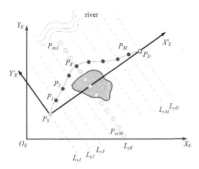

Fig. 1. Illustration of horizontal plane discretization modeling.

set $P_{xyz} = \{P_S, P_1, P_2, \ldots, P_D\}$ based on the principle of consistent coordinates on the X'_E axis. Thus, a three-dimensional discrete space model is established, and the path formed by connecting each node in set P_{xyz} is the optimal path in the three-dimensional space.

2.2 Local Path Planning for Underwater Glider Cluster

After global path planning, the optimized path from the starting point to the target point can be obtained for each underwater glider. Based on the corresponding path, the time when each underwater glider passes through each path point can be calculated, and this can be used to predict the path points where collisions between underwater gliders may occur.

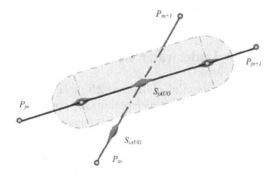

Fig. 2. Illustration of Collision Risk for Underwater Gliders.

As shown in Fig. 2, when the lower-priority i-th underwater glider appears in the yellow alert zone S_{jAUG}, collision danger may occur. The purpose of local adjustment is to use artificial potential field method to coordinate the local path when it is judged that collision may occur, so that the lower-priority underwater glider can adjust its heading in advance to avoid the higher-priority underwater glider, thereby preventing collision.

The traditional APF (Artificial Potential Field) method generates a repulsive potential field at the obstacle and an attractive potential field at the goal, so that the vehicle moves along the obstacle-free path under the combined force of attraction and repulsion. In this paper, we specifically consider the attraction from the desired path and the repulsion from other underwater gliders. The repulsive potential field can be represented as:

$$
\begin{cases}
U_{\text{rep}} = \begin{cases} \frac{1}{2} k_{\text{rep}} \left(\frac{1}{\rho(r_i, r_j)} - \frac{1}{\rho_0} \right) \rho(r_i, r_j)^2 \\ \qquad \text{if } \rho(r_i, r_j) < \rho_0 \\ 0, \text{ if } \rho(r_{i,j}) < \rho_0 \end{cases} \\
F_{\text{rep}} = -\nabla U_{\text{rep}}(\rho(r_{i,j}, r_j))
\end{cases}
\tag{1}
$$

r_i and r_j represent the positions of the ith and jth underwater gliders, respectively, $\rho(r_i, r_j)$ is the relative distance between the two underwater gliders, ρ_0 is the range of influence of the repulsive field, k_{rep} is the intensity coefficient of the repulsive field, and ∇ represents the gradient of the function.

The gravitational potential field can be expressed as:

$$
U_{\text{att}} = \frac{1}{2} k_{\text{att}} \rho(r_i, r_0)^2
$$
$$
F_{\text{att}} = \nabla U_{\text{att}}(\rho(r_i, r_0))
\tag{2}
$$

r_0 represents the coordinates of the reference position on the path, and k_{att} is the intensity coefficient of the attractive field.

The resultant force of repulsion and attraction is:

$$
F_{tol} = F_{rep} + F_{att}
\tag{3}
$$

The next position point can be calculated by the combined force of repulsion and attraction, until the force pulls the underwater glider back to the global path.

3   Path Planning for Underwater Glider Cluster

Due to the unique motion pattern of gliders, which makes it impossible for them to maintain horizontal motion, traditional path optimization problems in the plane are no longer applicable. Compared with agile and flexible unmanned aerial vehicle, gliders moving underwater require another parameter, the glide angle ξ, to describe their motion in three-dimensional space, in addition to the navigation speed V and curvature ρ. Moreover, due to the maximum working depth limit of gliders, their motion trajectory may consist of one or several incomplete Dubins paths, as shown in Fig. 3.

The glider starts diving from the horizontal plane, completes the Dubins motion in the xy plane, simultaneously performs a zigzag motion in the $\pi_1 \pi_3$ plane, and returns to the horizontal plane at the end point.

Due to the long working time of underwater gliders in the ocean, gliders that are typically battery-powered carry very limited energy. Therefore, it is crucial to maximize the efficiency of glider motion and minimize energy consumption for long-distance and long-duration navigation. The trajectory $P^* = [s_0, ..., s_f]$ is defined as the path of lowest

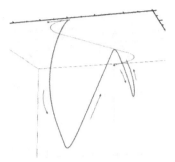

Fig. 3. Dubins path of glider in three-dimensional space

energy consumption from the initial state s_0 to the final state s_f during the motion of the glider.

$$P^* = P \in_{\min} \rho \sum\nolimits_{i=1}^{|P|} \text{energy}(s_i) \tag{4}$$

$s_i = (x_i y_i)^T$ represents the path node along the trajectory $P = [s_0, ..., s_f]$, and energy(s_i) denotes the energy consumption at path node s_i. Energy consumption is closely related to the power of each subsystem of the glider:

$$\text{energy}(s_i) = \sum\nolimits_{j=1} e_r(u_j) \tag{5}$$

u_j represents the control variable of the actuator, and $e_r(u_j)$ represents the power of the subsystem. The objective of this study is to find the optimal trajectory of the glider during Dubins motion in three-dimensional space, in order to minimize the energy consumption of the glider.

Unlike a glider moving at a constant velocity in a plane, underwater gliders experience changes in velocity during their ascent/descent due to changes in their motion state, which is usually slow. Additionally, due to the possibility of multiple cycles of motion, this velocity change process will repeat itself. Assuming the glider moves from point A to point B with a distance of S and a velocity and glide angle of V and ξ, respectively. The maximum operating depth of the glider is $D_{\max}$. The number of complete motion cycles n and the depth D_m of the last descent during the entire motion process can be approximated as:

$$n = \ln(\frac{S\tan\xi}{2D_{\max}}) \tag{6}$$

$$D_m = \text{Re}(\frac{S\tan\xi}{2D_{\max}}) * D_{\max} \tag{7}$$

In mathematics, $\ln(\cdot)$ represents the quotient obtained by dividing two numbers, while $\text{Re}(\cdot)$ represents the remainder obtained by dividing two numbers.

During the transition between descent and ascent, the glider loses some velocity, and the duration of this transition is closely related to the speed of the buoyancy adjustment mechanism. Assuming the buoyancy adjustment mechanism has a constant adjustment rate, and the mass of the volume of water that needs to be adjusted during each ascent/descent is constant, the time consumed during each adjustment process is also constant. The total time t during the entire operation can be expressed as:

$$t = \frac{S}{V\cos\xi} + 2(n+1)c_b\Delta m_b \tag{8}$$

In the equation, c_b is a constant and Δm_b is the constant change in mass of the displaced volume.

Suppose we ignore the time consumed by the glider during attitude changes. In that case, the path with the minimum time can be considered a combination of the maximum horizontal velocity motion and the shortest path on the horizontal plane. For a specific glider model, its maximum horizontal velocity is a constant value, while the shortest path on the horizontal plane results in the smallest turning radius, which corresponds to the maximum rotation angle of the glider's rotating mechanism.

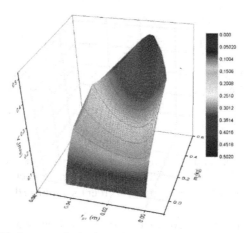

Fig. 4. The relationship between the horizontal movement speed of the glider and the control input of the actuator

Figure 4 shows the relationship between the input of the execution mechanism and the horizontal velocity of the glider during its zigzag motion. It can be seen that the maximum horizontal velocity of the glider is 0.5016 (m/s), with the control quantity of the execution mechanism being $r_p = 0.0107$ cm, $m_b = 0.494$ kg. The corresponding glider state variables are shown in Fig. 5.

According to the simulation results, the corresponding glide angle of the glider at maximum horizontal velocity is $\xi = -0.6269$ rad, and the glide speed is $V = 0.6193$ (m/s). After determining the control quantity of the linear execution mechanism and the buoyancy adjustment mechanism, the turning radius of the glider in the Dubins trajectory should be minimized, in order to obtain the shortest Dubins path of the glider in three-dimensional space. This path serves as the control group to compare with the optimized path with the lowest energy consumption.

During the actual operation of the glider, its motion speed is variable. On the other hand, the shortest time does not necessarily mean the lowest energy consumption, therefore a detailed analysis of the energy consumption during the glider's motion is needed.

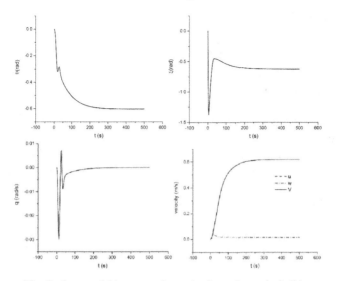

Fig. 5. State variables at maximum horizontal speed of glider

The system energy consumption of the glider can be divided into the energy consumption of the buoyancy adjustment mechanism E_b, pitch adjustment mechanism energy consumption E_l, roll adjustment mechanism energy consumption E_r, and circuit system/sensor basic energy consumption E_s, respectively. Thus, the total system energy consumption can be expressed as:

$$E = E_b + E_l + E_r + E_s \tag{9}$$

The energy consumption of the buoyancy control mechanism can be expressed as:

$$E_b = \frac{2\Delta m_b}{\eta_b \rho}(n\rho g D_{max} + \rho g D_m + n P_0) \tag{10}$$

Here, η_b represents the working efficiency of the buoyancy control system, ρ is the density of seawater, and P_0 is the standard atmospheric pressure.

The energy consumption of the pitch control mechanism can be expressed as:

$$E_l = \frac{m_p}{\eta_l} \int |\ddot{r}_{p1}| dr_{p1} \tag{11}$$

where, η_l represents the working efficiency of the pitch control system. Similarly, the energy consumption of the roll control mechanism can be expressed as:

$$E_r = \frac{m_p}{\eta_r} \int |\ddot{\gamma}| dr_\gamma \tag{12}$$

where, η_r represents the working efficiency of the roll control system.

To keep the position of the movable ballast fixed during gliding, the basic energy consumption of the system is related to the gliding angle ξ and the rotation angle of the slider γ:

$$E_s = \int P_{s1} + P_{s1}\sin\xi + P_{s3}\sin\gamma \, dt \qquad (13)$$

where, P_{s1}, P_{s2} and P_{s3} represent the basic power of the system.

The process of Dubins path planning in three-dimensional space is shown in Fig. 6. After the initial conditions are given, the initial motion trajectory is selected. Combined with the equilibrium state solution method, the control quantity of each actuator is obtained, and the corresponding energy consumption is calculated. Unlike in profile motion, there may be up to 6 Dubins paths with the same motion parameters, which are solved and the optimal option is selected.

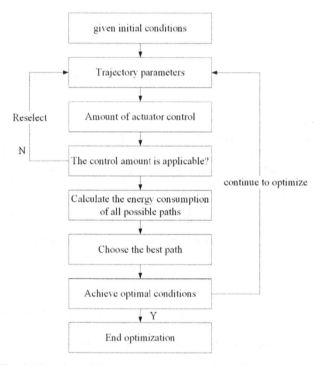

Fig. 6. Flowchart of Dubins path planning in three-dimensional space

4 Simulation Experiments and Result Analysis

MATLAB's Simulink module was used to simulate the glider and analyze the feasibility and efficiency of the optimization algorithm proposed in this paper for glider path planning. First, a verification simulation of path planning for a single glider was conducted, and then the algorithm was applied to the rendezvous and docking problem of multiple gliders.

The simulation used the mathematical model of the Seagull glider, and the other parameters involved were: $\eta_b = 0.7$, $\eta_l = 0.85$, $\eta_r = 0.85$, $a_l = 0.1$, $a_r = 1$, $a_s = 1.5$, iteration number $n_i = 10$, repeat calculation number $n_r = 100$, crossover probability $P_c = 0.7$, and mutation probability $P_m = 0.05$. The fitness function is:

$$E_s = \int P_{s1} + P_{s1}\sin\xi + P_{s3}\sin\gamma dt \tag{14}$$

where, E_i is the total energy consumption (total time) of the i-th individual in the population, and E_{min} is the minimum energy consumption of individuals in the population.

In the optimization process at a steady speed, the glide angle is $\xi = -0.6269$ rad and the glide speed is $V = 0.6193$ (m/s). In the energy-minimizing trajectory optimization process based on genetic algorithm, the range of the solution space is selected as: turning radius $R \in [50, 500]$, glide speed $V \in [0.2, 0.8]$, glide angle $|\xi| \in [0.2, 1]$, and maximum diving depth $D_{max} = 100$ m.

4.1 Optimization of Glider Path Given Final Heading Angle

Assuming the initial position of the glider is $A_1 = (0, 0)$, and the initial heading angle is $\Phi_A = 0°$. In some specific cases, the glider needs to move to the final point $B_1 = (500, 700)$ with a final heading angle of $\Phi_B = 270°$. The trajectory of the glider during the movement is optimized using both constant speed and the genetic algorithm-based optimization. In the genetic algorithm optimization process, the population size is selected as $S_p = [10, 20, 50, 100, 200]$.

Figure 7 shows the changes in the average energy consumption and standard deviation of individuals in each generation of the optimization process for different population sizes. When the population size is small ($S_p = 10, 20$), the optimization process does not converge, but maintains a converging trend, indicating that the number of iterations should be increased. As the population size increases ($S_p = 50, 100$), the optimization process performs well, and the standard deviation approaches 0 in the last iteration. When the population size continues to increase ($S_p = 200$), the average energy consumption curve has become very stable, and it is difficult to observe the standard deviation of individuals. With the same conditions, a larger initial population means faster convergence speed, but the computational complexity will also increase exponentially.

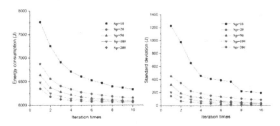

Fig. 7. Comparison of trajectory optimization results for different population sizes

Figure 8 shows the relationship between total energy consumption and velocity and glide angle when R = 50 m. The minimum energy consumption (E = 6050 J) occurs when V = 0.38 (m/s) and ξ = ±20.9°, indicating that energy consumption will increase regardless of changes in velocity or glide angle during the glider's movement. This result is consistent with the optimization results based on genetic algorithms proposed in this paper, which also verifies the correctness of the optimization results from another perspective.

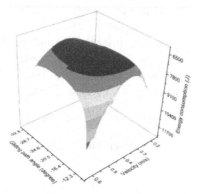

Fig. 8. The relationship between total energy consumption and velocity and glide angle when R = 50 m

Figure 9 compares the final glider trajectories obtained by the two optimization methods. In both optimization methods, the glider's turning radius converges to the minimum value of R = 50 m and performs three-dimensional Dubins motion in the LSR mode. At this point, the glider dives to the maximum depth twice under constant speed. Although the time consumption is less than that of the GA-optimized trajectory, the energy consumption significantly increases (E_c = 7675 J). On the other hand, the trajectory optimized by GA, although slower in speed and longer in time than the trajectory under constant speed, increases the steady-state energy consumption due to the longer duration of the mission. However, the energy consumption of the actuation system is significantly reduced, resulting in the lowest overall energy consumption of the glider.

4.2 Glider Path Optimization Under Uncertain Final Heading

In some practical scenarios, the final heading of the glider during the motion is uncertain, and therefore, it should also be considered in the optimization process. Let the initial position of the glider be A_2 = (0, 0), the initial heading angle be Φ_A = 0°, and the endpoint coordinate be B_2 = (500, 700), while the final heading angle is uncertain. Compared with the optimization in Sect. 3.1, the solution space of the optimization process is expanded from R = (R, V, ξ) to R = (R, V, ξ, Φ_{B2}).

Figure 10 shows the variation of the average energy consumption of each generation and the standard deviation of individuals in each generation for different population sizes (S_p = [100, 200, 400]) during the optimization process when the final heading

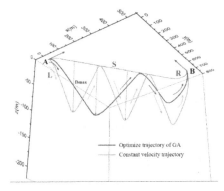

Fig. 9. Comparison of glider trajectory optimization results

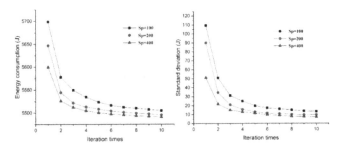

Fig. 10. Comparison of trajectory optimization results for different population sizes

angle of the glider during motion is uncertain. Due to the increased dimensionality of the solution space, the initial population size must also be correspondingly increased to ensure convergence within the number of iterations. According to the GA optimization results, the glider's motion parameters converged to $R_0 = 50$ m, $V_0 = 0.3839$ (m/s), $\xi_0 = \pm 20.91°$, and $\Phi_{B2o} = 55.54°$, and the energy consumption during the glider's motion was $E_c = 7122$ J, which is less than the optimization result under constant speed.

The GA optimization trajectory and the trajectory optimized under constant speed are shown in Fig. 11. It can be seen that the results obtained by both methods indicate that the glider starts with a left turn (L), then performs a zigzag motion in the longitudinal section (S), and does not turn left or right at the end stage. Therefore, it can be inferred that the optimal trajectory in Dubins motion with an uncertain final heading angle starts with a minimum radius left/right turn and ends with a zigzag motion. In this way, the solution space can be reduced to $R = (V, \xi)$ during the optimization process, reducing the computational cost while accelerating convergence.

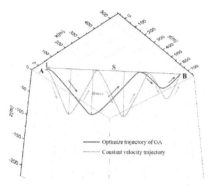

Fig. 11. Comparison of glider trajectory optimization results

5   Conclusion

An underwater glider is an autonomous underwater robot that can control its sinking and rising by changing its buoyancy and gravity. During underwater missions, the glider needs to autonomously detect and navigate in the underwater environment to complete its assigned tasks. Path planning is one of the key issues in glider missions because there are various complex environmental factors in the ocean that can directly affect the efficiency and accuracy of task completion. This paper first introduces a path planning model for underwater glider clusters. Then, based on cluster size optimization and path planning for underwater gliders, this paper proposes a method that combines the classic path optimization problem in a plane with the glider's unique motion patterns and extends it to three-dimensional motion optimization. Simulation results show that the optimization of the path of a single glider has been significantly improved, given a given final heading angle situation and an uncertain final heading angle situation. The optimization of heading angle, endpoint coordinates, and path when two gliders intersect have also been optimized, validating the effectiveness of the proposed method in providing better optimization results for underwater glider path planning.

References

1. Tuna, G., Gungor, V.C.: A survey on deployment techniques, localization algorithms, and research challenges for underwater acoustic sensor networks. Int. J. Commun. Syst. **30**(17), e3350 (2017)
2. Ullah, B., Ovinis, M., Baharom, M.B., Javaid, M.Y., Izhar, S.S.: Underwater gliders control strategies: a review. In: 2015 10th Asian Control Conference (ASCC), pp. 1–6. IEEE (2015)
3. Paley, D.A., Zhang, F., Leonard, N.E.: Cooperative control for ocean sampling: the glider coordinated control system. IEEE Trans. Control Syst. Technol. **16**(4), 735–744 (2008)
4. Jo, S.W., Jeong, S.K., Choi, H.S., Jeong, D.W.: Design of a new high speed unmanned underwater glider and motion control. In: OCEANS 2016, pp. 1–6. IEEE (2016)
5. Isa, K., Arshad, M.R.: Motion simulation for propeller-driven USM underwater glider with controllable wings and rudder. In: 2011 2nd International Conference on Instrumentation Control and Automation, pp. 316–321. IEEE (2011)

6. Rudnick, D.L., Davis, R.E., Sherman, J.T.: Spray underwater glider operations. J. Atmos. Oceanic Tech. **33**(6), 1113–1122 (2016)
7. Chen, B., Pompili, D.: Team formation and steering algorithms for underwater gliders using acoustic communications. Comput. Commun. **35**(9), 1017–1028 (2012)
8. Sitaba, A.I., Trilaksono, B.R., Hidayat, E.M.I., Sagala, M.F.: Communication system and visualization of sensory data and HILs in autonomous underwater glider. In: 2017 6th International Conference on Electrical Engineering and Informatics (ICEEI), pp. 1–6. IEEE (2017)
9. Stommel, H.: The slocum mission. Oceanography **2**(1), 22–25 (1989)
10. Rudnick, D.L.: Ocean research enabled by underwater gliders. Ann. Rev. Mar. Sci. **8**, 519–541 (2016)
11. Webb, D.C., Simonetti, P.J., Jones, C.P.: SLOCUM: an underwater glider propelled by environmental energy. IEEE J. Oceanic Eng. **26**(4), 447–452 (2001)
12. Eriksen, C.C., et al.: Seaglider: a long-range autonomous underwater vehicle for oceanographic research. IEEE J. Oceanic Eng. **26**(4), 424–436 (2001)
13. Sherman, J., Davis, R.E., Owens, W.B., Valdes, J.: The autonomous underwater glider "Spray." IEEE J. Oceanic Eng. **26**(4), 437–446 (2001)
14. de Fommervault, O., Besson, F., Beguery, L., Le Page, Y., Lattes, P.: SeaExplorer underwater glider: a new tool to measure depth-resolved water currents profiles. In: OCEANS 2019-Marseille, pp. 1–6. IEEE (2019)
15. Hussain, N.A.A., Arshad, M.R., Mohd-Mokhtar, R.: Underwater glider modelling and analysis for net buoyancy, depth and pitch angle control. Ocean Eng. **38**(16), 1782–1791 (2011)
16. Isern González, J., Hernández Sosa, D., Fernández Perdomo, E., Cabrera Gámez, J., Domínguez Brito, A.C., Prieto Marañón, V.: Obstacle avoidance in underwater glider path planning. **6**(1), 11–20 (2012)
17. Lan, W., et al.: Path planning for underwater gliders in time-varying ocean current using deep reinforcement learning. Ocean Eng. **262**, 112226 (2022)
18. Cai, J., Zhang, F., Sun, S., Li, T.: A meta-heuristic assisted underwater glider path planning method. Ocean Eng. **242**, 110121 (2021)
19. Ji, H., Hu, H., Peng, X.: Multi-underwater gliders coverage path planning based on ant colony optimization. Electronics **11**(19), 3021 (2022)
20. Lan, W., Jin, X., Wang, T., Zhou, H.: Improved RRT algorithms to solve path planning of multi-glider in time-varying ocean currents. IEEE Access **9**, 158098–158115 (2021)
21. Wen, H., Zhou, H., Fu, J., Zhang, X., Yao, B., Lian, L.: Consensus protocol based attitudes coordination control for Underwater Glider formation. Ocean Eng. **262**, 112307 (2022)
22. Ma, X., Wang, Y., Zhang, G., Yang, S., Li, S.: Discrete-time formation control of multiple heterogeneous underwater gliders. Ocean Eng. **258**, 111728 (2022)
23. Cococcioni, M., Lazzerini, B., Lermusiaux, P.F.: Adaptive sampling using fleets of underwater gliders in the presence of fixed buoys using a constrained clustering algorithm. In: OCEANS 2015-Genova, pp. 1–6. IEEE (2015)
24. Yilmaz, N.K., Evangelinos, C., Lermusiaux, P.F., Patrikalakis, N.M.: Path planning of autonomous underwater vehicles for adaptive sampling using mixed integer linear programming. IEEE J. Oceanic Eng. **33**(4), 522–537 (2008)
25. Mahmoudian, N., Geisbert, J., Woolsey, C.: Approximate analytical turning conditions for underwater gliders: implications for motion control and path planning. Oceanic Eng. **35**(1), 131–143 (2010)
26. Mahmoudian, N.: Efficient motion planning and control for underwater gliders. Virginia Polytechnic Institute and State University (Doctoral dissertation, Virginia Tech) (2009)

UUV Fault Diagnosis Model Based on Support Vector Machine

Lihua Wu[✉], Yu Liu, Zhenhua Shi, Zhenyi Ai, Man Wu, and Yuanbao Chen

Wuhan Second Ship Design and Research Institute, Wuhan 430025, China
wlhcheers@126.com

Abstract. Aiming at the most common power system faults in the UUV ancillary system, in order to diagnose related faults in a timely and accurate manner and prevent the occurrence of faults as early as possible, in this paper, we analyze four common faults based on the characteristic of the UUV power system, and obtain the fault mode of each fault. Then, according to the failure mode, the power system operation model and the failure model of the power system are established in Simulink under the Matlab platform, and four types of typical faults of the UUV power system are reproduced through the combination of experiment and simulation, and then all the necessary follow-up diagnosis process is obtained including training and validation data sets. Finally, the fault diagnosis model of UUV power system based on support vector machine algorithm is adopted, corresponding to different faults, we use genetic algorithms to optimize the selected diagnostic parameters and complete the fusion training verification process for multi-source parameter sets. The experimental results show that when the power system has a small number of fault samples, the use of support vector machine algorithm for fault diagnosis has good adaptability, and the diagnosis results have higher accuracy.

Keywords: Support Vector Machines · Failure model · Troubleshooting · Genetic algorithm

1 Introduction

Power system fault is the most common multiple fault in the UUV auxiliary system. When the power system is abnormal and cannot be solved in time, it is likely to cause the change of the operation state of the related equipment components or the auxiliary subsystem, and even cause new faults. Therefore, in order to diagnose the fault of UUV power system timely and accurately and prevent the occurrence of fault as far as possible in advance, it is very important to study a set of fault diagnosis method combining with the change law of various parameters of power system and gas engine.

In the study of fault detection and fault diagnosis for different systems and models, one of the more commonly used algorithms is Support Vector Machines [1–4] (SVM). On the basis of SVM, scholars have extended it. On the basis of SVM, the Least square method (LS) was used to extract the fault features, and then DEIWO was used to optimize the parameters to complete the establishment [5] of the fault diagnosis model. Liu Kaishi et al. proposed a fault diagnosis model [6] based on artificial bee colony (ABC) optimized SVM for four typical faults of photovoltaic arrays. Gong et al. proposed a recurrent neural network deep learning algorithm based on improved LSTM-SVM to solve the problem [7] of rapid fault diagnosis of UUV rotating machinery in multi-sensor monitoring environment. On the basis of SVM algorithm, Zhang Shaojie used fireworks algorithm to optimize support vector machine for the case that SVM parameters were difficult to determine, so as to build a fault diagnosis model [8] of support vector machine optimized by fireworks algorithm. CAI Bo et al. proposed a planetary gearbox fault diagnosis method [9] based on improved ensemble Empirical Mode Decomposition (MEEMD) multi-feature fusion and least squares support vector Machine (LS-SVM) in view of the nonlinear and non-stationary characteristics of planetary gearbox vibration signal and the problem that fault features are difficult to extract effectively.

In this paper, the UUV power system is taken as the research object. According to the composition of the power system, the working process and the working principle of each component, the causes of four common faults of the UUV power system (including filter blockage, filter screen corrosion and damage, fuel pump leakage and fuel valve leakage) are analyzed, and the engineering solutions are analyzed, as well as the changes of related component performance parameters when each single fault occurs. The fault modes of each fault in the UUV power system are obtained. Based on the combed FMEA results of UUV, according to the fault mode of each fault of the power system, the power system operation model and each fault model of the power system are established by Simulink under Matlab platform, and the docking debugging with the existing gas engine model is completed. Through the combination of test and simulation, the four types of typical faults of the UUV power system are reproduced, and then all the training and validation data sets required for the subsequent diagnosis process are obtained.

In this paper, SVM algorithm is used to establish the fault diagnosis model of UUV power system. Corresponding to different faults, Genetic Algorithm (GA) is used to optimize [10–14] the selected diagnostic parameters, and the fusion training and verification process of multi-source parameter sets is completed. The related verification results show that the use of support vector machine algorithm for fault diagnosis has good adaptability and high precision in the case of less fault samples of the power system.

2 SVM Theory

SVM is a new technology in data mining, which is proposed by Vapnik, and it is a new tool to solve machine learning by using optimization method based on the principle of structural risk minimization. It has great advantages in solving small sample, nonlinear and high-dimensional pattern recognition problems. At the beginning of its design, it is mainly used to solve data classification and regression problems. The core idea is to maximize [15, 16] the distance between positive examples and negative examples through the optimal hyperplane of components.

Suppose first that there is such a sample set:

$$Q = \{(x_i, y_i)\}_{i=1}^{n}, i = 1, \ldots, n, x_i \in \mathbb{R}^n, y_i \in \{\pm 1\} \tag{1}$$

where, $y_i = 1$ is the sample set of class 1; $y_i = -1$ is a 2-class sample set. According to Largerange duality principle, the decision function can be obtained as follows.

$$g(x) = \text{sgn}\left(\sum\nolimits_{i=1}^{n} c_i^* y_i (x \cdot x_i) + q^*\right) \tag{2}$$

where, c_i^* is the support vector; q^* is the offset. For solving nonlinear problems, we can use the method of constructing mapping to transform the data set of low-latitude space to high-latitude space, so as to maximize the distance between feature data. In this case, the radial basis function $K(x_i, y_i) = \phi(x_i)\phi(y_i)$ (RBF) is used to obtain the decision function:

$$g(x) = \text{sgn}\left(\sum\nolimits_{i=1}^{n} c_i^* y_i (x_i \cdot y_i) + q^*\right) \tag{3}$$

The main features of support vector machine are:

(1) Structure optimal classification hyperplane

Because of adopting the principle of structural risk minimization, support vector machine algorithm transforms the problem of constructing the optimal hyperplane into solving a quadratic optimization problem, and the solution of the quadratic optimization problem is the global optimal solution. In the case of non-separable samples, the relaxation factor is introduced, which makes support vector machine obtain high generalization under the premise of satisfying the classification condition. This is the core content of the support vector machine.

(2) VC (Vapnik-Chervonenkis Dimension) dimension theory

At present, there is no complete theory to calculate the VC dimension of a function, which is mainly used to describe the convergence speed and generalization of the learning process of a function. Under the premise of a certain number of samples, the larger the VC dimension is, the more complex the learning machine will be. Therefore, the support vector machine with structural risk minimization needs to reduce the VC dimension while ensuring the learning accuracy.

(3) Dealing with nonlinear problems

In practice, the classification problems are often nonlinear. SVM maps the sample space to a higher dimensional space, and the non-linear non-separable data becomes linearly separable in the higher dimensional space. After finding the classification surface in the higher dimensional space, the hyperplane corresponding to it in the lower dimensional space is the sought hyperplane.

(4) Kernel Function Method

When the samples are mapped to the high dimensional space, the amount of calculation will increase exponentially. In order to solve this problem, the support vector machine adopts kernel method to transform the inner product operation into the kernel function operation in the low dimensional space, which avoids the increase of calculation. From the mathematical point of view, support vector machines are proposed from the case that there is an optimal classification surface in the case that the function set is linearly separable. Consider the case where two classes are linearly separable in two dimensions, as shown in Fig. 1.

In Fig. 1, the hollow points and solid points represent the training samples of the two classes respectively, which are generally defined as normal data samples and fault

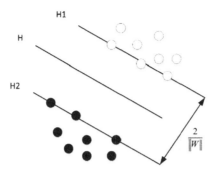

Fig. 1. Diagram of the optimal 2D linearly separable classification surface for support vector machines

data samples in fault diagnosis. H is the classification line that completely separates the two classes of samples, and H1 and H2 are the lines parallel to the classification line H and passing through the nearest point to the classification line respectively. The distance between H1 and H2 is called the classification interval or the classification gap. The optimal classification line is the one that can guarantee the largest classification gap between the two classes. When extended to high dimensional space, the concept of optimal classification line is generalized to the optimal classification surface, that is, the optimal hyperplane.

3   Fault Diagnosis Model

In the process of fault diagnosis model establishment of Simulink module under Matlab platform based on SVM algorithm and theory, it is difficult to deal with different faults, and the selection of fault diagnosis parameters. Aiming at the problem that it is difficult to select the appropriate parameters of SVM, this paper uses GA algorithm to determine the optimal values of SVM parameters.

3.1   Parameter Optimization of GA-SVM Based on Genetic Algorithm

Traditional SVM usually uses manual trial calculation method to set the key parameters of kernel function parameter r and penalty factor C, which is usually over-dependent on the subjective experience of experts. It is not only time-consuming, laborious and inefficient, but also difficult to achieve reasonable parameter setting for slightly complex problems, which hinders the promotion and application of SVM to some extent.

In view of the above problems, many scholars begin to transform the parameter setting of SVM into the parameter optimization problem, and then use intelligent algorithms to solve the optimal problem. Among the many intelligent algorithms, genetic algorithm, as an optimization intelligent algorithm based on the survival and natural selection mechanism in the theory of biological evolution, has the advantages of good applicability, high optimization efficiency and fast search speed. In this paper, genetic algorithm is introduced on the basis of SVM algorithm, and GA-SVM algorithm is constructed.

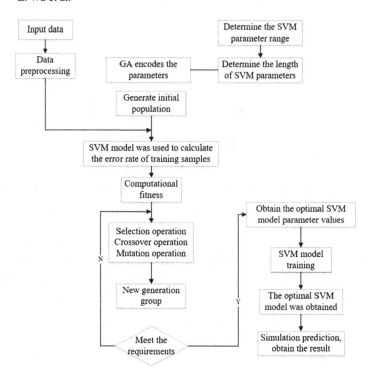

Fig. 2. Flowchart of GA-SVM algorithm

The flowchart of GA-SVM algorithm is shown in Fig. 2, and its algorithm steps are as follows.

Step 1: Parameter setting: Set the addressing range of GA running parameters and SVM main parameters reasonably.

Step 2: Encoding and formation of initial population: In this paper, the kernel function parameter r and penalty factor C of support vector machine are expressed as genotype data with genetic characteristics by using binary coding method, and the initial population is randomly generated.

Step 3: Calculation of individual fitness: calculate the fitness of all individuals, which is used to measure the quality of individuals.

Step 4: Judgment of stopping criterion: Judge whether the fitness of the individual meets the stopping criterion, if it does, perform step 6, otherwise continue to perform step 5.

Step 5: Genetic operation: use roulette wheel method, select the individual with high fitness for selection, crossover, mutation operation, and finally form a new population, return to step 3 and continue to execute until the stopping criterion is met.

Step 6: Decode and output the optimal solution of parameters.

Step 7: Put the optimal parameters into the SVM to solve the model.

3.2 Normalize the Data

After the fault diagnosis model is established, before the fault diagnosis of each fault in the power system, the obtained fault data need to be preprocessed first. Because the detected parameters include flow, pressure, speed, temperature, etc., and these parameters themselves are not the same in units and orders of magnitude, so the data need to be normalized before extracting samples for training. The purpose is to eliminate the influence of the dimension of each parameter on the diagnosis results. There are many methods to normalize the parameters. In this paper, we choose to normalize the parameters by dividing the corresponding parameters minus the minimum value of the parameters by the maximum allowable value of the parameters minus the minimum value.

$$\bar{x} = \frac{x_i - x_{\min}}{x_{\max} - x_{\min}}, i = 1, \ldots, n \tag{4}$$

4 Experiment and Result Analysis

Support vector machine is suitable for the fault diagnosis process in this paper because of its good diagnostic ability for small sample size. According to the 10 groups of parameters of the power system based on Simulink simulation and experimental process (see Attached Table 1, Attached Table 2 and Attached Table 3), the corresponding fault diagnosis of the UUV power system is carried out based on the characteristics of 3/4 samples selected as training samples and 1/4 samples selected as testing samples in the support vector intelligent algorithm.

Based on the classification principle of C-support vector classifier, according to the processing method of each fault data of UUV power system, five typical states are respectively analyzed: Including the normal state of the system (corresponding to classification 0), filter blockage (corresponding to classification 1), filter corrosion damage (corresponding to classification 2), fuel pump leakage (corresponding to classification 3) and fuel valve leakage (corresponding to classification 4), the fault data under the determined working condition 1.0 were diagnosed. The diagnosis results are as follows:

For the four types of fault diagnosis results in Table 1, the ROC chart of diagnosis evaluation and the statistical table of related fault diagnosis results are as follows (see Fig. 3 and Table 1):

For the four types of fault diagnosis results in Table 2, the ROC chart of diagnosis evaluation and the statistical table of related fault diagnosis results are as follows (see Fig. 4 and Table 2):

For the four types of fault diagnosis results in Table 3, the ROC chart of diagnosis evaluation and the statistical table of related fault diagnosis results are as follows (see Fig. 5 and Table 3):

According to Fig. 3, 4 and 5 and Table 1, 2 and 3, it can be seen that the diagnosis results of each fault diagnosis group are more than 90%, so using SVM to classify and identify the fault of UUV power system has high feasibility and effectiveness. For the fault diagnosis of the power system operation, there is a small difference between the training and fault diagnosis results of the three groups of test data, so it can be

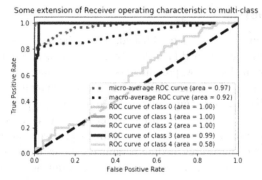

Fig. 3. ROC statistics of power system fault diagnosis results Fig. 1

Table 1. Statistics of fault diagnosis results Table 1

/	Operating condition 1.0
Training accuracy	93.7%
Diagnostic accuracy	92.8%
Classification results	935/1000

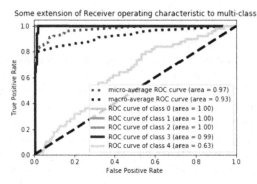

Fig. 4. ROC statistics of power system fault diagnosis results Fig. 2

Table 2. Statistics of fault diagnosis results Table 2

/	Operating condition 1.0
Training accuracy	94.5%
Diagnostic accuracy	92.4%
Classification results	940/1000

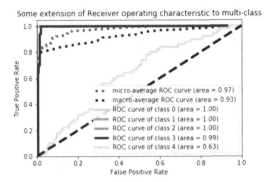

Fig. 5. ROC statistics of power system fault diagnosis results Fig. 3

Table 3. Fault diagnosis results Table 3

/	Operating condition 1.0
Training accuracy	90.3%
Diagnostic accuracy	91.6%
Classification results	906/1000

considered that the support vector machine is suitable for the fault diagnosis of the UUV power system under the working condition 1.0.

5 Conclusion

Aiming at the most common multiple faults in the UUV auxiliary system – power system fault, in order to timely and accurately diagnose the fault of the UUV power system and prevent the occurrence of fault as far as possible in advance, this paper takes the UUV power system as the research object, analyzes the common faults of the UUV power system according to the composition of the power system, the working process and the working principle of each component. The fault mode of each fault is obtained, and the fault diagnosis model of UUV power system is established by using the Simulink module of Matlab combined with SVM algorithm and GA algorithm. The diagnosis parameters are selected for different faults and the fusion training and verification process of multi-source parameter sets is completed. Through the test and result analysis, the use of support vector machine algorithm for fault diagnosis has good adaptability and high precision in the case of less fault samples of the power system.

Genetic algorithm is used to optimize the parameters of SVM. Because the optimal solution of genetic algorithm is found by repeated iteration, the speed of parameter optimization will be greatly limited when the number and quality of samples are too large. Therefore, the next step is to optimize the genetic algorithm and try to improve the efficiency of parameter optimization of genetic algorithm.

References

1. Liu, J.W., Luo, J., Zhang, X.C.: Statistics of new vision: big data and machine learning. J. Stat. Theory Pract. **2023**(10), 55–60 (2023)
2. Zhou, W.D.: Research on nuclear machine learning methods. Xidian University, Ph.D. dissertation (2003)
3. Xiao, R., Wang, J.C., Zhang, F.Y.: Overview of support vector machine theory. Comput. Sci. **27**(3), 1–3 (2000)
4. Guo, J.Y., Li, T., Li, Y.: SVM fault detection based on principal component augmented matrix. J. Shenzhen Univ. (Sci. Technol. Ed.) **38**(05), 543–550 (2021)
5. Zhou, W.: Simulation research of analog circuit fault diagnosis based on SVM parameters optimization. Microcomput. Appl. **37**(09), 70–72+76 (2021)
6. Liu, K.S., Li, T.Z., Liu, D., Wang, M.J., Xu, L.B.: Photovoltaic array fault diagnosis based on ABC-SVM algorithm. Power Supply Technol. **45**(09), 1171–1174 (2021)
7. Gong, W.F., Chen, H., Wang, D.W.: Fast fault diagnosis method of multi-sensor UUV rotating machinery based on improved LSTM-SVM. UUV Mech. **25**(09), 1239–1250 (2021)
8. Zhang, S.J.: Transformer fault diagnosis based on SVM optimized by fireworks algorithm. Mech. Electr. Inf. **2021**(22), 30–31 (2021)
9. Cai, B., Huang, J.Y., Du, J.B., Ma, J.C., Wang, Z.C.: Fault diagnosis of planetary gearbox based on MEEMD multi-feature fusion and LS-SVM. China Test **47**(09), 126–132 (2021)
10. Chen, L., Wang, P.Y., Chen, L., Wang, P.Y.: Prediction model of first feed water pipe leakage time based on genetic least squares support vector machine. J. Zhejiang Univ. Technol. **49**(05), 546–549 (2021)
11. Cheng, H.D., Liu, P.H., Gou, W., Yu, F.R., Chen, H.X., Wang, J.Y.: Building model identification and temperature prediction based on genetic algorithm-resistent-capacity model. Refrigeration Air Conditioning 1–6 (2021)
12. Liu, C.Y., Xu, H.R., Duan, F., Wang, T.S., Lu, Z.W., Yu, W.X.: Spectral identification of rabbit liver VX2 tumor based on genetic algorithm and support vector machine. Spectrosc. Spectral Anal. **41**(10), 3123–3128 (2021)
13. Walters, D.C., Sheble, G.B.: Genetic algorithm solution of economic dispatch with valve point loading. IEEE Trans. Power Syst. **8**(3), 1325–1332 (1993)
14. Ahn, C.W., Ramakrishna, R.S.: A genetic algorithm for shortest path routing problem and the sizing of populations. IEEE Trans. Evol. Comput. **6**(6), 566–579 (2003)
15. Fan, Y.Q., Cui, X.Y., Han, H., Wu, H., Xu, L.: Fault diagnosis of water chiller based on field sensor parameters and SVM. Energy Res. Inf. Technol. **37**(03), 147–152 (2021)
16. Yao, B., Li, M.C., Wang, F.Z.: Fault diagnosis and reliable configuration based on GA-SVM trapezoid region. J. Shenyang Normal Univ. (Nat. Sci. Ed.) **39**(03), 210–214 (2021)

Research on Airborne Radar Multi-target Continuous Tracking Algorithm on Sea Surface Based on Deep Kalman Filter

Zhisuo Xu[✉]

China Ship Research and Design Center, Wuhan 430070, China
xuzhisuo@126.com

Abstract. It is difficult for airborne radar to track multiple targets on the sea surface because of the large number of targets, high density and various types of targets. The application of traditional tracking algorithm is limited by operation, especially in the case of airborne radar tracking of sea target, the amount of tracking calculation will increase explosively with the increase of target track and radar echo number. In this paper, a multi-target continuous tracking algorithm based on deep Kalman filter is used to predict the state matrix through slicing recurrent neural network, combined with linear Kalman filter, which can improve the tracking accuracy of the target and improve the computing efficiency. Compared with the traditional tracking algorithm, the tracking accuracy of the proposed method is improved by about 10 m, and the convergence time is reduced by about 25 s. Simulation results verify the effectiveness of the proposed multi-target continuous tracking algorithm, and it has good performance.

Keywords: Airborne Radar · Target Tracking · Deep Kalman Filter

1 Introduction

Airborne radar Multi-Target Tracking [1] (MTT) is the process of estimating the number and state (motion characteristics, such as position, velocity and acceleration, etc.) of multiple targets on the sea surface through the echo information containing noise obtained by radar scanning in the presence of false alarm, clutter or interference [2]. With the increase of the number of targets that airborne radar can deal with, the simultaneous tracking of multiple targets has become an indispensable system. For example, when airborne radar intercepts targets in the open sea, it should be able to track multiple targets at the same time, and select some targets with the greatest threat or the highest hit rate to attack [3]. When attacking a single target, it is often necessary to monitor the surrounding targets at the same time. In addition, surface ship and submarine surveillance and sea battlefield intelligence situation analysis also raise the problem of multi-target tracking.

The concept of multi-target tracking was formally proposed back in 1955. In the subsequent development, theories such as information fusion, random set theory, sequential

L. Pan et al. (Eds.): BIC-TA 2023, CCIS 2062, pp. 331–341, 2024.
https://doi.org/10.1007/978-981-97-2275-4_26

Monte Carlo, particle filter and dynamic programming [4] were successively involved, providing new ideas for the further development of multi-target tracking [5].

While the academic achievements of multi-target tracking are booming, its role in the application level is also becoming more and more prominent. The National Missile Defense (NMD) and Territorial Missile Defense (TMD) systems of the United States take multi-target tracking as the core technology [6, 7]. Up to now, Universities and research institutions, such as George Mason University, University of New Orleans, University of Connecticut, Georgia Institute of Technology, University of Cambridge, University of Melbourne, Boeing Avionics Flight Laboratory, Naval Operations Center, and French Institute of Computer Science and Control, are at the forefront of multi-target tracking research.

In this paper, a multi-target continuous tracking algorithm based on deep Kalman filter is proposed, which solves the problem of target tracking difficulty and poor tracking accuracy in the process of airborne radar tracking multi-target at the sea surface.

2 Deep Kalman Filter Network Framework

In the field of airborne radar multi-target tracking, the Kalman filter is widely used, but it is only suitable for linear modeling and cannot filter nonlinear systems [8]. Extended Kalman filter is an extended form of nonlinear modeling of Kalman filter, and its filtering effect is stable in not very violent nonlinear environment. However, the extended Kalman filter also has limitations, it needs to solve the Jacobian matrix in the modeling process, and the calculation and complexity are relatively high. With the increase of the number of targets, the amount of calculation increases exponentially, and the model error is introduced when the local linearization is performed.

With the continuous update of computer technology and the rapid development of deep learning and artificial intelligence based on neural network, neural network has powerful nonlinear processing ability, self-learning ability, infinite approximation to the fitting function and other advantages, and its application field has expanded to all aspects. In recent years, the algorithm fusion of neural network and Kalman filter has become a research hotspot at home and abroad [9]. At present, there are many literatures on the fusion of Kalman filter using error back propagation and Elman Neural network, but there are few literatures using Recursive Neural Networks (RNN) [10, 11]. Moreover, compared with ordinary fully connected neural networks, RNN adds a hidden layer to save the input information. With the input of data, the information stored in the hidden layer is constantly updated iteratively, that is, the memory of historical information is formed. In the target following problem, the historical state information can be used for state prediction at the present or future time. Moreover, the parallel processing ability of RNN can be improved after slicing, which is suitable for processing a large amount of data and efficient calculation. These characteristics are very crucial for the real-time solution of radar multi-target continuous tracking on the sea surface.

As shown in the Fig. 1, a tracking algorithm framework combining deep neural network and Kalman filter is proposed to solve the problem of multi-target tracking in the sea surface environment. The framework combines deep neural network, state equation of motion and Kalman filter to infer the motion state of the target, which mainly includes

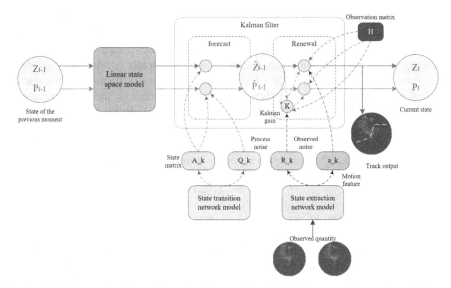

Fig. 1. The framework of the tracking algorithm combined with deep neu-ral network and Kalman filter

three parts: state extraction network, state transition network and Kalman filter. Among them, the state extraction network is used to extract useful features from radar points. The state transition network is used to dynamically update the state matrix of the system. The Kalman filter combines the system state equation to estimate and predict the system state.

3   Continuous Multi-target Tracking Algorithm on Sea Surface Based on Deep Kalman Filter

3.1   State Extraction Network Model

Consider a time-varying dynamic model, denoted by

$$z_t = f(z_{t-1}, w_t) \tag{1}$$

where, $z \in \mathbb{R}^d$ represents the d - dimensional state vector, t represents the current moment, and w represents the random vector of system noise and measurement noise. The function f is a Markov process and represents the state z_{t-1} change to z_t. The model (1) can be written in the form of the following linear structure:

$$z_t = A_t z_{t-1} \tag{2}$$

where, A is the time-varying system state transition matrix.

In general, the system state containing useful information is often located in a different state space from the original measurement. For example, given a set of radar points (radar measurements), the key system states needed are the speed, direction, and position of the target. This project does not explicitly specify the physical state of the target motion model as in the classical state space model, but uses a deep neural network to automatically extract the latent state features, and dynamically acquire the relationship state matrix that follows the similar linear state matrix in (2). It can be automatically realized by optimizing the model with stochastic gradient descent and back propagation algorithm. Through this dynamic linearization, the Kalman filter can be directly used for state feature inference in the subsequent process.

In the state feature extraction model, the encoder $f_{encoder}$ is used to extract features a_t and estimate errors σ_t^a from observations X_t at time t:

$$a_t, \sigma_t = f_{encoder}(x_t) \tag{3}$$

where, the feature a denotes the observed quantity of the feature state space. σ denotes the measurement confidence that can be converted into the observation error matrix R in the Kalman filter. a and σ can be used in the update stage of the Kalman filter.

Features a may sometimes not be fully informative in the predicted state z of a dynamic system. For example, the occasional loss of sensing data. In this case, the compensation matrix H can be used to transform a into a matrix with the same dimension as the state z.

$$a_t = Hz_t \tag{4}$$

In practice, when the extracted feature a contains all the states of the dynamic system, H is an identity matrix I_d of d - dimension; When a can represent only $m(m < n)$ features, $H = \begin{bmatrix} I_m, 0_{m \times (d-m)} \end{bmatrix}$.

3.2 State Transition Network Model

Equation (2) can be used to carry out the state transition of the system through the state matrix $\mathbf{A}$, but in practical applications, when the system is changing, it is difficult to directly specify the state matrix $\mathbf{A}$. Given the method shown in Fig. 2, the state matrix $\mathbf{A}$ can be regenerated based on the previous system state and the current system information through the Dirichlet distribution combined with the long short-term memory network, and the parameters of the long short-term memory network can be pre-trained by the historical track sets of different targets.

Deep neural networks can be trained to approximate dynamic system models through data training. However, these trained "black box" models are difficult to interpret or adjust. The lack of behavioral indicators of the model and the lack of system reliability largely limit the practical application of these learning models. Stability of dynamical systems is crucial for autonomous systems, as it guarantees that the predictions of dynamical models will not change abruptly in the presence of slightly perturbed inputs. However, most pure deep learning network models do not guarantee this desirable property. Therefore, this study aims to ensure the stability of the system state model by resampling the state matrix from a specific Dirichlet distribution.

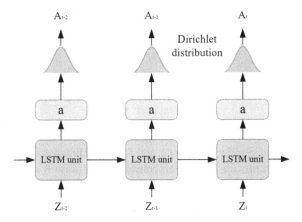

Fig. 2. LSTM-based extraction of the state transition matrix

The concentration parameters α of Dirichlet distribution in the framework algorithm proposed in this study can be generated from the system history state by long short-term memory network:

$$\alpha = \mathrm{LSTM}(z_{t-1}, h_{t-1}) \tag{5}$$

where, z_{t-1} denotes the state of the system at time $t-1$, and h_{t-1} is the hidden state of the long short-term memory network. At any time, the state matrix A can be obtained from the constructed Dirichlet distribution by Eq. (6).

$$A_t \sim \mathrm{Dirichlet}(\alpha) \tag{6}$$

3.3 Kalman Filter-Based Object Tracking

The state extraction network model estimates the system state from noisy sensor measurements, while the generated system state matrix describes the evolution of the system and predicts the system state based on the state at the previous time. However, there is uncertainty in both of them, which can be further filtered and smoothed by the Kalman filter. The Kalman filter can recursively process the uncertainty and produce a state prediction and a new weighted average of the observations.

Further, the Kalman filter consists of two stages: the prediction stage and the update stage. In the prediction phase, prior estimates of the mean and covariance $\left(z_{t|t-1}, P_{t|t-1}\right)$ at the current time are derived from posterior state estimates at the previous time:

$$
\begin{aligned}
Z_{t|t-1} &= A_t Z_{t-1} | P_{t-1} P_{t|t-1} \\
&= A_t P_{t-1|t-1} A_t^T + Q_t
\end{aligned}
\tag{7}
$$

When the observations a_t are available, the update process can produce the posterior mean and covariance $(z_{t|t}, P_{t|t})$ of the states as follows:

$$
\begin{aligned}
r_t &= a_t - H_t Z_{t|t-1} \\
S_t &= R_t + H_t P_{t|t-1} H_t^T \\
K_t &= P_{t|t-1} H_t^T S_t^{-1} \\
z_{t|t} &= z_{t|t\ 1} + K_t r_t \\
P_{t|t} &= (I - K_t H_t) P_{t|t-1}
\end{aligned}
\tag{8}
$$

where, r is the residual, S is the covariance, and K is the Kalman gain. Compared with the manual adjustment of process noise Q and measurement noise R in traditional Kalman, these two values are jointly learned by the proposed deep neural network model.

4 Simulation Experiments

Python simulation is used to simulate the movement of sea surface target in the sea clutter environment, and the tracking algorithm proposed in this paper is compared with the traditional Kalman filter tracking. Assume that the motion state of the target i at time k is $[x_k^i, \dot{x}_k^i, y_k^i, \dot{y}_k^i]^T$, where x_k^i, y_k^i denotes the position information of the axis X, Y, $\dot{x}_k^i, \dot{y}_k^i$ denotes the velocity information of the axis X, Y, the process noise $w_k \sim N(0, Q)$, and the process co-noise covariance matrix is:

$$
Q = \begin{bmatrix}
T^3/3 & T^2/2 & 0 & 0 \\
T^2/2 & T & 0 & 0 \\
0 & 0 & T^3/3 & T^2/2 \\
0 & 0 & T^2/2 & T
\end{bmatrix} \sigma_w^2
\tag{9}
$$

The measurement state equation is as follows.

$$
z_{k+1}^i = \begin{bmatrix} 1 & 0 & 0 & 0 \\ 0 & 0 & 1 & 0 \end{bmatrix} x_{k+1}^i + v_{k+1}
\tag{10}
$$

Measurement noise is $v_{k+1} \sim N\left(0, \text{diag}\left[\sigma_x^2, \sigma_y^2\right]\right)$. Sampling time is $T = 1s$, $\sigma_w^2 = 25$, $\sigma_x^2 = \sigma_y^2 = 9$. The target detection probability is 0.5, and the number of targets can be set according to different scenes.

1) Multiple targets on the sea surface move in a straight line

In this simulation example, the actual trajectory of the target movement is shown in Fig. 3. There are 20 targets in the simulation, and the size of the target movement area is 1200 * 2500. Figure 4 and Fig. 5 are the tracking effects of the multi-target intelligent tracking algorithm and the extended Kalman filter algorithm proposed in this paper when the target moves in a straight line, respectively. Figure 6 and Fig. 7 respectively show the target tracking errors of the two tracking algorithms.

As can be seen from the figure above, the simulation experiment results of the two algorithms for tracking linear moving targets are shown in the following table.

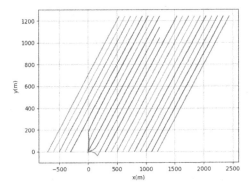

Fig. 3. Actual trajectories of linear moving targets of multiple targets.

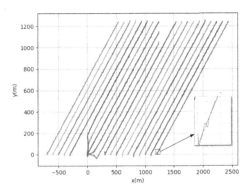

Fig. 4. Multi-target linear motion tracking trajectory based on deep Kalman filter.

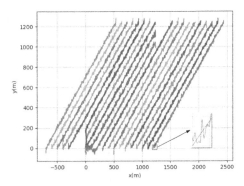

Fig. 5. Multi-target linear motion tracking trajectory based on extended Kalman filter.

As can be seen from Table 1, the average tracking error of classical Coleman filter is 14.25 m, and the simulation time is 34.5 s. The average tracking error of the proposed algorithm is 4.537 m, and the running time is 12.6 s. The tracking effect of the proposed

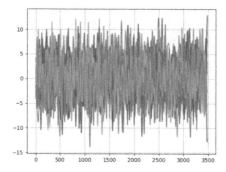

Fig. 6. Multi-target linear motion tracking error based on deep Kalman filter

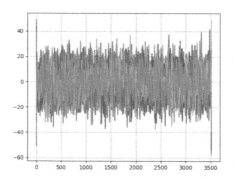

Fig. 7. Multi-target linear motion tracking error based on extended Kalman filter.

Table 1. Experimental results of tracking multi-target linear motion by two algorithms.

Algorithm	Simulation run time (s)	Average distance error (m)
Deep Kalman filter algorithm	12.6 s	4.537
Extended Kalman filtering algorithm	34.5 s	14.25

algorithm is much better than that of the extended Kalman filter algorithm, and the running efficiency is higher.

2) curvilinear motion of multiple targets on the sea surface

In this simulation example, the actual trajectory of the curve movement of the target is shown in Fig. 8. There are 10 targets in the simulation, and the size of the target movement area is 600 * 600. Figure 9 and Fig. 10 are the target tracking effects of the multi-target intelligent tracking algorithm and the extended Kalman filter algorithm respectively proposed in this paper. Figure 11 and Fig. 12 respectively show the results of tracking errors of each target under the two tracking algorithms.

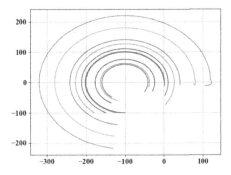

Fig. 8. Actual trajectory of multi-target curve moving target.

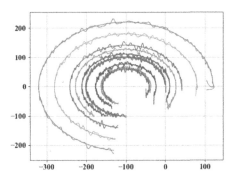

Fig. 9. Multi-target curvilinear motion tracking trajectory based on deep Kalman filter.

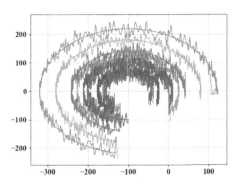

Fig. 10. Multi-target curvilinear motion tracking trajectory based on extended Kalman filter.

As can be seen from the figure above, the simulation experiment results of the two algorithms for tracking curvilinear moving targets are shown in the following table.

It can be seen from Table 2 that the average tracking error of the classical Coleman filter is 15.829 m, and the simulation time is 41.3 s. The average tracking error of the proposed algorithm is 4.483 m, and the running time is 14.1 s. The tracking effect of

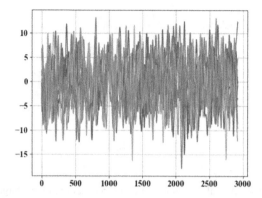

Fig. 11. Multi-target curvilinear motion tracking error based on deep Kalman filter.

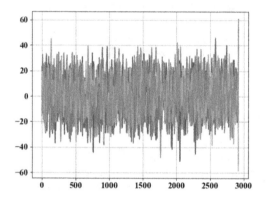

Fig. 12. Multi-target curvilinear motion tracking error based on extended Kalman filter.

Table 2. Experimental results of tracking multi-object curve motion by two algorithms.

Algorithm	Simulation run time (s)	Average distance error (m)
Deep Kalman filter algorithm	14.1	4.483
Extended Kalman filtering algorithm	41.3	15.829

the proposed algorithm is much better than that of the extended Kalman filter algorithm, and the running efficiency is higher.

5 Conclusion

In order to solve the problems of strong sea clutter background and large amount of calculation in the process of airborne radar continuous tracking of multiple targets on the sea surface, a multi-target tracking algorithm based on deep Kalman filter is designed.

In order to overcome the shortcomings of traditional Extended Kalman filter (EKF) and other tracking algorithms, such as large amount of computation, inaccurate target association and lack of real-time performance, and realize airborne radar multi-target continuous tracking on sea surface in complex sea conditions. The feasibility of the proposed method is verified by the simulation study of two typical routes: straight line and curve. Based on the simulation results, the proposed method has good performance and is expected to be used in practical applications in the future.

References

1. Liu, Z., et al.: Robust multi-drone multi-target tracking to resolve target occlusion: a benchmark. IEEE Trans. Multimed. (2023)
2. Fang, W., Chen, X.H.: Research and simulation of airborne radar tracking method. Comput. Simul. **3**, 71–73 (2012)
3. Li, J.: Research on Airborne Radar tracking Filter Algorithm Based on Interpolation. Hebei University of Science and Technology (2020)
4. Fiorini, P.: Motion planning in dynamic environments using velocity obstacles. Int. J. Robot. Res. **17**(7), 760–772 (1998)
5. Xiong, J., Zuo, Z.Y., Xiong, J.: An improved target tracking algorithm for airborne Doppler radar. Telecommun. Technol. **59**(09), 1026–1030 (2019)
6. Jaganath, S., Steve, F.: Defending the United States: revisiting National Missile Defense against North Korea. Int. Secur. **46**(3), 51–86 (2022)
7. Zhang, H.G., He, Q.: Current situation and development trend of anti-jamming technology of airborne radar. Mod. Radar **43**(03), 1–7 (2021)
8. Lu, S., Zhang, S.Y.: Characteristic analysis and filtering algorithm design for UNGM model. J. Northwestern Polytechnical Univ. **41**(2), 293–302 (2023)
9. Zeng, G.R., Yao, J.M., Yan, Q., Lin, Z.X., Guo, T.L., Lin, C.: Real-time hand tracking method based on neural network and Kalman filter. Chin. J. Liquid Crystals Displays **35**(5), 464–470 (2020)
10. Ding, X.: Prediction of GSM-R field strength based on error backpropagation neural network. Electrified Railways **33**(1), 67–70 (2022)
11. Zhang, X.Q., Jiang, R.H., Wang, T., Wang, J.X.: Recursive neural network for video deblurring. IEEE Trans. Circuits Syst. Video Technol. **31**(8), 3025–3036 (2021)

Application-Aware Fine-Grained QoS Framework for 5G and Beyond

Xi Liu[1(✉)] and Yongwei Zhang[2]

[1] School of Computer Science, Beijing University of Posts and Telecommunications,
Beijing 100876, China
xiliu@bupt.edu.cn
[2] School of Information and Communication Engineering, Beijing University of Posts
and Telecommunications, Beijing 100876, China

Abstract. In the face of diverse traffic types, existing 5G standards have adopted a flow-based Quality of Service (QoS) framework and introduced dual mapping to enhance resource allocation flexibility. However, the cognitive ability of these standards, particularly in terms of application traffic recognition (based on network features, eg. IP five-tuples), remains relatively rudimentary. This limitation results in coarse-grained QoS flow classification. Consequently, it's unable to cater to the diverse requirements of priority requirements or service quality needs of applications. To overcome this limitation, we extend our previously proposed cognition-driven core network architecture by developing a cognition-based fine-grained QoS control framework. This framework employs service tags to accurately disclose the QoS requirements of application traffic, enabling 5G networks and beyond to provide more precise services. We provide an in-depth discussion of the framework's internal processes, service tag definitions, and implementation. Through experiments conducted on both the core network and Radio Access Network (RAN) sides, our framework effectively enhances the user's quality of experience (QoE).

Keywords: Quality of service · 5G and beyond · QoS framework

1 Introduction

The 5G QoS framework [1,2], more advanced than 4G, incorporates QoS flows and introduces the Service Data Adaptation Protocol (SDAP) for New Radio (NR) support. It features a two-fold mapping process, mapping IP flows to QoS flows and then to Data Radio Bearers (DRBs). This enhances flexibility but does not address the coarse granularity issue. It maps single or multiple IP flows to one QoS flow, failing to differentiate sub-services within the same IP flow. This limitation becomes prominent in practical applications, especially under poor network conditions.

Consider a scenario where users stream live video content through a 5G network. Under ideal network conditions, the NR can satisfy the requirements of all video traffic, preventing video buffering. However, when the number of users surges and resources become scarce, video buffering may occur. This buffering is **not content-aware** but is only dependent on network conditions. Imagine such buffering happens in crucial moments of a live video stream (e.g., a goal scored during a World Cup match). In the current 5G QoS framework, this video stream is typically mapped as a single QoS flow (due to identical IP five-tuples [3]), making it impossible to prioritize those critical moments. Users' experience will be significantly degraded due to this coarse-grained QoS management.

This motivates us to construct a content-aware fine-grained QoS framework, enabling the core network to recognize traffic more precisely.

The key idea involves the network's ability to discern the varying priorities and QoS demands associated with different content streams. This understanding allows for a more granular division of traffic, which in turn facilitates the allocation of suitable resources. Although previous literature [4] has mentioned similar ideas, their work is primarily conceptual, and realizing their design is non-trivial. First, regarding the most important component, the service tags, they did not mention how to define them, which directly affects the efficiency of the sender and the core network; Second, they did not introduce the signaling processing flow of service tags in the core network; Third, their definition of the insertion point for service tags is inappropriate and lacks an injection position for the IPv4 type.

In this paper, we aim to design and implement this fine-grained QoS control framework, which enables 5G networks to more effectively adapt to diverse application needs, offering highly tailored quality of service for different scenarios. However implementing a truly cross-layer QoS control framework is not easy, and its challenges mainly include the following points:

1. Integration of 5G network and service tags. From the core network's perspective, the challenge is to incorporate service tags into the existing 5G network with minimal changes to the 5G QoS control framework, while considering the costs of storing and processing this information in the control plane and during signaling.
2. Efficiently embedding service tags in data packets challenges balance and fairness. From the perspective of service tags, the issue is defining and inserting these tags into data packets in a way that utilizes limited space efficiently to represent the QoS requirements of service providers. This insertion should be straightforward for receivers and perceptible to the core network's User Plane Function (UPF). Additionally, given the need for global traffic scheduling, there's a challenge in balancing and ensuring the fairness of unified service tags.

Therefore, we have thoroughly designed the entire framework in detail. This work is based on our proposed cognition-driven core network architecture [5]. This architecture draws inspiration from the decentralized neural system structure of an octopus, analogizing the cloud core network to the octopus's brain and the edge core network to its tentacles, thereby forming a cognitive network. The edge core network is brought closer to the user side, with cognitive services deployed on them, guided by the AI of the cloud core network, aiding in the coordination among edge core networks.

The main contributions of this paper are as follows:

1. We have proposed a fine-grained QoS architecture, designed and implemented the integration process of service tags with the existing 5G network, proposing a new signaling process for handling service tags.
2. We developed a mechanism for service tags, including the definition methods and insertion points, and also integrated DSCP [6] and standard 5QI [7] values into the tags for ease of use, universality, and measurement solutions.
3. Simulation experiments were carried out to validate the feasibility and effectiveness of the method proposed in this paper.

2   Background and Related Work

2.1   5G QoS Control Framework

5G QoS framework [1,2] is built on top of QoS Flows, which serve as the smallest granularity unit for delivering differential services. Each QoS Flow is identified by a unique QoS flow ID (QFI), which remains unique within the Packet Data Unit (PDU) session. In the 5G network, data flows with the same QFI values are treated equally, resulting in the inability to fulfill demands for different priorities or QoS.

In a typical scenario of downstream traffic transmission, data from the Data Network (DN) to UPF is mapped from IP flows to QoS flows based on the IP five-tuple, with the QFI value carried via the GTP-U protocol on N3 and N9 interfaces (Fig. 1). At the NR, QoS Flows are mapped to DRBs, a process handled by the SDAP layer. The standard framework's main issue is coarse QFI granularity and inflexibility, hindering differentiated services for application subservices.

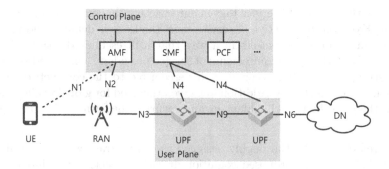

Fig. 1. 5G Core Network.

2.2   Fine-Grained QoS Control Framework

This concept, first introduced by Zhang *et al.* [4,8], intended to resolve the issue of poor user experience caused by the coarse granularity of QoS flows. Its main innovation is using semantic tags[1] in the QFI decision process for finer QoS Flow management. It involves pre-communication between application service providers and the core network for understanding semantic tags and QoS profiles. Thereby achieving the transition from One-to-Many to Many-to-One of IP Flows to QoS Flows (Fig. 2). With the finer-grained QoS Flows, resource allocation on the NR becomes more precise, enabling improved differential services.

Although this offers a potential framework, it remains a concept not yet fully realized, lacking specific design details such as signaling processes and service tag configurations. Furthermore, there is a lack of effective experimental evidence to demonstrate the practical effectiveness of the finer-grained QoS framework.

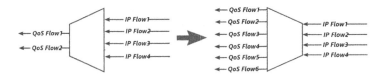

Fig. 2. Transition from One-to-Many to Many-to-One of IP Flows to QoS Flows.

Nightingale *et al.* [10] utilized H.265 encoded video, embedding information about video layering and additional details into the stream using the Supplemental Enhancement Information (SEI) format. Then the 5G network uses the SEI and current network conditions to control the adaptation of video layers. However, SEI information is exclusive to H.265 encoded videos and is not universal. Zhang *et al.* [11] describes a video streaming system, they observe that quality sensitivity is inherent to video content and modify existing ABR [12,13] algorithms, significantly enhancing user QoE. However, no one has yet integrated this insight into the existing 5G QoS framework.

3   Implementation

3.1   QoS Control Process

The overall process of QoS service tag handling is illustrated in Fig. 3, involving various components such as application service provider, Application Function (AF), Policy Control Function (PCF), and more. Our framework exhibits a high degree of flexibility and customizability.

The overall process is that application service providers negotiate with the core network to clarify the exact QoS needs of their applications, encompassing

---

[1] To avoid confusion with semantic communication [9], we propose renaming semantic tags to service tags.

aspects like bandwidth, latency, reliability, and other key metrics. These requirements are then converted into service tags and embedded in data streams, signifying their desired service quality. The core network, considering these tags and IP five-tuple information (source and destination IP addresses and ports, and transport protocol), collaboratively assigns appropriate QFI. Traffic with varying QFI values receives differential treatment across UPFs and NR. This approach enables 5G networks to more effectively adapt to diverse application needs, offering highly tailored quality of service for different scenarios.

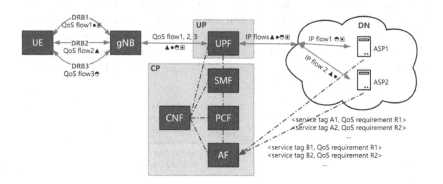

Fig. 3. The Overall QoS Service Tag Processing Flow.

Specifically, application service providers can predefine specific QoS service tags and assign corresponding QoS requirements to each tag. Subsequently, the application service providers transmit this QoS configuration information in JSON format to the AF within the core network through a specific interface for configuration. This configuration allows the downstream traffic to use the QoS service tags.

Now, we will provide a detailed description of the signaling process for internal core network handling of QoS configuration with service tags. This process involves signaling for QoS information, as shown in Fig. 4:

1. Initially, the AF receives the QoS configuration information from the service providers, which includes the definition of service tags. Since this AF is a third-party AF and cannot directly communicate with internal network functions, it needs to forward this information to the Network Exposure Function (NEF).
2. Through the N33/Nnef interface, this information is transmitted to the NEF, responsible for executing the transfer and translation of external information to meet the invocation requirements of internal network functions.
3. The NEF sends the received QoS configuration information via the N5/Rx interface to the PCF. In the standard 5G process, this configuration information primarily includes application-related details (such as service type and bit rate requirements) and traffic parameters (like IP five-tuples). However,

in our fine-grained framework, it is evident that multiple sets of application-related information need to be processed and stored in the UDM.

4. The PCF may request service-related information from the UDR and formulate Policy and Charging Control (PCC) rules by considering factors such as the previously defined QoS configuration. Unlike the standard 5G definitions, these PCC rules also include the definition of service tags to enable finer-grained decision-making.

5. The PCF delivers the PCC rules to the Session Management Function (SMF) through the N4 interface. The SMF is responsible for determining the mapping and allocation of IP flows to QoS flows, which includes the establishment and modification of QoS flows. For more detailed information, refer to Sect. 3.3.

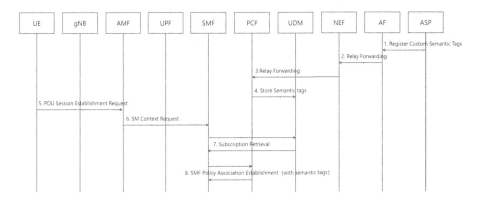

Fig. 4. Flowchart of Service Tag Processing and Partial PDU Session Establishment in Core Network.

This signaling process ensures the effective handling of QoS service tags to meet the specific QoS requirements of different services, providing the network with increased flexibility and customization. This contributes to improving network performance and user experience, especially in scenarios that demand more precise QoS control.

3.2 Design of Service Tag

Service tags are the core elements of the fine-grained QoS framework, with each service tag representing a specific set of QoS requirements. It is these service tags that allow us to break down the originally coarse-grained IP flows into finer-grained sub-IP flows and map them to different QoS flows on the user data plane, thereby providing more nuanced services.

To efficiently embed service tags into data flows, we propose an innovative service tag design scheme, which includes the structure, management, application, and integration with existing network architectures of service tags. This section provides a detailed overview of this scheme.

Insertion Position of Service Tag. The first challenge is identifying an optimal position for inserting service tags, ensuring ease of communication and core network recognition. A straightforward method is to insert tags in the application layer's message headers or body, akin to Deep Packet Inspection (DPI) [14], but differing in real-time processing. This requires alterations to application layer protocols for both senders and receivers, complicating positioning and detection in the core network's user plane.

When considering moving down to lower layers, the diversity of transport layer protocols (such as UDP, TCP, QUIC, etc.) makes adding service tags within these protocols challenging. In contrast, at the network layer, the IP protocol is predominantly used, offering a relatively uniform protocol with existing QoS configuration methods. Therefore, we believe that inserting service tags within the IP protocol is a more suitable choice.

Previous proposal [4] suggested inserting service tags into the IPv6 extension headers but did not propose an insertion location applicable to IPv4. However, due to the comparatively inflexible and less universal nature of IPv4 extension headers, in contrast to IPv6, inserting service tags into IPv4 extension headers is complex and would significantly degrade performance when processing the headers [15].

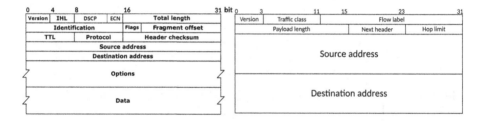

Fig. 5. DSCP field (for IPv4) and the Traffic Class field (for IPv6).

DiffServ [6] is a network architecture developed for traffic classification and management in IP networks to enhance QoS. It differentiates packet forwarding treatments at each hop based on the Differentiated Services Code Point (DSCP), a 6-bit categorization in IP packets. In IPv4, DSCP is in the DS field (replacing the ToS field) of the packet header, while in IPv6, it is in the 8-bit Traffic Class field. For the purpose of unifying service tags, we set the service tags to 6 bits.

We recommend inserting service tags into the DSCP field (for IPv4) or the Traffic Class field (for IPv6) in the IP protocol header, as shown in Fig. 5. It offers several advantages:

1. Applications can easily set the DSCP field directly. There are existing use cases, such as WebRTC [16] shown in Table 1, which utilize this field to provide more suitable services for different voice and video traffic.

2. Modifications at the IP layer do not directly affect the upper layers, and the receiver can receive these service tags without any awareness.
3. Network switching devices experience minimal performance impact when processing the DSCP field. Previous research [17] has shown that DSCP field settings have a negligible effect on latency.

This method has clear advantages for the insertion of service tags, making the application of service tags more convenient and efficient.

Table 1. Recommended DSCP Values for WebRTC Applications [16].

Flow type	Very Low	Low	Medium	High
Audio	LE (1)	DF (0)	EF (46)	EF (46)
Interactive Video with or without Audio	LE (1)	DF (0)	AF42, AF43 (36, 38)	AF41, AF42 (34, 36)
Data	LE (1)	DF (0)	EF (46)	EF (46)

Definition of Service Tag. In this section, we will provide a detailed explanation of our service tag definition. In the previous section, we decided to place service tags in the DSCP field (for IPv4) or the TrafficClass field (for IPv6).

If application service providers opt not to use standard DSCP values [6,18], they can define up to 64 unique service tags using the 6-bit DSCP field, matching specific QoS needs. However, this approach has limitations in universality. Service tags, while unique within a domain, don't carry over between different domains. For instance, two providers might independently create similar tags, leading to duplication in the core network. At the network level, the User Data Manager (UDM) would need to store up to 64 <service tag, QoS configuration> pairs per IP five-tuple, leading to data redundancy and inefficient QFI determination by the SMF.

On the other hand, we can also choose to use some RFC-recommended DSCP values directly as our service tag definitions. For generic traffic types, recommended DSCP values can be used, while for specific cases where generic DSCP values cannot represent QoS requirements, customization can be performed. This approach allows the UDM to only store the IP five-tuple and their corresponding service tag meanings, reducing the workload for application service providers to customize service tags and effectively solving the core network's storage redundancy issue.

While generic (RFC-recommended DSCP values) and custom service tags have been introduced, there's a need to incorporate specific 5G QoS features to define the QoS requirements of these tags. Standard DSCP values like AF41/42 for "Multimedia" and CS4 for "Real-Time" are broadly defined and lack detailed quantitative metrics. Moreover, for custom tags, different application service providers might have varying standards for <service tag, QoS requirements>.

Given that the network is a global entity, coordinating service tags across multiple providers is complex. Even within a single domain, aligning QoS requirements between generic and custom tags poses a significant challenge.

Therefore, **quantifying** and **standardizing** service tags across different application service providers are essential. We suggest mapping standard DSCP values to 5G QoS characteristics [7] to quantize tags within and across application domains. Key 5G QoS characteristics include Resource Type (GBR or Non-GBR), Priority, Packet Delay Budget, Packet Error Loss Rate, Maximum Data Burst Size (for GBR), and Data Rate Average Window (for GBR). These QoS characteristics are extended to standard DSCP values for user guidance. The generic service tags are detailed in Table 2, with CS0 to CS7 representing Non-GBR types, AF11 to AF32 for GBR types, and AF33 to AF43 for "Low-Latency GBR" types.

For custom service tags, DSCP Pool1 (0bxxxxx0) and Pool3 (0bxxxx01) are predefined, while Pool2 is reserved for local use or experimentation. We recommend that application service providers utilize custom Pool2 (0bxxxx11) for setting their tags, offering 16 combination options. This allows an IP five-tuple to correspond to 16 distinct custom service tag schemes, in addition to 48 predefined generic tags, adequately refining traffic. Providers can customize their service tags based on the mapping of QoS requirements to these generic tags.

In Table 3, the capacity of different methods to express QoS requirements is shown, where N represents the count of IP five-tuples. Without service tags, N QoS types can be expressed. Using service tags, there are 48 generic and 16 custom QoS types, providing 17 options per IP five-tuple, totaling $48 + N * 17$. This significantly surpasses the QoS subdivision capacity in the current 5G using only DSCP.

3.3   IP Flow to QoS Flow Mapping

With the incorporation of service tags, we now have a more precise way of determining the generation of QFI. This involves the synergy of IP five-tuples and service tags. In the case of traffic that possesses service tags, we can directly determine the corresponding QFI based on these tags, thus providing more fine-grained services for this traffic. For traffic without service tags, we continue to employ the same processing approach as existing 5G, which is to determine the appropriate QFI based on their IP five-tuples. The overall process is illustrated in Fig. 6, demonstrating the collaborative interaction of service tags and IP five-tuples to determine the QFI.

Table 2. Definition of generic service tags (combined with 5QI)

DSCP name	DSCP value	Description (RFC 4594)	Our Suggestions	5QI	Resource Type	Priority	Packet Delay Budget	Packet Err Loss Rate
CS0	000000	Standard	Any flow that has no BW assurance	10	Non-GBR	1100	300	10^{-6}
CS1	001000	Low-Priority	Any flow that has no BW assurance	9	Non-GBR	90	300	10^{-6}
CS2	010000	OAM&P	Video (Buffered Streaming) TCP-based	8	Non-GBR	80	300	10^{-6}
CS3	011000	Broadcast Video	Voice, Video (Live Streaming)	7	Non-GBR	70	100	10^{-3}
CS4	100000	Real-Time Interactive	Video (Buffered Streaming) TCP-based	6	Non-GBR	60	300	10^{-6}
CS5	101000	Signaling	IMS Signaling	5	Non-GBR	10	100	10^{-6}
CS6	110000	Network Control	V2X messages	79	Non-GBR	65	50	10^{-2}
CS7	111000	Reserved for future	Low Latency eMBB applications (AR)	80	Non-GBR	68	10	10^{-6}
AF11	001010	High-Throughput Data	"Live" Uplink Streaming	76	GBR	56	500	10^{-4}
AF12	001100	High-Throughput Data	Conversational Voice	1	GBR	20	100	10^{-2}
AF13	001110	High-Throughput Data	Conversational Voice (Live Streaming)	2	GBR	40	150	10^{-3}
AF21	010010	Low-Latency Data	Real Time Gaming, V2X messages	3	GBR	30	50	10^{-3}
AF22	010100	Low-Latency Data	Non Conversational Video (Buffered Streaming)	4	GBR	50	300	10^{-6}
AF23	010110	Low-Latency Data	Mission Critical user plane	65	GBR	7	75	10^{-2}
AF31	011010	Multimedia	Non-Mission Critical user plane	66	GBR	20	100	10^{-2}
AF32	011100	Multimedia	Mission Critical Video user plane	67	GBR	15	100	10^{-3}
AF33	011110	Multimedia	Discrete Automation	82	Delay Critical GBR	19	10	10^{-4}
AF41	100010	Multimedia Conferencing	Discrete Automation	83	Delay Critical GBR	22	10	10^{-4}
AF42	100100	Multimedia Conferencing	Intelligent transport systems	84	Delay Critical GBR	24	30	10^{-5}
AF43	100110	Multimedia Conferencing	Electricity Distribution – high voltage	85	Delay Critical GBR	21	5	10^{-5}

Table 3. Comparison of QoS Requirements Expressible by Different Approaches.

Current 5G	Only DSCP	Ours
N	48 + N	**48 + N * 17**

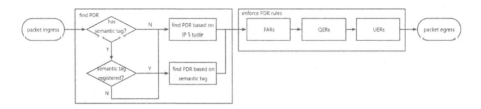

Fig. 6. Collaborative QFI Generation with Service Tags and IP Five-Tuples.

4 Experiments

In this chapter, we present our experimental design and results to validate whether our fine-grained QoS framework can achieve a higher level of granularity for traffic that currently relies on IP five-tuple sub-division, enabling more targeted service delivery and improving users' QoE. The testbed of the entire fine-grained QoS control framework involves two parts. One part is within the core network modules responsible for registering, modifying, identifying, and making QFI generation decisions based on service tags. The other part is on the RAN side, responsible for allocating appropriate DRB resources based on the QFI values.

4.1 Experimental Setup

Firstly, the core network portion is based on free5GC [19] (an open-source 5G core network framework that supports the SDAP layer). We made modifications to the GTP5G kernel module, adding the logic for setting QFI based on service tags. By examining kernel logs, we can verify that our system can successfully set QFI based on service tags, providing support for our framework.

Secondly, the RAN-side implementation was accomplished through the simulation of the NR environment using Mininet [20]. While we initially planned to use UERANSIM [21] (an open-source NR and UE simulator), unfortunately, UERANSIM's QoS support capabilities are still limited and do not allow dynamic resource switching for QoS flows. Therefore, we decided to use Mininet to simulate the NR environment.

Our experiment, depicted in Fig. 7, has three parts:

1. 5G Network: Focused on the RAN and excluding the core and transport networks, we incorporated five switches (s1, s2, s3, s4, s5). We simulated different DRBs using links between these switches, specifically three DRBs: s2–s5, s3–s5, and s4–s5, with bandwidth limits of 1 Mbps, 1 Mbps, and 2 Mbps,

respectively. Other links had unlimited bandwidth. By configuring S1's flow tables, we directed packet paths to simulate the mapping of QoS flows to various DRBs, akin to different NR resources.

2. DN: This part consists of multiple application service providers. We set up three application service providers to simulate traffic transmission. These application service providers are connected to the core network via the S1 switch.

3. User Equipment (UE): We set up three UEs, which are connected to the RAN through the S5 switch.

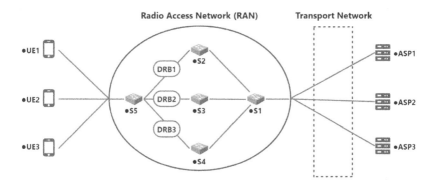

Fig. 7. Network Topology Diagram Constructed with Mininet.

Our fine-grained QoS framework is designed to be universal. However, to simplify experiments and better illustrate its functionality, we chose a typical video streaming scenario as an example: users streaming 1080P and 4K resolution videos on demand. In both resolution videos, there are two types of video segments: some are regular video segments, while others are highly engaging segments that greatly impact the user viewing experience.

To manage these two different priority traffic types, we assigned two distinct values in the service tags: 011000 and 001010. The specific configurations are shown in the Table 4:

Table 4. Service Tag Configurations for Four Traffic Types.

Traffic type	Service tag	Resource type	DSCP value
Regular 1080P video	011000	Non-GBR	CS3
High-priority 1080P video	001010	GBR	AF11
Regular 4K video	011000	Non-GBR	CS3
High-priority 4K video	001010	GBR	AF11

Regarding the traffic generator, we selected D-ITG [22] to simulate and generate realistic network traffic on the application service provider side.

Our experiments use three different sessions: ASP1-UE1, ASP2-UE2, and ASP3-UE3. In these sessions, we consider the transmission of two different video resolutions, 1080P, and 4K, using the UDP protocol. Each video stream contains segments of high user interest, with a segment length of 1 MB. The specifics of the video content are illustrated in Fig. 8 We set the traffic sending rate for 1080P video at 1 Mbps and 4K video at 2 Mbps. The total maximum bandwidth available on the NR (s2–s5, s3–s5, and s4–s5) is 4 Mbps. However, the bandwidth demand at this point is 5 Mbps which means at any given moment, the QoS requirements of at least one session cannot be met, leading to video buffering or other issues.

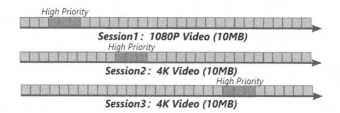

Fig. 8. Video Content Information for Three Sessions.

4.2   Experimental Result

We simultaneously initiated three sessions, and Fig. 9 illustrates the variations in UE-side receiving rates over time for these sessions, with segments of user-interest video content marked with red circles (indicating high priority). In Session 1, video stream transmission is consistently allocated to DRB1, precisely meeting its bandwidth requirements. However, for Session 2 and Session 3, since the total required bandwidth exceeds what is available, it results in a situation where their transmission rate requirements cannot be simultaneously satisfied. Without fine-grained QoS control, users might experience buffering or video packet loss, ultimately diminishing their experience.

Nevertheless, through our framework and the traffic granularity introduced by service tags, a single QoS flow can be further divided into multiple QoS flows. For example, here, we separated user-interest video segments from regular video content into two distinct flows. This enhanced flexibility allows us to dynamically allocate the appropriate DRBs and prioritize meeting the bit rate requirements of high-priority user-interest traffic. Consequently, this significantly elevates the user experience.

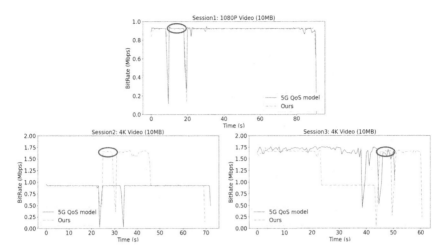

Fig. 9. QoS Performance of three sessions. (Color figure online)

5   Conclusion

In this paper, we have further refined the previously proposed cognitive-driven core network architecture to realize an application-oriented, universal, fine-grained QoS cognitive framework. This framework leverages service tags to expose the QoS requirements of application traffic, thereby providing more precise and high-quality services. In particular, we have discussed in detail the design of internal processes, as well as the definition and implementation of service tags within the framework. Through experiments consisting of both the core network and RAN sides, we have demonstrated that this framework can enhance the user experience. We believe that the application of this framework will play a vital role in future 5G networks, offering improved support and services for various applications.

References

1. 3GPP: System architecture for the 5G system (5Gs); stage 2 (release 17): TS 23.501 (2021)
2. 3GPP: Procedures for the 5G system; stage 2: TS 23.502 (2019)
3. Rommer, S., Hedman, P., Olsson, M., Frid, L., Sultana, S., Mulligan, C.: 5G Core Networks: Powering Digitalization. Academic Press (2019)
4. Zhang, Z.L., Dayalan, U.K., Ramadan, E., Salo, T.J.: Towards a software-defined, fine-grained QoS framework for 5G and beyond networks. In: Proceedings of the ACM SIGCOMM 2021 Workshop on Network-Application Integration, pp. 7–13 (2021)
5. Li, Y., Huang, J., Sun, Q., Sun, T., Wang, S.: Cognitive service architecture for 6G core network. IEEE Trans. Ind. Inform. **17**(10), 7193–7203 (2021). https://doi.org/10.1109/TII.2021.3063697

6. Babiarz, J., Chan, K., Baker, F.: RFC 4594: configuration guidelines for DiffServ service classes (2006)
7. 3GPP: TS 23.203 v17.2.0, technical specification group services and system aspects, policy and charging control architecture (2022)
8. Dayalan, U.K., Fezeu, R.A.K., Salo, T.J., Zhang, Z.L.: Prototyping a fine-grained QoS framework for 5G and NextG networks using POWDER. In: 2022 18th International Conference on Distributed Computing in Sensor Systems (DCOSS), pp. 416–419 (2022). https://doi.org/10.1109/DCOSS54816.2022.00075
9. Luo, X., Chen, H.H., Guo, Q.: Semantic communications: overview, open issues, and future research directions. IEEE Wirel. Commun. **29**(1), 210–219 (2022)
10. Nightingale, J., Salva-Garcia, P., Calero, J.M.A., Wang, Q.: 5G-QoE: QoE modelling for ultra-HD video streaming in 5G networks. IEEE Trans. Broadcast. **64**(2), 621–634 (2018). https://doi.org/10.1109/TBC.2018.2816786
11. Zhang, X., Ou, Y., Sen, S., Jiang, J.: SENSEI: aligning video streaming quality with dynamic user sensitivity. In: 18th USENIX Symposium on Networked Systems Design and Implementation (NSDI 2021), pp. 303–320 (2021)
12. Mao, H., Netravali, R., Alizadeh, M.: Neural adaptive video streaming with pensieve. In: Proceedings of the Conference of the ACM Special Interest Group on Data Communication, pp. 197–210 (2017)
13. Arunruangsirilert, K., Wei, B., Song, H., Katto, J.: Pensieve 5G: implementation of RL-based ABR algorithm for UHD 4K/8K content delivery on commercial 5G SA/NR-DC network. In: 2023 IEEE Wireless Communications and Networking Conference (WCNC), pp. 1–6 (2023). https://doi.org/10.1109/WCNC55385.2023.10118834
14. El-Maghraby, R.T., Abd Elazim, N.M., Bahaa-Eldin, A.M.: A survey on deep packet inspection. In: 2017 12th International Conference on Computer Engineering and Systems (ICCES), pp. 188–197. IEEE (2017)
15. Fransson, P., Jonsson, A.: End-to-end measurements on performance penalties of IPv4 options. In: IEEE Global Telecommunications Conference, 2004. GLOBECOM 2004, vol. 3, pp. 1441–1447. IEEE (2004)
16. Jones, P., Dhesikan, S., Jennings, C., Druta, D.: RFC 8837 differentiated services code point (DSCP) packet markings for WebRTC QoS (2021)
17. Welzl, M., Islam, S., Barik, R., Gjessing, S., Elmokashfi, A.: Investigating the delay impact of the DiffServ Code Point (DSCP). In: 2019 International Conference on Computing, Networking and Communications (ICNC), pp. 612–616. IEEE (2019)
18. Nichols, K., Blake, S., Baker, F., Black, D.: RFC2474: Definition of the differentiated services field (DS field) in the IPv4 and IPv6 headers (1998)
19. free5GC: an open-source 5G core network (2023). https://www.free5gc.org
20. mininet: an open-source network emulator (2023). https://mininet.org/
21. Güngör, A.: UERANSIM: open source 5G UE and RAN (gNodeB) implementation (2023). https://github.com/aligungr/UERANSIM
22. Avallone, S., Guadagno, S., Emma, D., Pescape, A., Ventre, G.: D-ITG distributed internet traffic generator. In: First International Conference on the Quantitative Evaluation of Systems 2004, QEST 2004. Proceedings, pp. 316–317. IEEE (2004)

Research on UUV Rudder Angle Control Method Based on Sliding Mode Control

Lin Li[1], Fengyun Li[1(✉)], Zihao Zhan[1], Kaiyu Li[1], Shuang Huang[1], and Haike Yang[2]

[1] Wuhan Second Ship Design and Research Institute, Wuhan 430025, China
lfynevermore@126.com
[2] Wuhan University of Technology, Wuhan 430070, China

Abstract. It is difficult to control the rudder Angle of UUV due to its nonlinear characteristics, dynamic matching, and many interference terms. With the increase of the number and type of uncertainty items, the corresponding system interference of traditional PID controller increases sharply, which limits its application in engineering. In this paper, the rudder Angle control method based on sliding mode control is adopted. By establishing the UUV mathematical model and combining with the backstepping method, the control efficiency can be improved while having strong robustness. The simulation results verify the effectiveness of the proposed control method.

Keywords: UUV · Sliding mode control · Rudder Angle controller

1 Introduction

The difficulty of UUV rudder Angle control mainly lies in the control of head and tail parallel steering, which requires the dynamic matching of head and tail rudder Angles [1, 2]. In order to achieve the minimum Angle of attack control, it is necessary to ensure that the vector sum of the UUV's trim torque is approximately 0 in the steady state, and the disturbance can be compensated in time. However, due to the errors in the UUV mathematical model, the nonlinear fluid resistance on the rudder surface, and the difficulty in accurately obtaining the dynamic proportional relationship between the head and tail rudder effect, it is difficult to obtain good control results by using conventional PID control methods, especially under disturbance conditions [3].

For the uncertainty of UUV rudder Angle system and the external disturbance of the system, the sliding mode variable structure control makes the system have strong robustness and anti-disturbance ability through the switching function in the control quantity [4]. In the application research of UUV, the sliding mode variable structure control combined with other control algorithms has been applied in the control of UUV rudder Angle holding, dynamic positioning and so on. For example, sliding mode variable structure control is combined with fuzzy control, neural network control and other control algorithms to estimate the upper bound of the uncertainty or the coefficient of the switching function [5].

L. Pan et al. (Eds.): BIC-TA 2023, CCIS 2062, pp. 357–367, 2024.
https://doi.org/10.1007/978-981-97-2275-4_28

Therefore, considering the uncertain terms such as the load disturbance of the ocean current has a direct impact on the performance of the UUV motion system, reducing the sensitivity of the system to the uncertain terms, and requiring strong robustness of the controller. In this paper, the control technology of rudder Angle holding mode is introduced to design a UUV rudder Angle controller based on sliding mode control. The final control law is obtained by stepwise recursive backstepping method [6], which has strong robustness to system uncertainties [7, 8].

2 Mathematical Model of UUV Control System

2.1 Load Disturbance Mathematical Model

When UUV is sailing underwater, the load is disturbed by the external environment. Therefore, it is necessary to model the main interference factors, starting from the physical nature of the interference term, and through certain simplification, it is more suitable for engineering applications.

(1) Ocean current interference [9]

The disturbance force of the ocean current is a main external force encountered by the UUV when sailing underwater. The ocean current has a certain speed, which makes the speed of the UUV relative to the inertial coordinate system and the speed relative to the current vary, so that the force and torque of the UUV change. Under the influence of current, the speed of UUV is:

$$\begin{cases} u = u_r + V_c \cos(\psi_c - \psi) \\ v = v_r + V_c \sin(\psi_c - \psi) \end{cases} \tag{1}$$

where, u_r, v_r is the projection of the relative velocity of the UUV motion on the water flow, V_c is the flow velocity, and ψ_c is the direction of flow. In the case of incomplete experimental data, a constant steering Angle can be used to simulate the constant interference of the current in engineering applications, and zero mean white noise can be used to simulate the random interference of the sea breeze and the current.

(2) The steering Angle model [10]

The steering torque generated by the UUV thruster is nonlinear coupled with the steering Angle of the thruster. If the Norribin nonlinear model is directly used, the accuracy of the UUV heading mathematical model is poor. Therefore, the slewing characteristics should also be considered. The slewing motion control of the thruster is similar to the steering servo system. The actual slewing Angle cannot reach the given Angle immediately, so it can be expressed as a first-order inertial link:

$$T_E \dot{\delta} + \delta = K_E \delta_E \tag{2}$$

After transformation, we can obtain that,

$$\dot{\delta} = \frac{K_E \delta_E - \delta}{T_E} \tag{3}$$

where, δ represents the actual rotation Angle; δ_E is the expected rotation Angle; K_E is the control gain, usually set to 1; And T_E is the time constant, usually 2.5–3 s.

When the thruster performs the rotary motion, it is limited by the frequency, and its rotary speed is also limited to a certain extent, so its rotary speed has the saturation characteristic of: $|\dot{\delta}| \leq 4.5°/s$, and for the general heading control, the limit condition of the rotary Angle is $|\delta| \leq 30°$.

2.2 The Mathematical Model of UUV Rudder Angle is Maintained

2.2.1 UUV Heading Motion Model

The steering motion of UUV is to purposefully adjust the rudder Angle of UUV according to the actual needs of UUV in the course of navigation. In the control of UUV rudder Angle, the speed and heading control of UUV is the most basic. In this paper, the rudder Angle of UUV is controlled by adjusting the steering Angle. The interaction between the hull and the water during the UUV sailing is affected by the external uncertain interference, which makes the system behave as a time-varying, nonlinear, and large disturbance dynamic process [11].

In this paper, based on the mathematical model of UUV steering motion, the line-motion equation of UUV is obtained. The established mathematical model of UUV steering motion is decoupled, and the state variables related to the course of UUV steering motion model are extracted, and the irrelevant state variables are set as uncertainties. Because of the nonlinear coupling of steering torque and steering Angle, the mathematical model of steering Angle is added, and the UUV heading motion equation is obtained as shown in the formula:

$$\begin{cases} \dot{\psi} = r \\ \dot{r} = [x_p(1 - t_p)T\sin\delta + N_H + N_E]/(I_Z + J_Z) \\ \dot{\delta} = (-\delta + \delta_E)/T_E \end{cases} \rightarrow \begin{cases} \dot{\psi} = r \\ \dot{r} = \theta_2 f(\delta) + \varphi_2 r + d_N \\ \dot{\delta} = \theta_3 \delta + \varphi_3 \delta_E \end{cases} \quad (4)$$

where, T is a time-varying uncertain quantity; N_H is the hydrodynamic torque, including the time-varying uncertainty related to r and the time-varying uncertainty unrelated to r; N_H is the external interference of wind and wave currents, and is the uncertain quantity, which can be written as d_N.

2.2.2 The Rudder Angle Maintains the Performance Index

The UUV should try to maintain the given heading during the course to reduce the change of heading. However, UUV will inevitably be disturbed by the external environment such as wind, wave and current. If the heading accuracy requirement is set high, UUV will frequently change the steering Angle, which will cause the wear of mechanical equipment and increase the loss of fuel. Therefore, in the UUV heading control system, it is necessary to find an optimal balance index between heading maintenance and steering Angle rotation frequency [12].

Course keeping requires that the UUV has certain robustness in its own parameter changes, has strong adaptability to the external environment interference, and can eliminate the constant value interference. In general sea conditions, the heading accuracy is required to be 0.5–1°, and in bad sea conditions, the heading accuracy is 1–3°.

3   Design and Principle of UUV Rudder Angle Controller

3.1   Sliding Mode Control

The basic problem of sliding mode control is to determine the sliding switching surface and solve the control law [13].

$$\dot{x} = f(x, u, t), \quad x \in R^n, \quad u \in R^m, \quad t \in R \tag{5}$$

The switching function needs to be determined.

$$s(x), \quad s \in R^m \tag{6}$$

Solving the control function,

$$u_i(x) = \begin{cases} u_i^+(x) s_i(x) > 0 \\ u_i^-(x) s_i(x) > 0 \end{cases} \tag{7}$$

where, $u^+(x) \neq u^-(x)$ and satisfies the following three conditions,

(1) Existence of sliding modes;
(2) All the state variables outside the switching surface $s(x) = 0$ can reach the switching surface in finite time.
(3) The stability of sliding mode motion is guaranteed.

When the system reaches the sliding surface, for the nonlinear system, the sliding mode motion equation of the system at this time should be satisfied as follows.

$$\begin{cases} s(x) = 0 \\ \frac{\partial s}{\partial x} f(x, u, t) = 0 \end{cases} \tag{8}$$

In the case that the actual motion of the system is the same as the mathematical model of the system and there is no external disturbance, the equivalent control quantity u_{eq} can be obtained according to the derivative of the switching surface in the equation is zero. The actual control quantity is the sum of the equivalent control and switching control:

$$u = u_{eq} + u_c \tag{9}$$

Among them, the switching control of u_c realizes the robust control of uncertainty and external interference in the system.

3.2   Basic Principle of Backstepping Method

Sliding mode variable structure control requires the controlled system to have a standard regular form, or a regular form after a certain transformation. However, most systems do not have this form, so the design of sliding mode controller for high-order nonlinear systems becomes more complex. And for nonlinear systems with parameter uncertainties and structural uncertainties, it is particularly difficult to transform into a regular form [14].

The inverse method uses a step-by-step recursive calculation method, introduces the concept of virtual controller, and deduces the design method of the whole system controller based on the Lyapunov stability principle. Since the mathematical model of UUV heading motion is of third order, a third order nonlinear system is taken as an example to explain the principle of backstepping method.

Let the third-order nonlinear system be:

$$\begin{cases} x_1 = \theta_1 f_1(x_1) + \varphi_1 g_1(x_1)x_2 \\ x_2 = \theta_2 f_2(x_1, x_2) + \varphi_2 g_2(x_1, x_2)x_3 \\ x_3 = \theta_3 f_3(x_1, x_2, x_3) + \varphi_3 g_3(x_1, x_2, x_3)u \\ y = x_1 \end{cases} \tag{10}$$

where, x_i is the system state variable, θ_i, φ_i is the system parameter, f_i, g_i is a smooth function and is not zero, and u is the control variable. Then, define the function

$$\begin{cases} z_1 = x_1 - \alpha_1 \\ z_2 = x_2 - \alpha_2 \\ z_3 = x_3 - \alpha_3 \end{cases} \tag{11}$$

where, $\alpha_1 = y_d$, α_2, α_3 is the virtual control quantity.

Then, due to Lyapunov stability principle, the final control quantity is,

$$u = \frac{1}{\varphi_3 g_3}(-k_3 z_3 - \theta_3 f_3 - \varphi_2 g_2 z_2 + \dot{\alpha}_3) \tag{12}$$

According to the Lyapunov stability principle, the system is asymptotically stable, and the controller can be designed according to the backstepping method mentioned above when each parameter of the system is determined. The actual controller designed according to this method has u coupled with each virtual controller, and all of them contain parameters θ, φ.

3.3 Design of Sliding Mode Controller for UUV Rudder Angle

3.3.1 Selection of Nonlinear Functions

The UUV heading motion system not only contains uncertain parameters, but also has uncertain structure, and the upper and lower bounds of the uncertain structure are known. Sliding mode control is the control method for this situation. However, the control law of sliding mode control contains sgn function, which is a discontinuous function. However, in the design step of backstepping method, the derivative needs to be controlled virtually, and the calculation of its derivative becomes complicated by directly using the sgn function in sliding mode control. Therefore, nonlinear derivable function is selected instead of discontinuous function.

Consider the following heterobounded model with uncertain structure,

$$\dot{x} = f(x) + g(x)u + \Delta \tag{13}$$

Let the sliding mode surface be

$$s = (x - x_d) \tag{14}$$

And assume that

$$|\Delta| \leq d \tag{15}$$

According to sliding mode control theory, its control law is,

$$u = \frac{1}{g(x)}\left[-f(x) - ks + x_d - d\mathrm{sgn}(s)\right] \tag{16}$$

However, this control law will cause serious chattering phenomenon, and its derivation is cumbersome, and it is difficult to realize the backstepping principle to obtain the control law of the high-order system. Therefore, it is necessary to construct a continuous derivable function h(s) instead of sgn function.

By Lyapunov function, h(s) satisfies:

$$sh(s) \geq \frac{\Delta s - \sigma}{d} \geq |s| - \frac{\sigma}{d} \tag{17}$$

3.3.2 Specific Design of Sliding Mode Controller

The UUV rudder Angle control model includes uncertain parameters and uncertain structure. The sliding mode controller provides a systematic method to maintain stability under the condition of modeling inaccuracy. However, the UUV course motion mathematical model can't be transformed into a regular form, so the sliding mode controller is designed by using the idea of backstepping.

$$\begin{cases} \dot{\psi} = r \\ \dot{r} = (T_0 + \Delta T)\sin\delta + N_{r0}r + \Delta N_r \\ \dot{\delta} = -\frac{\delta}{T_E} + \frac{\delta_E}{T_E} \end{cases} \tag{18}$$

Among them,

$$T_0 = x_p(1 - t_p)T(n_0, u_0, 0)/(I_Z + J_Z) \tag{19}$$

The initial value of thrust is a fixed value that is not zero;

$$\Delta T = x_p(1 - t_p)T(n_0, u_0 + \Delta u, v)/(I_Z + J_Z) \tag{20}$$

The change value of thrust along its own longitudinal direction during the course of UUV course change is an uncertain value.

$$N_{r0} = N_r/(I_Z + J_Z) \tag{21}$$

The hydrodynamic coefficient associated with r is a non-zero fixed value.

$$\Delta N_r = (N_v v + N_{r|r|}r|r| + N_{v|v|}v|v| + N_{vvr}v^2r + N_{vrr}vr^2 + N_E)/(I_Z + J_Z) \tag{22}$$

The nonlinearity associated with r is an uncertain quantity, considering the external disturbance.

According to the design method of backstepping method, when the parameters associated with the next state variables are uncertain, the corresponding virtual control law cannot be calculated. Therefore, the linearization method is adopted in this section, except for the equation of the highest order, the parameter uncertainties in the other order equations are attributed to the uncertain structure to facilitate the calculation of the control law. After the uncertain parameters and structures are processed accordingly, the state transformation is performed.

$$\begin{cases} x_1 = \psi \\ x_2 = r \\ x_3 = N_{r0}r + T_0 \sin \delta \end{cases} \tag{23}$$

Then its state space expression is

$$\begin{cases} \dot{x}_1 = x_2 \\ \dot{x}_2 = x_3 + \Delta_2 \\ \dot{x}_3 = N_{r0}\dot{r} + T_0(\sin \delta)' = (f_0 + \Delta f) + g_0 u \end{cases} \tag{24}$$

where

$$\Delta_2 = \Delta N_r + \Delta T \sin \delta \tag{25}$$

$$f_0 = -T \cos \delta \cdot \frac{\delta}{T_E} + N_{r0}{}^2 r + N_{r0}T_0 \sin \delta \tag{26}$$

$$\Delta f = N_{r0} \cdot \Delta N_r + N_{r0}\Delta T \sin \delta, g_0 = T_0 \cos \delta / T_E \tag{27}$$

Using a combination of backstepping and sliding mode, define a function,

$$\begin{cases} s_1 = x_1 - \alpha_1 \\ s_2 = x_2 - \alpha_2 \\ s_3 = x_3 - \alpha_3 \end{cases} \tag{28}$$

where $\alpha_1 = \psi_d$. α_2, α_3 is the virtual control quantity.

Using the stability principle of Lyapunov function, the equivalent control can be obtained.

$$u_{eq} = -\frac{(f_0 - \dot{\alpha}_3)}{g_0} \tag{29}$$

For systems with uncertainties, switching control should be added.

$$u_{sm} = -\frac{D_3}{g_0} \frac{s_3}{\sqrt{s_3{}^2 + \sigma}}, |\Delta \alpha| \leq D_3 \tag{30}$$

$$u_3 = u_{eq} + u_{sm} \tag{31}$$

The system is guaranteed to be asymptotically stable.

According to the derivative of Lyapunov function, u_3 is modified, and the final control law is:

$$u = -\frac{1}{g_0}\left((f_0 - \dot{\alpha}_3) + D_3\frac{s_3}{\sqrt{s_3{}^2 + \sigma}} + s_2 sgn(s_2 s_3)\right) \tag{32}$$

4   Simulation and Analysis of UUV Rudder Angle Keeping

To simulate the designed controller, it is necessary to select the appropriate parameters to make it reach the sliding mode surface and reduce the chattering as much as possible. For the estimation of the maximum value of the uncertainty term, the maximum value of the uncertainty value in the state without the controller is used. In the simulation process, the appropriate parameters are adjusted according to the simulation results.

According to the formula in Sect. 2 above, the calculation results of dimensionless hydrodynamic coefficients are obtained as follows (Table 1).

Table 1. Calculation results of dimensionless hydrodynamic coefficient

X	Y	N
m = 0.327	Ydv = −0.1916	Izz = 0.0204
Xdu = −0.0218	Yv = −0.3786	Ndr = −0.0099
Xuu = −0.0431	Yr = 0.1107	Nv = −0.1042
Xvv = −0.0268	Yvv = −3.4795	Nr = 0.0449
Xvr = −0.0563	Yrr = −0.0140	Nvv = −0.0013
Xrr = 0.0058	Yvrr = −0.4158	Nrr = −0.0124
	Yvvr = −0.4078	Nvrr = −0.028
		Nvvr = −0.1293

According to the simulation test and operation standards, namely Federal Development and Execution Process (FEDEP) and Test and Training Enabling System (TENA), the UUV rotation speed should not be less than 100 r/min, and the simulation time should not be less than 40 s.

Therefore, in this simulation, the thrust of UUV at constant speed = 120 r/min is selected, the above hydrodynamic coefficient calculation results are selected, and the other control parameters are selected as follows: $k_1 = 2.4, k_2 = k_3 = 0.5, D_2 = 0.45, D_3 = 5, \sigma = 0.01$. In the case of no interference, the simulation results are shown in Figs. 1 and 2.

It can be seen from the simulation results that, given a constant speed of 120 r/min, the UUV heading responds within 5 s, increases at a speed of 0.36°/s from 0 to 50 s, slows down from 50 s to 75 s, and stabilizes at 20° at 75 s. Then the heading does not change, and

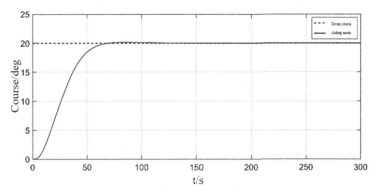

Fig. 1. UUV heading change curve.

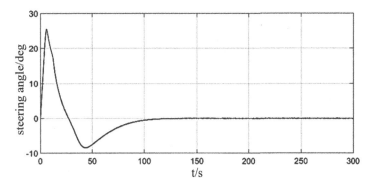

Fig. 2. Change curve of steering Angle.

continues to stabilize at 20°. The steering Angle responds within 1 s at a given constant speed of 120 r/min, reaches a peak of 25° at 8 s at a speed of 3.125°/s, then decreases irregularly at a speed of about 0.825°/s, reaches 0 at 25 s, continues to decrease, and then reaches a valley value at 48 s, that is −8°/s. The corresponding heading slowdown tends to 20°, which means that when the steering Angle drops to the lowest value, the heading corresponding acceleration decreases. Between 48 s–100 s, the steering Angle gradually returns to 0, and the corresponding heading does not change, which means that the system tends to be stable. We can see that the sliding mode controller can achieve stable rudder Angle within 100 s under a given constant speed.

The actual course of UUV can well maintain the desired course, and the overshoot is small, and the response time is short. The steering Angle has small chattering and the UUV roll Angle is small, which is conducive to the stable operation of the UUV. In the case of no disturbance, the sliding mode controller makes the UUV have a good rudder Angle retention performance.

When the control parameters are unchanged, the disturbance of wind and wave current is added, and the simulation results as shown in the figure are obtained (Figs. 3 and 4):

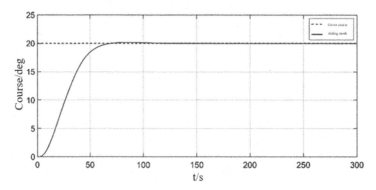

Fig. 3. UUV heading change curve under disturbance.

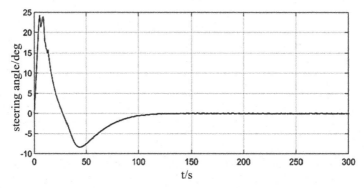

Fig. 4. Change curve of steering Angle under interference.

It can be seen from the simulation results that compared with the above situation without interference, the UUV heading change curve remains quantitatively unchanged, and the steering Angle is about to reach the peak but still at 24°. At this time, the corresponding time is 8 s, the chattering phenomenon occurs, and the peak value of 25° is not reached directly, and then in the descent process, the steering Angle is 15°. Then in the descent process, at the steering Angle of 15°, the slight chattering occurs at 23 s. In the case of adding disturbance, the steering Angle failed to reach the peak value, and the chattering phenomenon appeared twice.

In the case of interference, the actual course of UUV can still track the desired course well, with smaller overshoot and shorter response time. After reaching the given heading, the UUV can maintain stable operation in this heading, and has good heading retention performance. The steering Angle has no reverse rotation for many times, and has a small chattering phenomenon. Because the maximum value of external interference is taken into account in the parameter selection process, the stability of UUV rudder Angle will not be affected by adding interference without changing the parameters.

In summary, it shows that under appropriate control parameters, the designed sliding mode controller has good control ability for the nonlinear UUV rudder Angle keeping system with uncertainty. When the uncertain parameters and uncertain structure change, that is, when the internal parameters of the UUV heading system change and are disturbed by wind and wave currents, the sliding mode controller has good control ability for the nonlinear UUV rudder Angle keeping system with uncertainty. When the control parameters are unchanged, the proposed controller still has good performance of rudder Angle retention and strong robustness.

5   Conclusion

In this paper, a rudder Angle keeping controller based on sliding mode control is designed for the nonlinear UUV heading motion model. A sliding mode heading controller based on reverse step method is designed for UUV heading motion mathematical model with uncertain terms. By using the design idea of backward step method, the control law can be calculated by the sliding mode control in the high order system without regularization. A method is proposed to replace the original switching function with continuous differentiable function for the discontinuous term in the virtual control law. The simulation results show that the designed sliding mode controller has good rudder Angle retention performance without changing the control parameters, and the Angle and buffeting are small, so it has strong robustness.

References

1. Yan, Z.P., Li, Z., Chen, T., Zhao, Y.F., Du, P.J.: UUV course fault-tolerant control based on GA-BP neural network. Chin. J. Sens. Technol. **9**, 1236–1242 (2013)
2. Duan, Y.S., Gong, S.H.: Research on the design of underactuated UUV head-to-tail parallel steering controller. Ship Sci. Technol. **41**(5), 65–69 (2019)
3. Zhao, X., Gong, S.H., Yang, J.: Interference compensation control for unmanned underwater vehicles based on LMI. J. Unmanned Underwater Syst. **28**(3), 271–277 (2020)
4. Ma, Y.: Rudder angle control strategy of full rotary propeller based on angular velocity and acceleration correction. Tianjin Navigation **2**, 12–16 (2015)
5. Tuo, G.J.: Research on Multi-Model Switching Control Method of Unmanned Underwater Vehicle. Harbin Engineering University (2015)
6. Jiang, Y.B., Guo, C., Yu, H.M.: Depth control of underwater unmanned vehicles based on reverse sliding mode algorithm. Ship Eng. **40**(2), 83–87 (2018)
7. Li, Q.M.: Research on Multi-UUV formation Control with time Delay. Harbin Engineering University (2020)
8. Pei, H.R.: Research on compliant guidance and Robust Control Method for UUV retracting unmanned craft. Harbin Engineering University (2022)
9. Hou, S.P., Bai, R., Yan, Z.P., Mou, C.H.: Path tracking of multi-UUV formation under ocean current interference. Shipbuilding China **54**(4), 126–136 (2013)
10. Xu, H.: Research on Energy Saving Path Tracking Control Technology of X-rudder UUV. Harbin Engineering University (2021)
11. Xu, H.X., Hu, C., Yu, W.Z., Yao, G.Q.: Buoyancy regulated UUV depth control with input constraints. Syst. Eng. Electron. **44**(11), 3496–3504 (2022)
12. Zhang, W., Teng, Y.B., Wei, S.L., Hu, S.Y., Zhang, J.N.: Underactuated UUV adaptive RBF neural network backward step tracking control. J. Harbin Eng. Univ. **39**(1), 93–99 (2018)
13. Zhao, H.T., Zhu, D.Q.: UUV Underwater model prediction sliding mode tracking control algorithm. Control. Eng. **29**(7), 1195–1203 (2022)
14. Yan, Z.P., Yu, H.M., Lo, B.Y., Zhou, J.J.: Underactuated UUV terrain tracking control based on integral sliding mode. J. Harbin Eng. Univ. **37**(5), 701–706 (2016)

Research on Distributed Control Technology of Ship Pod Propeller Turning Small Angle

Ningjun Xu[1,2], Zhangsong Shi[1], and Xu Lyu[3](✉)

[1] College of Weapons Engineering, Naval University of Engineering, Wuhan 430033, China
[2] Jiangsu Automation Research Institute, Lianyungang 222062, China
[3] Department of Precision Instrument, Tsinghua University, Beijing 100084, China
lvclay@163.com

Abstract. At present, most of the control systems of the full swing ship adopt the classical PI control or open loop transfer method, which has low control accuracy and slow response speed. In this paper, the idea of intelligent distributed control is introduced into the control of the propulsion device of the full swing ship pod, and the physical model and mathematical model of the hydraulic propulsion system of the pod are established. The PID control strategy based on fuzzy reasoning is designed for the full rotary propulsion device, which can accurately and reliably control the rotation Angle of the pod. Finally, the modeling and simulation of the full rotary hydraulic propulsion system of the pod are carried out under small Angle conditions. The results show that the rotation Angle error is small, and the proposed method has application value.

Keywords: fuzzy reasoning · distributed PID · ship control

1 Introduction

The ship full swing pod propulsion device is one of the hot points of ship equipment today. Because of its high flexibility, high propulsion efficiency and low maintenance cost, it is widely used in various types of tugs, engineering ships and special ships [1]. However, the control of ship full gyro has not been deeply discussed, and its nonlinear characteristics also determine that the open-loop control can't achieve its control accuracy [2]. Therefore, it is of great practical significance to study the control of ship full gyro push pod inlet device.

At present, most of the full sleeving ship control systems use the classical PI link, and some manufacturers also use the open-loop transmission method [3, 4]. The result is that the control accuracy is not high, the response speed is not fast, and so on [5].

Compared with the traditional control method, the intelligent control method has many advantages. Taking the common fuzzy control as an example, the linguistic logic control mode based on the operator's experience in the control process or similar reasoning is used. Compared with the classical PID algorithm, the fuzzy control needs to establish the mathematical model of the controlled object, and has certain adaptability to the nonlinear, time-varying and large inertia of the controlled object [6]. And it has the characteristics of fast response, good stability and good robustness. In many fields, fuzzy control has been successfully applied. Fuzzy-pid combines the two technologies, and uses fuzzy logic to query and adjust the parameters of PID controller online. It can meet the changes of the parameters of the control system, effectively control the output of the controlled object, shorten the response time and improve the control characteristics [7].

In addition, there are still some shortcomings in the electric propulsion of ship pod thruster, such as low propulsion efficiency, poor overload protection ability from thermal and system, low safety and reliability. Compared with electric propulsion, hydraulic propulsion has good system overload protection ability, high safety and reliability. Applying the hydraulic propulsion technology to the pod thruster, the pod-type hydraulic propulsion system is formed, which combines the maneuvering flexibility of the pod thruster with the safety and overload protection ability of the hydraulic propulsion, and can make up for some shortcomings of the current electric thrusters to a certain extent.

In summary, this paper will use the hydraulic drive mode, design a PID control strategy based on fuzzy control to realize the system control, and set up in the small Angle condition to simulate the system to verify its performance.

2 Control System Model

2.1 Overall System Design

The hydraulic system of the full slewing pod designed in this paper is composed of the hydraulic system according to the performance requirements of the classification society for the slewing system, that is, when the ship is advancing at the fastest speed, the time from $35°$ on one side to $35°$ on the other side of the pod can not exceed 28 s (small Angle response). According to the design requirements, it is set that the pod can deflute at a speed of $12°$/s (rotation speed of 2 rpm) under the normal operation of the hydraulic pump of the pod rotating mechanism. The following figure is the schematic diagram of the designed pod rotating hydraulic drive system. The slewing system mainly includes the following parts: the main circuit of the hydraulic system, the oil replying circuit, the flushing circuit, the overload protection circuit, and the variable pump servo control system (Fig. 1).

The servo valve of the hydraulic control system can be approximately regarded as an oscillating link when the natural frequency of the system execution element is high (>50 Hz), that is, when the dynamic characteristics of the system have a large impact, and its transfer function can be described as follows.

$$\frac{Q_{out}(s)}{Q_{in}(s)} = \frac{1}{\frac{s^2}{w^2} + \frac{2\zeta_{sv}}{w_{sv}}s + 1} \tag{1}$$

where, $Q_{out}(s)$ is the output flow rate (m^3/s); $Q_{in}(s)$ is the output flow (m^3/s); ζ_{sv} is the damping ratio of the valve dimensionless; And w_{sv} is the natural frequency of the valve (rad/s).

The valve port flow equation is as follows.

$$Q_L = Q_{out}(s) - K_c p_L(s) \tag{2}$$

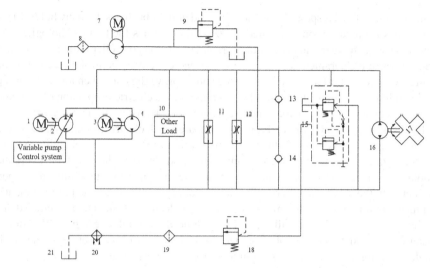

Fig. 1. The overall design of the hydraulic propulsion system of the full slew pod. 1, 3, 7- Motors; 2-bidirectional variable pump; 4-bidirectional quantitative pump; 5-variable pump control mechanism; 6- filling pump; 8, 19- filter; 9, 18- Overflow valve; 10- Other loads; 11, 12- pilot relief valve; 13, 14- check valve; 15- Flush valve; 16- bidirectional quantitative motor; 17- propeller; 20- cooler; 21- Fuel tank

where, Q_L is the motor load flow (m³/s); $Q_{out}(s)$ is the output flow (m³/s); K_c is the valve port flow pressure coefficient (m³/(s·Pa)); p_L is the load pressure (Pa).

The load flow equation of hydraulic motor can be expressed as follows.

$$Q_L = Q_1 + Q_2 + Q_3 \tag{3}$$

where, Q_1 is the flow required for motor rotation (m³/s); Q_2 is motor internal and external leakage loss flow (m³/s);

Additional flow (m³/s) Q_3 is generated for oil compression. According to this idea, its load equation can be expressed as

$$Q_L(s) = D_m\theta(s) \cdot s + C_{tm}p_L + \frac{V_t}{4\beta_e}p_L \tag{4}$$

where, D_m is the motor displacement in radians (m³/rad); θ is motor angular displacement (rad); C_{tm} is the total leakage coefficient of motor (m³/(s·Pa)); V_t is the total volume of motor, valve cavity and connecting pipe (m³); β_e is the effective volume elastic modulus of oil (Pa).

Ignoring the nonlinear factors of the load and the quality factor of the oil, the force balance equation of the motor and the load and its Laplace transformation can be expressed as

$$\begin{cases} D_m p_L = J_t\ddot{\theta} + B_m\dot{\theta} + G\theta + T_L \\ D_m p_L(s) = J_t\theta(s) \cdot s^2 + B_m\theta(s) \cdot s + G\theta(s) + T_L(s) \end{cases} \tag{5}$$

where, J_t is the moment of inertia of the motor and its bearing (kg·m2); B_m is the viscous damping coefficient (N·m/(rad/s)); G is the stiffness of the load spring (N·m/rad); And T_L is the load moment (N·m).

According to the analysis of (1)–(5), the overall transfer function of the hydraulic motor power can be obtained as

$$\theta(s) = \frac{\frac{Q_{out}(s)}{D_m} - \frac{K_{ce}}{D_m^2}\left(1 + \frac{V_t}{4\beta_e K_{ce}}s\right)T_L(s)}{\frac{J_t V_t}{4\beta_e D_m^2}s^3 + \left(\frac{J_t K_{ce}}{D_m^2} + \frac{B_m V_t}{4\beta_e D_m^2}\right)s^2 + \left(1 + \frac{B_m K_{ce}}{D_m^2} + \frac{GV_t}{4\beta_e D_m^2}\right)s + \frac{GK_{ce}}{D_m^2}} \tag{6}$$

At this time, it can be considered that the motor and the load are rigidly connected, so $G = 0$, and in general, $\frac{B_m K_{ce}}{D_m^2} \ll 1$, the load $T_L(s)$ does not change with time, and the hydraulic flow rate is constant due to the relief valve of the servo hydraulic pump, so $Q_{out}(s)$ remains constant. In summary, the above equation can be simplified as

$$\theta(s) = K_R \frac{Ts + 1}{s\left(\frac{s^2}{w_h^2} + \frac{2\xi_h}{w_h}s + 1\right)} \tag{7}$$

where, K_R is the hydraulic motor gain, dimensionless, $K_R = \frac{Q_{out}D_m - T_L K_{ce}}{D_m^2}$; T is the hydraulic motor frequency (rad/s), $T = \frac{T_L V_t}{4\beta_e(Q_{out}D_m - T_L K_{ce})}$; w_h is undamped hydraulic natural frequency (rad/s), $w_h = \sqrt{\frac{4\beta_e D_m^2}{J_t V_t}}$; ξ_h is hydraulic damping ratio, dimensionless, $\xi_h = \frac{K_{ce}}{D_m}\sqrt{\frac{J_t \beta_e}{V_t}} + \frac{B_m}{4D_m}\sqrt{\frac{V_t}{J_t \beta_e}}$.

2.2 System Transfer Function

If the rod reducer is regarded as a rigid joint and all viscous resistance and friction are ignored, it can be regarded as a proportional link with gain K; And ignoring the total viscous damping coefficient mB of the motor shaft, the block diagram of the transfer function of the hydraulic propulsion control system of the full rotation pod can be seen in Fig. 2 below.

On this basis, it can be considered that the natural frequency of the electro-hydraulic three-way four-way valve is far greater than that of the hydraulic motor, so the servo valve can only behave as a through-through action, that is, the transfer function of the three-position four-way valve is $\frac{Q_{out}(s)}{Q_{in}(s)} = 1$, and the hydraulic frequency $T = \frac{T_L V_t}{4\beta_e(Q_{out}D_m - T_L K_{ce})} \gg 1$ of the system, so the system can be simplified as

$$G(s) = K \frac{s}{s\left(\frac{s^2}{\omega_h^2} + \frac{2\xi_h}{\omega_h}s + 1\right)} \tag{8}$$

where, K is the system gain, dimensionless, $K = \frac{T_L V_t}{4\beta_e D_m^2}$; ω_h is undamped hydraulic natural frequency (rad/s), $\omega_h = \sqrt{\frac{4\beta_e D_m^2}{J_t V_t}}$; ξ_h is hydraulic damping ratio, dimensionless, $\xi_h = \frac{K_{ce}}{D_m}\sqrt{\frac{J_t \beta_e}{V_t}} + \frac{B_m}{4D_m}\sqrt{\frac{V_t}{J_t \beta_e}}$.

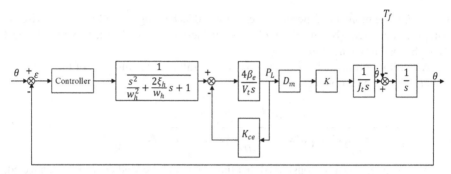

Fig. 2. Transfer function block diagram of hydraulic propulsion control system of full slewing pod.

3   Research on Control Algorithm

3.1   Principle of PID Control

PID control, as one of the most common control methods in the production process, has been widely used in industry. It is the given data and feedback data produced by the deviation, proportional (P), integral (I), and differential (D) linear combination, constitute a new control quantity, the controlled object is controlled, so called PID controller [8]. Since the digital computer was introduced into the control field, the control system composed of the traditional regulator is replaced by the numerical calculation of the computer, which can not only facilitate the use of software PID control, but also make use of the logic function of the computer, so that the PID control is more flexible [9]. The following Fig. 3 is the principle block diagram of PID control.

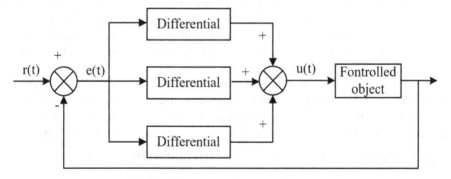

Fig. 3. PID control schematic diagram.

The given value $r(t)$ and the actual output value y (t) constitute the control deviation:

$$e(t) = r(t) - y(t) \tag{9}$$

Through linear combination of deviation $e(t)$ and proportion (P), differential (I), integral (D), the deviation control quantity is formed to control the controlled object, and its control law is as follows:

$$u(t) = k_p e(t) + k_i \int_0^t e(t)dt + k_d \frac{de(t)}{dt} \tag{10}$$

where, k_p is the proportional coefficient, k_i is the integral coefficient and k_d is the differential coefficient. If is in the frequency domain, its transfer function is considered as.

$$G(s) = \frac{U(s)}{E(s)} = k_p \left(1 + \frac{1}{T_i s} + T_d s \right) \tag{11}$$

where, k_p is the scaling coefficient, T_i is the integral time constant, and T_d is the differential time constant.

Since the system is fixed, the three parameters k_p, k_d, k_i are always unchanged, the static error of the system will not be adjustable, and the large inertia link and nonlinearity of the system are not fundamentally solved, so people put forward the fuzzy control theory into PID algorithm, forming the fuzzy PID algorithm.

3.2 The Principle of Fuzzy Control

The basic idea of fuzzy control is based on some control experience of human beings, using the way of computer to achieve, and the human experience is expressed by fuzzy human language to express the control rules. The core of Fuzzy control is Fuzzy Controller (FC) [10].

A large number of practices have proved that, compared with the classical PID algorithm, fuzzy control has faster response speed, smaller overshoot, strong robustness to the change of process parameters, and all nodes can be effectively controlled [11]. Especially in some difficult to establish mathematical model or with a large delay of the control process, the effect of fuzzy controller is obvious.

3.3 Design of PID Control System Based on Fuzzy Inference

During the control process, the characteristic parameters or structure of the controlled object often change with the influence of load changes or various interference factors, which increases the difficulty of control. The fuzzy controller improves the control strategy of the classical PID controller that the constant is used as the parameter, and puts forward the method of changing the PID controller parameters online according to the actual feedback signal, so as to improve the response and expand the scope of application.

In this paper, the fuzzy PID controller takes error and error change rate as input, and meets the requirements of PID parameters at different times. EeceecUsing fuzzy control rules to modify PID parameters online, a fuzzy PID controller is constructed. Its structure is shown in Fig. 4.

The PID controller of fuzzy logic is expressed as

$$\begin{cases} k_p(k) = k_p(k-1) + \gamma_p(k)\Delta k_p \\ k_i(k) = k_i(k-1) + \gamma_i(k)\Delta k_i \\ k_d(k) = k_d(k-1) + \gamma_d(k)\Delta k_d \end{cases} \tag{12}$$

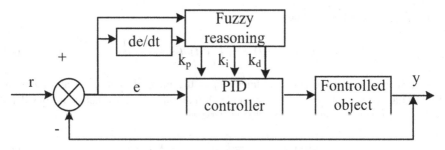

Fig. 4. PID control principle block diagram.

where, $\gamma_p \gamma_i \gamma_d$ is the correction speed parameter, the more times of correction, the smaller the parameter will be. It can be seen from the above equation that the parameters of the controller in the next step are obtained by weighting the parameters of the controller in this step with the parameters obtained from the fuzzy reasoning generated in this step. By taking the classical formula of PID into, we can obtain:

$$u(k) = k_p(k)e(k) + k_i(k) \sum_{j=0}^{k} e(j) + k_d(k)(e(k) - e(k-1)) \tag{13}$$

Since it is relatively difficult to solve when $\sum_{j=0}^{k} e(j)$ as an accumulation term, the state variable $x(k) = \sum_{j=0}^{k} e(j)$ is introduced, and has:

$$x(k+1) = x(k) + e(k) \tag{14}$$

Therefore, Eq. 13 can be rewritten as

$$u(k) = k_p(k)e(k) + k_i(k)x(k) + k_d(k)(e(k) - e(k-1)) \tag{15}$$

Analyzing the control system designed above, it can be seen that there is 1 state variable and 1 output variable. Considering the change of coefficient k_p, k_i, k_d, the number of outputs is chosen to be 4 and the input signal is $e(k), e(k-1)$. In this way, the basic content of fuzzy PID controller can be expressed. After that, the fuzzy PID control module can be constructed and encapsulated.

According to the relevant experience of fuzzy PID, the parameters change table can be constructed and fuzzy reasoning can be carried out in MATLAB environment. The system input is $e(k), ec(k)$, and the output is $\Delta k_p, \Delta k_i, \Delta k_d$. According to the control rule table, the fuzzy system can be established in MATLAB, and the system has the characteristics of 2 input and 3 output.

4   System Simulation

4.1   Overall System Simulation Model

According to the above modeling analysis, the simulation model of the hydraulic propulsion control system of the full slewing pod will be built through the Simulink toolbox in MATLAB. Figure 5 below shows the overall simulation model of the system.

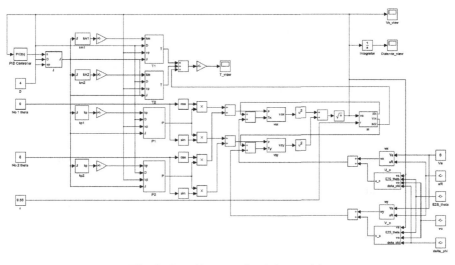

Fig. 5. Overall system simulation model.

4.2 Simulation Parameter Setting

Table 1. Simulation parameter Settings.

Heading level	Example	Font size and style
Prime mover speed	1750	r/min
Variable hydraulic pump speed	1750	r/min
Variable hydraulic pump displacement per revolution	500	ml/r
Ration hydraulic pump speed	1750	r/min
Quantitative hydraulic pump displacement per revolution	500	ml/r
Hydraulic motor displacement per revolution	320	ml/r
Pinion radius	150	mm
Large gear radius	900	mm
Load torque	19098	N·m
Relief valve set pressure	40	Mpa
Hydraulic motor rated torque	889.97	N·m
Drive efficiency of reducer	0.9	-
Maximum slewing load torque	84852.60	N·m
Slewing mechanism gear ratio	8	-

The simulation parameters of this paper are set according to the data in Table 1.

Since the rotation speed of the pod designed in this paper is $N=12°/s = 2$ r/min, Then the simulation setting of hydraulic motor speed n_m is:

$$n_m = i_h i_G N \tag{16}$$

where, i_G is the reduction ratio of the reducer, as follows:

$$i_G = \frac{M_m}{2T_m \eta_G} \tag{17}$$

where, T_m is the rated torque of hydraulic motor (N·m); η_G is the transmission efficiency of reducer; And M_m is the maximum value of slewing torque; i_h is the slewing mechanism transmission ratio.

The simulation parameters set by Table 1 are substituted into the corresponding data:

$$n_m = i_h i_G N = i_h \frac{M_m}{2T_m \eta_G} N = 8 \times \frac{84852.60}{2 \times 889.97 \times 0.9} \times 2 = 847.52 r/min \tag{18}$$

It can be seen that the simulation parameters can design the hydraulic motor speed as 847.52 r/min to meet the design requirements of the rotation speed 2 rpm.

4.3 Simulation of Slewing Motion Under Small Angle Condition

Based on the above simulation model, a linear model with two input and multiple output under small Angle condition is designed. The input control variables are command rudder Angle and command speed, and the output variables are six variables reflecting the ship motion state, including longitudinal velocity, lateral velocity, yaw angular velocity, lateral displacement, longitudinal displacement and heading Angle.

By using the simulation model, the propulsion motor model and the thrust and torque of the pod propeller are relatively independent and complete, so they are modularly packaged in the implementation. The initial value of the state variables and the basic parameters of the mother ship are set in the function, the hull hydrodynamic force/torque and the pod thrust vector are calculated respectively, and the resultant force/torque is obtained. Finally, the state variables are solved according to the pod propulsion ship steering motion equations.

In order to verify the correctness of the performance and control strategy of the full slewing hydraulic propulsion control system, the simulation model of the traditional propulsion and the pod hydraulic propulsion simulation model introduced into the control system in this study are simulated by setting the parameters of Simulink, and the computer simulation program is written by MATLAB to realize the slewing control. The simulation environment was that the ship was sailing at a straight and uniform speed of 8 m/s without sea wind or wave interference, the propulsion motor speed was 125 r/min, and the maximum pod rotation Angle was 35° (small Angle condition).

It can be seen from the comparison of Fig. 6 and Fig. 7 that under the small Angle condition, the simulation results show that the steady rotating diameter of the traditional propulsion ship is 788 m, and the steady rotating diameter of the ship using the full rotating pod hydraulic propulsion control system is 575 m. Through comparison, it can be seen that the ship using the hydraulic propulsion control system of the full swing pod has far better gyroscopic characteristics than the traditional propulsion ship, and all of them are less than the MO (International Maritime Organization) standard value, indicating that the control strategy designed in this study can correctly control the rotation of the pod propeller under the small Angle condition, and the effect is better than the traditional propulsion device.

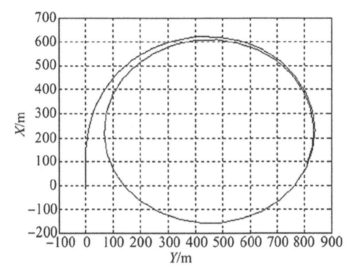

Fig. 6. Simulation results of traditional propulsion rotation under small Angle condition.

4.4 Simulation of Control Command Response Under Small Angle Condition

According to the response effect of the hydraulic propulsion control system of the full slewing pod to the control instructions under the condition of small Angle, the following simulation is made. The load-control system modeling method is adopted. After compiling the overall simulation model of the hydraulic propulsion control system of the full slewing pod, in the Simulink environment, with the help of Matlab's powerful data calculation and processing capabilities. The load model of the slewing mechanism of the pod is established. Through the slewing control system, the speed can be controlled to make the actual azimuth Angle close to the command azimuth Angle, and the Angle deviation value is set to 0.2°. When the actual azimuth Angle and the command azimuth Angle error is within the set range, the main loop of the system is locked to complete the rotation Angle command.

In the small Angle condition, the azimuth Angle change range is set to 35°~−35°, the simulation time is 40 s, the 35° Angle command is given at 0 s, and the −35° Angle command is given at 20 s.

The figure shows the comparison between the actual azimuth Angle and the command azimuth Angle change curve.

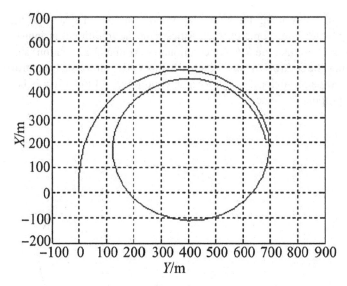

Fig. 7. Simulation results of hydraulic propulsion control system of full slewing pod under small Angle condition.

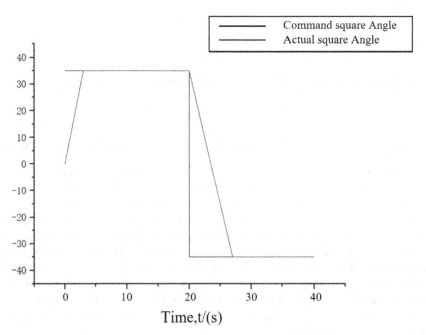

Fig. 8. Change curve of actual and instructed azimuth Angle of pod under small Angle condition (35°– −35°).

As can be seen from Fig. 8, the pod is given an azimuth Angle command of $35°$ at 0 s, and the pod completes the rotation Angle command from $0°$ to $35°$ at t = 3 s (the pod's rotation speed is set to $12°/s$). After the system adjustment and the commanded azimuth Angle error are within the allowable range, the system remains stable, and the rotation system is locked. At 20 s, the pod's azimuth Angle command of $-35°$ is given, and the actual azimuth Angle reaches $-35°$ at 27 s. After the system adjustment and the instructed azimuth Angle are within the allowable range of error, the slewing system is locked.

Analysis of the above simulation results: the system takes 3 s to complete the 0–$35°$ Angle conversion, which is represented as the rotation Angle $\Delta\theta_1 = 35°$, the rotation time $t_1 = 3\,s$, and the rotation angular speed $N_1 = \frac{\Delta\theta_1}{t_1} = \frac{35°}{3s} = 11.67°/s = 1.94\,r/min$;After the $-35°$ Angle instruction is given, the system takes 7 s to complete the Angle conversion from $35°$ to $-35°$, which is expressed as $\Delta\theta_2 = 70°$, Angle time $t_2 = 7\,s$, and angular velocity $N_2 = \frac{\Delta\theta_2}{t_2} = \frac{70°}{7s} = 10°/s = 1.67\,r/min$. The average value of the two calculations is $N = \frac{N_1+N_2}{2} = \frac{1.94+1.67}{2} = 1.8\,r/min$.

In summary, this result satisfies the simulation technical standard that the rudder can turn the rudder from any side to another side at $35°$ in no more than 28 s. And according to the simulation data calculation, in the hydraulic propulsion control system of the full slew pod designed in this study, the average slew speed of the pod thruster is 1.8 rpm, which is 0.2 rpm different from the 2 rpm specified in the design standard, and the error is 10%, within 20%, in line with the standard.

5   Conclusion

Based on the analysis and full understanding of the podded propulsion system, this paper chooses the full slewing podded propulsion system for design, and the power module chooses hydraulic drive. Then the system model is fully discussed, and the physical model and mathematical model of the hydraulic propulsion system of the full slew pod are established. On the basis of analyzing the theory of PID control algorithm and fuzzy control algorithm, an intelligent algorithm of PID control system based on fuzzy reasoning is introduced. The appropriate simulation parameters are selected according to the design indexes.

Finally, the simulation of rotary motion under small Angle condition is carried out to verify the correctness and superiority of the control strategy designed in this paper. The response of the control command to the rotation Angle is simulated under the small Angle condition. After the numerical calculation of the simulation results, the rotation Angle reaches 1.8 rpm, which is in line with the standard of 2 rpm in the design standard, and the error is 10%.

References

1. Cao, F.Y.: Research on Key Technologies of counter-rotating propeller Full rotating pod Propeller. Tongji University (2019)
2. Fu, Z.B.: Research on ship turning control based on simulink. Ship Sci. Technol. **2**, 31–33 (2017)

3. Shang, L.B., Wang, W., Liu, Z.H.: Optimization of working area of full rotary propeller for dynamic positioning ship. Chinese Ship Res. **15**(2), 104–110 (2020)
4. Xu, Y.Y.: Research on rudder Servo Controller of Full Rotary Thruster based on DSP. Wuhan University of Technology (2014)
5. Wen, H.B., Liu, Z.Z., Chen, N.: Research on vibration and noise prediction and control of a full-turning tug. J. Ship Mech. **22**(10), 1292–1299 (2018)
6. Li, H.L., Yang, R.N., Li, Q.N., Han, H.Y.: Consistency of arbitrary order linear multi-agent systems based on distributed PID Control. Control Decis. **32**(5), 899–905 (2017)
7. Yuan, P., Xu, C.F., Zhou, J., Yang, Z.B.: Research on improved anti-interference fuzzy PID for AGV control system. Mach. Des. Manuf. **385**(3), 212–220 (2023)
8. Ting, S., Cheng, L., Wang, X.G.: Distributed anti-windup NN-sliding mode formation control of multi-ships with minimum cost. ISA Trans. **138**, 49–62 (2023)
9. Zhang, G.Q., Yao, M.Q., Zhang, W.J., Zhang, W.D.: Event-triggered distributed adaptive cooperative control for multiple dynamic positioning ships with actuator faults. Ocean Eng. **242**, 110124 (2021)
10. Jeon, H., Kim, J.: Application of reference voltage control method of the generator using a neural network in variable speed synchronous generation system of DC distribution for ships. J. Mar. Sci. Eng. **8**(10), 802 (2020)
11. Jin, Z., Meng, L., Guerrero, J.M., Han, R.: Hierarchical control design for a shipboard power system with DC distribution and energy storage aboard future more-electric ships. IEEE Trans. Industr. Inf. **14**(2), 703–714 (2017)

Distributed Intelligence Analysis Architecture for 6G Core Network

Wen Sun$^{(\boxtimes)}$ and QiBo Sun

State Key Laboratory of Networking and Switching Technology,
Beijing University of Posts and Telecommunications, Beijing 100876, China
{wensun,qbsun}@bupt.edu.cn

Abstract. To achieve automation and intelligence in 5G networks, the 3rd Generation Partnership Project (3GPP) introduced the Network Data Analysis Function (NWDAF) as a novel network function. However, in the traditional 5G core network architecture, NWDAF relies on fixed configurations for data collection, lacking support for user customization and flexibility. Additionally, the current deployment of NWDAF is predominantly centralized, failing to provide real-time and reliable analysis services for the massive data in future 6G systems. Moreover, it is incapable of ensuring user privacy, making it incompatible with emerging scenarios like federated learning in 6G. Therefore, this paper proposes a user-customizable data collection approach and introduces a distributed NWDAF deployment based on the Raft algorithm, where the master node assigns data collection, analysis, and inference tasks to multiple worker NWDAFs. Our work and experimental results demonstrate that the proposed architecture effectively addresses these challenges and further achieves closed-loop network automation in 6G systems.

Keywords: Core network · Network data analysis · Raft algorithm

1 Introduction

To address the requirements of automation and intelligence in core networks, the Third Generation Partnership Project (3GPP) has introduced a new fifth-generation (5G) system architecture [1]. Unlike the fourth-generation (4G) Long Term Evolution (LTE) system, 5G defines a range of modular network functions (NFs) as depicted in Fig. 1, enabling flexible network management and operation. This flexibility is crucial for supporting the diverse needs of new 5G services. However, as the number of connected devices rapidly increases and 5G encompasses diverse applications, manual network management and operations become increasingly challenging [2]. In response to this context, 3GPP has defined a new NF called the Network Data Analytics Function (NWDAF). NWDAF collects data from other NFs and Operations, Administration, and Maintenance (OAM) systems, and utilizes machine learning (ML) and artificial intelligence (AI) techniques to generate analytics. These analytics can be applied to decision-making and automating the network. For example, NWDAF can obtain the status data of NF instances and their resource utilization from the Network Repository Function (NRF). By analyzing this data, NWDAF can build NF load analysis and provide it to NF instances. Based on these analysis results, the capacity of NF instances can be automatically adjusted, either increased or decreased [3].

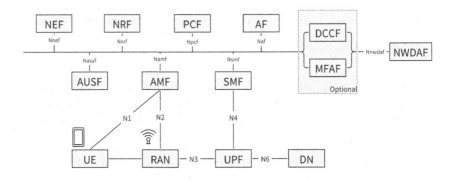

Fig. 1. 5G network function architecture

NWDAF performs four main functions: data collection, model storage and training, model inference, and providing specific analysis for network functions. Based on the analysis results generated by NWDAF, such as path selection analysis in the user plane, network functions can make decisions and generate new data to be reported back to NWDAF. The entire process forms a closed-loop of network automation and cognition.

However, the current data collection process of NWDAF lacks flexibility, thus unable to provide customized services to users. Additionally, the predominantly centralized deployment of NWDAF fails to ensure data privacy and is not suitable for scenarios such as federated learning and distributed artificial intelligence. Moreover, the centralized deployment lacks resilience capabilities. Therefore, optimizing the implementation of each NWDAF functionality and providing a universal distributed deployment approach for future 6G networks would be beneficial in delivering higher-quality services to internal network functions and users in scenarios involving massive data. It would also cater to various emerging business requirements and better realize the overall goal of network automation.

1.1 Shortage of Prior Architecture

NWDAF currently plays a crucial role in providing analysis of various network parameters, including slice load levels, service-related data, NF load, network performance, and user equipment (UE) information. The entire process can be divided into three key parts: data collection, model training and inference, and analysis result subscription and notification.

However, the current NWDAF proposed by 3GPP relies on the EventExposure mechanism for data collection. The entire process is shown in Fig. 2. Firstly, NWDAF needs to retrieve user consent information from UDM (User Data Management). If the corresponding data permissions are granted by the user, NWDAF will subscribe to event exposure from network functions. When the corresponding events are triggered, data collection takes place. Once the data collection is completed, subscriptions to events from the target network functions and UDM are terminated. The entire data collection process is initiated proactively by NWDAF based on configuration. As a result, the flexibility of the data collection scope is relatively limited. Furthermore, due to the exclusive initiation of the entire process by NWDAF, users are unable to customize their data set preferences and permission scopes. This inherent limitation creates challenges in offering data analysis services to third-party AFs, such as game servers.

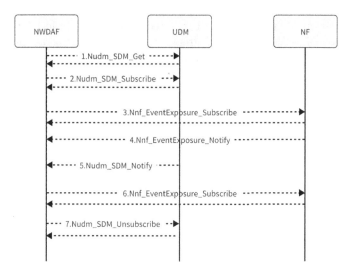

Fig. 2. Data collection based on event exposure [3]

Simultaneously, while 3GPP has proposed the possibility of deploying NWDAF in either a centralized or distributed manner, the centralized deployment of NWDAF presents several challenges. Processing all service requests and model training tasks within a single NWDAF instance can be time-consuming, leading to potential issues with service stability, data security, and privacy. Furthermore, there is a lack of widely adopted and standardized approaches for implementing distributed NWDAF to address various emerging businesses and scenarios. Therefore, this is a crucial problem that needs to be solved.

1.2   Technical Challenges

In response to the existing issues with NWDAF, we will optimize the data collection process and design a universal distributed NWDAF deployment approach to achieve network automation more effectively.

Challenge of Optimizing the Data Collection Process. Due to the lack of flexibility and customization in the current configuration-based data collection process, a new data collection approach needs to be designed to address the issues. However, achieving efficient collaboration between NWDAF and other network functions and addressing data permissions and privacy concerns are crucial considerations that require careful balancing.

Challenge of Propose the Universal Distributed NWDAF Architecture. There are numerous distributed algorithms available, and the selection and optimization of appropriate distributed algorithms, tailored to the specific scenarios of NWDAF, are of great importance. This strategic approach will ensure the provision of reliable services by NWDAF, effectively addressing critical challenges related to fault tolerance and scalability within the distributed architecture.

1.3   Proposed Approach

Drawing inspiration from the remarkable octopus, which possesses a central brain and eight highly developed arms, each with its intricate neural network to perceive the world, this paper introduces a deployment approach for distributed NWDAF and introduces a new data collection method to achieve octopus-like cognitive capabilities.

First, we propose a process where network functions actively report data and subscribe to analysis to address the limited scalability of configuration-based data collection and the inability to instantly perceive critical business data beyond the subscription scope. Additionally, We propose a distributed deployment approach for NWDAF based on the Raft consensus algorithm to enhance the service quality provided by NWDAF and achieve fault tolerance and scalability. Furthermore, we also optimize the architecture to reduce network signaling interaction and achieve more real-time network decision-making, enabling an octopus-like distributed cognitive decision-making capability.

1.4   Summary of Simulations Result

Based on the YAML documents provided by 3GPP, we implemented a simple version of centralized NWDAF using Golang. Additionally, we implemented our proposed distributed NWDAF deployment based on the Raft algorithm. Besides, we simulated the process of data collection and model training using the Lumos5G dataset and LSTM network.

The experimental results show that our distributed deployment approach significantly improves performance compared to the previous centralized NWDAF deployment. Our implementation is able to make decisions more promptly (%29 faster) and recover services more quickly (%21 faster) in the event of node failures.

2   Related Work

This section will be divided into three parts to introduce the related work: research on data collection process, research on distributed NWDAF and research on consensus algorithms.

Research on Data Collection Process. The research on the data collection part of NWDAF mainly originates from the 3GPP standard TS23.288, in which it is specified that when NWDAF collects data, it relies on the event exposure mechanism of 5GC, specifically that NWDAF subscribes to a set of event IDs (marking a certain type of event information generated at the network function, such as establishing a connection, node access, etc.), and when the event is triggered, NF will report relevant data to NWDAF [3,4], but the biggest problem of such a data collection process is that the network is not flexible enough, and how to determine the granularity of data collection is also a problem. If the granularity is too low, the transmission pressure will be too high, and if the granularity is too high, the response will not be timely, and there is a lack of awareness of abnormal states outside the subscription range. Regarding the implementation of data collection in the core network, Ahmad F proposed a new monitoring framework based on cloud-native principles for designing data collection in 5G networks [5]. On the other hand, Li proposed an implementation approach for data collection from the hardware perspective of the 5G Core (5GC) [6]. However, these proposals have limited reference value for the data collection implementation of NWDAF in an SDN architecture.

Research on Distributed NWDAF. At present, there is some work to explore the deployment of NWDAF in the 5G+ and 6G core network, P. Li presents a solution for native intelligence in 6G networks based on NWDAF [7], however, the paper only considers what functionality the distributed NWDAF needs to implement to support intrinsic intelligence, without proposing specific methods for implementing distributed deployment. Lu Yu introduces an intelligent communication network framework that combines 5G Network Data Analysis Function (NWDAF) and federated learning to solve data isolation and privacy protection [8]. However, he only proposed the functionalities of each node to realize federated learning based on distributed deployment of NWDAF, without addressing how to achieve distributed task collaboration. C. Chou further proposed an encryption method to address data privacy concerns and enable federated learning in the core network [9]. However, the introduction of this encryption method will lead to additional resource consumption. Youbin Jeon of Korea University proposed a tree-like structure to relieve the pressure on the central NWDAF node, and at the same time, in further work, experimental verification was carried out, but in the implementation, only the Anlf function (model inference) was delivered to the leaf node NWDAF, and the Mtlf (model training) function was still in the central NWDAF [2,10]. However, this implementation only improves the analysis response speed but does not solve the problem of serious NWDAF load on the central NWDAF during model training. Abdelkader from Bell Labs proposed that NWDAF be bound to each NF to ensure data privacy and that the central NWDAF should implement aggregation to achieve federated learning [11], but the problem of simple binding a specific type of NF with one NWDAF is that there will still be uneven load and still have task state loss for different NWDAFs. Mohammad Arif Hossain proposed a master-slave architecture using NWDAF in the 5G+ core network architecture in his paper. In this architecture, the master node assigns similar tasks to multiple nodes of the same type and waits for the completion of model training. The master node then aggregates the results. However, this approach does not consider node failures, and there can be differences in load among the network functions. The uneven task assignment can result in degraded overall system performance [12].

Research on Consensus Algorithms. The selection of a consensus algorithm for distributed deployment of NWDAF is a crucial problem that needs to be addressed to achieve high availability and scalability. Besides, maintaining data consistency between master and slave nodes for fault detection and recovery also needs to be considered. Currently, research on consensus algorithms in distributed systems is mainly based on Paxos, Raft, PBFT, PoW, and PoS, with PBFT, Pow, and PoW being primarily used in blockchain applications. The Paxos consensus algorithm was initially proposed to address the consensus problem in distributed systems [13]. However, due to its complexity and difficulty in understanding and implementation, the Raft algorithm designed to be easier to understand and implement was later introduced. The Raft algorithm achieves fault-tolerance distributed consensus through log replication and leader election process [14]. Paxos and Raft have commonly used consensus algorithms in mainstream distributed systems, and they both support CFT (Crash Fault Tolerance). However, Raft, being an algorithm that is easier to understand and implement, has gained more extensive application than Paxos. On the other hand, PBFT not only supports Crash Fault Tolerance but also considers the presence of malicious nodes, resulting in significant performance differences compared to Raft [15]. PoW and PoS algorithms impose high computational and communication overhead on nodes, and they do not provide support for data consistency [16]. They only determine the nodes' right to record transactions. These algorithms are not suitable for distributed consensus in NWDAF due to large resource consumption and significant differences in the internal core network scenario.

3   Intelligence Analysis Architecture

First, in this section, we will present the diagram (referred to as Fig. 3) illustrating our proposed distributed deployment approach for NWDAF based on the Raft algorithm. Subsequently, we will provide detailed explanations regarding the data collection process and the specific implementation of the Raft algorithm.

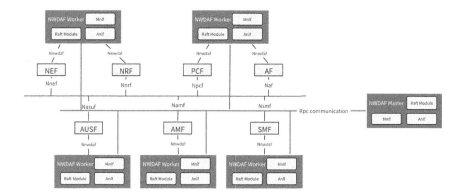

Fig. 3. Proposed architecture

3.1   Custom Data Collection Scheme Defined by Service Consumers

Because the data collection process of the core network depends on configuration, which fails to provide users with customizable data collection services and has limited awareness of events outside the configured scope. We propose a user-customized data collection approach and a data collection method based on local models at NF for anomaly detection to better establish a closed loop of network automation.

Custom Data Collection Scheme. Especially for the data collection of third-party AF, because there is a gap between the data format of AF and the core network status data, the conventional subscription-based approach may not efficiently and accurately retrieve the required data from the AF's internal business and status data. Therefore, we propose that the responsibility for initiating analysis subscription requests be shifted to the third-party AFs themselves. These requests can be tailored to the specific analysis requirements, and the data content and format can be agreed upon in advance. This proactive approach shown in Fig. 4 ensures a more effective and precise data acquisition process.

1. The consumer network functions of the data analytics service subscribe to the format of data collection and information about the analysis models from NWDAF.

2. Based on the Raft Module, task assignment logs are replicated and task assignments are carried out between the master and slave nodes of NWDAF. The entire process will be described in detail in the next subsection.

3. NWDAF workers determine the format of data collection and other related information and inform the service consumers that the services will be provided by the local node.

4–6. NWDAF provides services to consumer NFs.

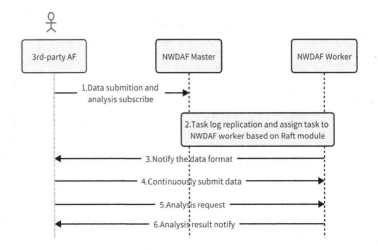

Fig. 4. Custom data collection by the AFs

Data Collection Based on Local Model. In particular, in scenarios where there are significant changes in the NF status and anomalies need to be detected, such as a sudden surge in load on the User Plane Function (UPF), responsible for routing and forwarding, having the model and inference module, namely Anlf, within the UPF, allows for direct detection and reporting of anomalies to NWDAF. Simultaneously, NWDAF can train the model (model update), generate the decision results, and notify the Service Management Function (SMF) to avoid using the UPF with the current load surge as a path node. Additionally, if necessary, traffic migration can be performed to alleviate the pressure on this UPF. The process is shown in Fig. 5. This deployment approach ensures timely decision-making for abnormal events, reporting recognized abnormal information to NWDAF for further decision-making, thereby achieving a networked automatic cognitive closed-loop system resembling an octopus.

Overall, adopting this deployment method enables efficient decision-making for abnormal events, facilitating information exchange between NFs and NWDAF, and achieving a closed-loop cognitive network architecture.

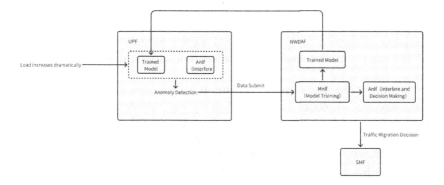

Fig. 5. UPF's automatic cognition closed-loop realization

3.2   Implementation of Distributed Deployment for NWDAF

In our proposed architecture shown in Fig. 3, we introduce a hierarchical structure for NWDAF nodes, consisting of worker nodes and a master node. This division allows for efficient coordination and management of data collection and analysis tasks through the utilization of a consensus algorithm.

The master node assumes the critical responsibility of task allocation within the NWDAF system. By leveraging its centralized authority, the master node effectively assigns tasks to worker nodes based on their capabilities and availability. This ensures optimal utilization of resources and promotes efficient task execution.

Furthermore, the Raft consensus algorithm employed in our architecture facilitates the synchronization and coordination of data collection and analysis tasks across the worker nodes. This algorithm ensures that all nodes reach a consensus on the status and progress of tasks, enabling seamless and reliable task coordination throughout the NWDAF system.

Log Replication and Task Assignment Process. In our proposed Architecture, we have also implemented a well-defined and efficient signaling process based on the Raft algorithm shown in Fig. 6 to ensure a seamless and reliable data collection and analysis task. Let's break down the steps involved:

1. Task Assignment Broadcast: The master NWDAF takes the lead by broadcasting task assignment logs to all worker nodes. These logs contain crucial information such as task identifiers and the corresponding network functions (NF) assignments. This ensures that all worker nodes are aware of their assigned tasks and the associated NFs.

2. Worker Node Response: Upon receiving the task assignment, each worker node promptly writes its own log and responds to the master node. This step confirms the worker node's acknowledgement of the assigned task and its readiness to proceed.

3. Consensus Confirmation: The master node awaits responses from more than half of the worker nodes. Once it receives these responses, it indicates that the majority of nodes have recognized the task allocation information. At this point, the master node broadcasts an allocation success message to all worker nodes.

4. Committing Task Information: With the allocation success message received, the worker nodes are instructed by the master node to commit the task information to their respective logs. This ensures that the task details are securely recorded and readily available for further processing and analysis.

5. Task Assignment to Target Worker Node: Finally, the master node assigns a specific task to the intended target worker node. It is worth noting that when a particular network function establishes a one-to-one relationship with a specific worker NWDAF, subsequent tasks are more likely to be assigned to that NWDAF. This assignment approach offers several advantages in terms of optimizing data transmission time, ensuring data privacy, and improving the overall performance of the NWDAF systems.

This refined signaling process reflects a well-coordinated and robust framework for data collection and analysis tasks. By implementing clear communication channels, consensus confirmation, and precise task assignment, our proposed approach enhances the efficiency, reliability, and scalability of the NWDAF architecture.

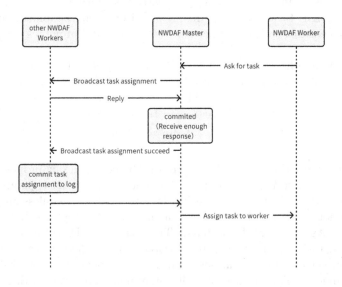

Fig. 6. Process of log replication and task assignment among nodes

Fault Tolerance for Master and Worker Node. Utilizing the Raft algorithm, we can achieve state detection of the slave nodes and provide fault tolerance by migrating tasks to new nodes when a slave node (service provider) fails. As shown in the diagram below, when a slave node is serving consumer network elements and suddenly fails, the system generates continuous log replication information (which degrades into heartbeat signals when no new logs are produced within the heartbeat period). This allows us to detect all the abnormal nodes in the distributed architecture and migrate their services to new nodes, thereby resuming service provision for the consumers. The fault tolerance process is shown in Fig. 7.

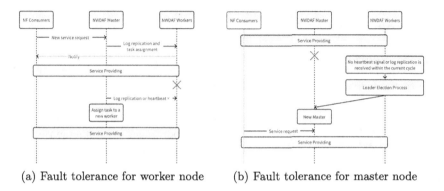

(a) Fault tolerance for worker node   (b) Fault tolerance for master node

Fig. 7. Fault tolerance process

Signal Reduction by Placing the Model Within the NF. It is important to highlight that the Network Data Analytics Function (NWDAF) encompasses various functions, including model training and inference. By directly placing the inference function within the service consumer NF of NWDAF and delivering the model to the NF, the delay involved in sending an analysis request to NWDAF and receiving the decision-making result can be eliminated. This enables more timely decision-making.

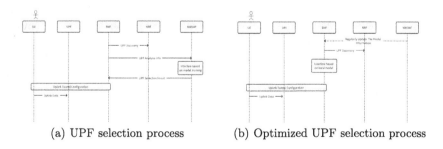

(a) UPF selection process   (b) Optimized UPF selection process

Fig. 8. Compared UPF selection process

In addition, based on the optimized distributed deployment approach of NWDAF proposed in this paper, we are also able to achieve a significant reduction in network signaling interaction for internal decision-making scenarios within the core network. For instance, in the session establishment process during the UPF selection phase, the SMF first requests UPF service discovery from the NRF, then requests UPF analysis results from NWDAF, and proceeds with decision-making, the process is shown in Fig. 8.a.

However, in our proposed architecture, the SMF itself possesses a model related to the UPF topology, enabling direct inference based on the Anlf and generating decision results. As a result, this process shown in Fig. 8.b reduces the signaling for analysis request transmission and response retrieval.

4   Experiment

4.1   Simulation Setup

Due to the lack of an open-source NWDAF project at present, this paper simulated the implementation of NWDAF as a control group based on Golang and according to the interface definitions provided by 3GPP. Meanwhile, the proposed NWDAF implementation in this paper was also implemented using Golang. RPC communication was employed internally within the NWDAFs, while HTTP communication was used for requests initiated by different network functions to NWDAF.

4.2   Data Collection Time

Based on the NWDAF data collection method we proposed, we simply carried out experimental verification to compare the transmission delay (the time taken from the sender NF to the NWDAF to receive the data), firstly, for the case of third-party AF, when the NWDAF does not describe the subscription event, the data transmission will be transmitted in the form of a complete set (assuming that the Lumos 5G dataset is transmitted and for the prediction of throughput, in fact, the UE location and other information have nothing to do with the model results). Therefore, we compare the time of transmission of the full set with the time of actively describing the time of sending and transmitting the data. In the data collection method we proposed, the data transmission takes 23.4 s (after filtering irrelevant data), while the full data transmission consumes 27.4 s (in full accordance with the NWDAF implementation method, when a single event is triggered, the full data is transmitted), and after further optimization of the data transmission, only 14.4 s are consumed, of course, the data transmission process can be further optimized (according to the size limit of a single transmission according to the relevant protocols).

4.3   Service Provision Time

In the experiment, we simulated requests and compared the performance differences of model training and analysis requests among the centralized single-node architecture in NWDAF, the tree-based NWDAF structure proposed by Youbin, and the Raft distributed architecture proposed in this paper. The simulation result is shown in Table 1.

Table 1. Service provision time

Architecture	Training time	Analytics provision time
Centralized NWDAF	61.7 s	Average 41 ms
Tree-based NWDAF	61.7 s	Average 41 ms
Our proposed architecture	17.4 s	Average 16 ms

4.4   Optimized Decision Making Process

Based on the further optimized NWDAF deployment approach proposed by us, we conducted experiments on the time cost required to make decisions in the user plane path selection process (UPF selection) within the core network, as depicted in Figs. 8 and 9. The experimental results are shown in Fig. 9. According to the experimental results, it can be demonstrated that deploying the model and Anlf module inside the network functions where decisions are made significantly reduces the time cost for decision-making and also reduces signaling interaction.

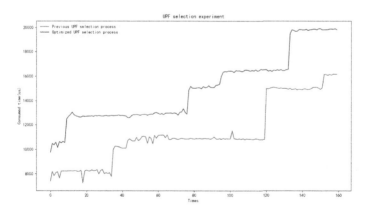

Fig. 9. Decision making time result

4.5   Fault Tolerance Experiment

First, we set the retry count for failed requests due to network reasons to 3 times (the actual count may vary in different scenarios). When the retry attempts fail, the network element needs to request service from the NWDAF master node, which then assigns a slave node to serve the network element. This process can be understood as a total of 5 signaling exchanges. However, with our proposed Raft-based distributed deployment approach, during the heartbeat detection process, if a node is found to be faulty, the relevant tasks can be directly handed over to a new worker node. This process involves log replication, as shown in Fig. 6, with a total of three signaling exchanges. The new service provider network element then sends a notification message to inform the service consumers that it will take over the service, requiring a total of 4 signaling exchanges. The conducted experiment result is shown in Fig. 10.

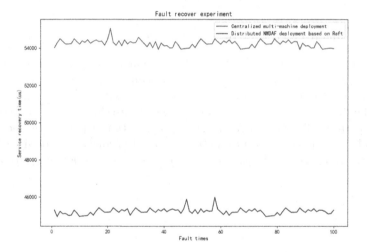

Fig. 10. Worker fault experiment

5   Conclusion

In this paper, we propose a distributed NWDAF architecture utilizing octopus cognitive capabilities. We introduce a user-customizable data collection approach to enhance the flexibility of data collection and analysis services. By leveraging the Raft consensus algorithm, we achieve distributed deployment of NWDAF, providing fault tolerance, scalability, and suitability for distributed federated learning scenarios. To validate the superiority of our proposed architecture, we establish a control group based on the YAML file provided by 3GPP and compare it against our experimental group using the proposed architecture. Evaluation results demonstrate that our implementation method enables scalable and faster data collection and analysis compared to traditional architectures, while also providing faster error recovery capabilities. In the future, we plan to further extend the application of this architecture, including incorporating AI algorithms for anomaly detection at the network element level and intelligent decision-making within the core network.

References

1. 3GPP: System architecture for the 5G System (5GS). Technical Specification (TS) 23.501, 3rd Generation Partnership Project (3GPP)
2. Jeon, Y., Jeong, H., Seo, S., Kim, T., Ko, H., Pack, S.: A distributed NWDAF architecture for federated learning in 5G. In: 2022 IEEE International Conference on Consumer Electronics (ICCE), pp. 1–2 (2022). https://doi.org/10.1109/ICCE53296.2022.9730220
3. 3GPP: Study of enablers for network automation for 5G. Technical Specification (TS) 23.791, 3rd Generation Partnership Project (3GPP)
4. 3GPP: Architecture enhancements for 5G system (5GS) to support network data analytics services. Technical Specification (TS) 23.288, 3rd Generation Partnership Project (3GPP)
5. Ahmad, F.: Data collection for Machine Learning in 5G Mobile Networks. Master's thesis, Oslomet-storbyuniversitetet (2023)
6. Li, J., et al.: 5GC network and MEC UPF data collection scheme research. In: 2021 International Conference on Information and Communication Technologies for Disaster Management (ICT-DM), pp. 80–85. IEEE (2021)
7. Li, P., Xing, Y., Li, W.: Distributed AI-native architecture for 6G networks. In: 2022 International Conference on Information Processing and Network Provisioning (ICIPNP), pp. 57–62. IEEE (2022)
8. Yu, L., Xin, L., Guo, M.: A safe architecture of 5G network intelligence based on federated learning and NWDAF. In: Proceedings of the 4th International Conference on Advanced Information Science and System, pp. 1–5 (2022)
9. Zhou, C., Ansari, N.: Securing federated learning enabled NWDAF architecture with partial homomorphic encryption. IEEE Network. Lett. (2023)
10. Jeon, Y., Pack, S.: Hierarchical network data analytics framework for B5G network automation: design and implementation. arXiv preprint arXiv:2309.16269 (2023)
11. Rajabzadeh, P., Outtagarts, A.: Federated learning for distributed NWDAF architecture. In: 2023 26th Conference on Innovation in Clouds, Internet and Networks and Workshops (ICIN), pp. 24–26. IEEE (2023)
12. Hossain, M.A., Hossain, A.R., Liu, W., Ansari, N., Kiani, A., Saboorian, T.: A distributed collaborative learning approach in 5G+ core networks. IEEE Netw., 1–8 (2023). https://doi.org/10.1109/MNET.133.2200527
13. Lamport, L.: Paxos made simple. ACM SIGACT News (Distrib. Comput. Column) 32, 4 (Whole Number 121, December 2001), 51–58 (2001)
14. Ongaro, D., Ousterhout, J.: In search of an understandable consensus algorithm. In: 2014 USENIX Annual Technical Conference (USENIX ATC 14), pp. 305–319 (2014)
15. Castro, M., et al.: Practical byzantine fault tolerance. In: OsDI, vol. 99, pp. 173–186 (1999)
16. Bamakan, S.M.H., Motavali, A., Bondarti, A.B.: A survey of blockchain consensus algorithms performance evaluation criteria. Expert Syst. Appl. **154**, 113385 (2020)

Research on Local Collision Avoidance Algorithm for Unmanned Ship Based on Behavioral Constraints

Junjun Wang[✉]

Marine Equipment Major Project Center, Beijing 100841, China
z18336265177@163.com

Abstract. Aiming at the problem that the global planning path of the unmanned vehicle is difficult to meet the requirements of the motion and maneuvering characteristics of the unmanned vehicle, a local collision avoidance algorithm based on behavioral constraints is designed. Through the analysis of the collision avoidance scene and motion characteristics of the unmanned boat, the behavioral constraints that meet the motion characteristics of the unmanned boat, dynamic obstacle collision avoidance, and maritime collision avoidance rules are constructed. The optimization objective function of heading and speed is established, and the optimal heading and speed values for collision avoidance are obtained by solving the objective optimization problem. Combining the above path planning and local collision avoidance algorithm, a collision avoidance scheme is given. The local collision avoidance behavior is constrained by the global path, so that the unmanned vehicle can return to the optimal path. The simulation shows the effectiveness of the algorithm proposed in this paper.

Keywords: Behavioral constraints · USV · local collision avoidance · maritime collision avoidance rules

1 Introduction

Local collision avoidance of Unmanned Surface Vessel (USV) is to perform correct avoidance behavior to avoid dangerous obstacles/vessels when it suddenly encounters an unknown obstacle in the global planning based on perception information. The advantages of this method are that the real-time performance of collision avoidance is good, and the safety is better than that of the global planning method. Therefore, the local collision avoidance algorithm has a good performance in the USV platform. At present, many local collision avoidance methods for USV have been proposed at home and abroad.

In the research of local collision avoidance in the field of USV, on the one hand, some scholars consider the influence of the motion control characteristics of USV on the collision avoidance constraints, and propose a behavior-based collision avoidance algorithm without dynamic hazards, which transforms the multi-objective problem into a single-objective optimization problem, and improves the efficiency of the algorithm. On the other hand, the interference of the maritime environment also has an impact on

© The Author(s), under exclusive license to Springer Nature Singapore Pte Ltd. 2024
L. Pan et al. (Eds.): BIC-TA 2023, CCIS 2062, pp. 396–408, 2024.
https://doi.org/10.1007/978-981-97-2275-4_31

the collision avoidance of USV [1]. Some scholars have proposed the collision avoidance strategy of USV based on the characteristics of the maritime environment and the change of the interference level of the sea state. Considering the constraints of the USV by the motion model, environment and maritime rules, the multi-objective optimization algorithm is introduced into the local collision avoidance research [2]. In addition, for the collision avoidance problem of USV under multiple navigation constraints, some scholars have established a behavior control framework and navigation constraints to achieve behavior synchronization, and proposed a method based on multi-objective optimization and interval programming to solve the problem [3]. This method realizes the behavior synchronization of USV, such as tracking and collision avoidance.

Although many collision avoidance methods for USV have been proposed at home and abroad, they all have good performance. However, these methods usually simplify the research object to a particle for analysis, and do not consider the motion and control characteristics of the USV and the influence of environmental constraints, so there are dangers in the actual navigation. To solve these problems, this paper adds constraints such as the motion characteristics of USV and dynamic obstacle collision, and optimizes the local path through the obstacle avoidance behavior constraints of USV. In order to meet the requirements of local collision avoidance of USV, this paper designs an USV collision avoidance algorithm based on behavior constraints.

2 The Construction of Behavioral Constraints

2.1 Motion Characteristics Constraints of USV

When an obstacle is encountered in an emergency, the unmanned surface vehicle (USV) will have the risk of hull capsize if it changes the large course by steering control at high speed [4–6]. Therefore, it is necessary to establish the behavior constraints based on the handling and motion characteristics of USV in this situation. The plane motion coordinate system of USV is shown in Fig. 1.

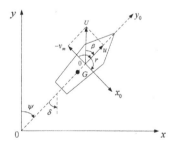

Fig. 1. Plane coordinate system for unmanned craft.

In Fig. 1, xoy is the inertial coordinate system of unmanned craft, which is the fixed coordinate system (referred to as "fixed system"), o is the origin of coordinates, the axis y is the direction due north of the earth, and the axis x is the direction due east of the earth; $x_0 G y_0$ is the attached body coordinate system of USV, G is the center of gravity of USV, y_0 axis is the heading direction of USV, x_0 axis is the starboard direction of USV; u is the forward speed of the USV, r is the yaw angular velocity, v is the transverse velocity; β is the drift Angle.

According to the handling and motion characteristics of USV, the behavior constraints are constructed, which can be divided into two aspects: motion characteristics constraints and maneuverability constraints.

(1) Constraints of motion characteristics.

This paper determines the motion state of unmanned boat according to its motion characteristics, adopts the dynamic window method in literature [7] to analyze the motion state of unmanned boat, and establishes the constraint conditions of the motion characteristics of unmanned boat such as dynamic window, linear velocity window and heading window [8]. The expression is as follows:

$$v_a = \{(v, \omega) \mid v \in [v_t - a_{max}\Delta t, v_t + a_{max}\Delta t] \wedge \omega \in [\omega_t - \omega_{max}\Delta t, \omega_t + \omega_{max}\Delta t]\} \tag{1}$$

where, v_a represents the dynamic window of the unmanned boat, and v_t represents the current speed of the USV, that is, the linear speed. ω_t Represents the current turning Angle speed of the USV, that is, the angular speed. v, ω Represents the solution of the linear velocity and angular velocity of the USV respectively. a_{max} Represents the current maximum acceleration of the USV, $a_{max} \geq 0$. ω_{max} is the maximum angular acceleration of the drone, $\omega_{max} \geq 0$. Δt Represents the shortest collision time of the USV, that is, the time window.

According to the speed v_t and yaw angular velocity ω_t of the USV, the state space after the time window Δt is calculated and denoted as the speed window and the heading window of the USV, respectively. The speed window expression is as follows.

$$v_{speed} = \{v \mid v \in [v_t - a_{max}\Delta t, v_t + a_{max}\Delta t]\} \tag{2}$$

where, v_{speed} represents the speed window of USV, and the other expressions are the same as above. The expression for the heading window of USV is as follows:

$$\varphi_{heading} = \left\{\varphi \mid \varphi \in \left[\varphi_{usv} + \omega_t\Delta t - \frac{1}{2}\omega_{max}\Delta t^2, \varphi_{usv} + \omega_t\Delta t + \frac{1}{2}\omega_{max}\Delta t^2\right]\right\} \tag{3}$$

where, $\varphi_{heading}$ is the heading window of USV, φ is the solution of the heading Angle of USV, and φ_t is the heading Angle of USV at the current moment.

(2) maneuverability limitation.

The maneuverability behavior constraint conditions of unmanned boat were established, and $K - T$ maneuverability equation was introduced for analysis [9]. The steering range, the rate of change of rudder Angle and the rate of change of heading are taken as the constraints of the maneuverability of USV. If the planned path satisfies the constraints, the output will be maintained, otherwise the path output of the previous moment will be maintained. The rudder Angle range of the small USV is $-35°\sim35°$. It is necessary to limit the left and right maximum rudder Angle of the rudder Angle control to avoid rudder damage caused by excessive rudder Angle output. The rudder Angle limit is as follows:

$$\delta(\theta) = \begin{cases} \delta_{min} & \theta \leq \theta_{min} \\ \delta_{max} & \theta \geq \theta_{max} \end{cases} \tag{4}$$

where, $\delta(\theta)$ is the rudder Angle output, δ_{min} is the maximum Angle limiting value of the left rudder, δ_{max} is the maximum Angle limiting value of the right rudder, $\theta_{min} = -35°$, and $\theta_{max} = 35°$.

According to the literature [10], when the speed of the USV is too fast, if the rudder Angle range is too large, the driving power of the servo will be insufficient and the mechanical structure of the servo will be damaged. Therefore, when the speed of the USV is too fast, the limit value of the left and right rudder should be adjusted accordingly. The expression for the right rudder Angle limit is as follows:

$$\delta_{max} = \frac{2r_{max}L}{Kv} \tag{5}$$

where, L is the length, v is the speed, K is the slewing parameter, and r_{max} is the maximum yaw angular velocity of the USV. Similarly, the left rudder Angle limit δ_{min} can also be obtained by the above formula.

The function relationship between rudder Angle and course Angle is established by $K - T$ maneuverability equation of USV, and the course Angle corresponding to the change value of rudder Angle at every moment is changed. According to the ship maneuverability index K, T of the unmanned boat, the transfer function expression of the whole ship can be obtained.

$$G(s) = \frac{K}{Ts^2 + s} = \frac{\varphi}{\delta} \tag{6}$$

where, K is the gyrating parameter and T is the stability parameter, namely the maneuverability index of the USV, which can be obtained through the Z-shape test. φ is the heading Angle of the USV, δ is the rudder Angle of the USV, and represents a derivative relation on the time state, for example: $r = \varphi \cdot s$. Then, the range limit of heading Angle change rate per unit time can be deduced from the above range of rudder Angle limit as follows:

$$\Delta\varphi_{zt}(\theta) = \begin{cases} \frac{K\delta_{min}}{Ts^2+s} & \theta \leq \theta_{min} \\ \frac{K\delta_{max}}{Ts^2+s} & \theta \geq \theta_{max} \end{cases} \tag{7}$$

2.2 Dynamic Obstacle Collision Constraint

In the actual navigation, there is a danger of collision between the USV and the dynamic ship if the static obstacle modeling method is used, so the collision constraint condition of the dynamic obstacle is established [11].

As shown in the figure below, the unmanned boat is regarded as v_{usv} particle located at point O, the unmanned boat moves in the direction of $\overrightarrow{v}_{usv}$ at a constant speed, the dynamic ship is regarded as an obstacle circle with radius R at point P, the dynamic ship moves in the direction of $\overrightarrow{v}_{obs}$ at a constant speed of v_{obs}, and the motion state of the two boats remains unchanged within time E.

The relative speed of unmanned boat and dynamic ship along the line OP direction is v_r, and the relative speed along the direction perpendicular to the line is v_θ, which is expressed as follows.

$$\begin{cases} v_r = v_{obs}\sin(\beta - \theta) - v_{usv}\sin(\alpha - \theta) \\ v_\theta = v_{obs}\cos(\beta - \theta) - v_{usv}\cos(\alpha - \theta) \end{cases} \tag{8}$$

where, α denotes the Angle between the speed of the USV and the horizontal axis, β denotes the Angle between the speed of the dynamic ship and the horizontal axis, and θ denotes the Angle between the line OP direction and the horizontal axis.

The collision geometry of the dynamic vessel is shown in Fig. 2.

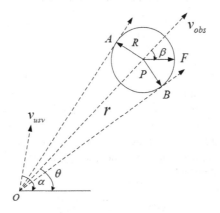

Fig. 2. Collision geometry of dynamic ships.

When the USV and dynamic ships meet the following two conditions, the two boats will collide. When $v_\theta = 0$ and $v_r < 0$, the USV and the dynamic ship will collide in a straight line, and the collision constraint conditions at this time are established. The expression of the feasible solution set for USV to avoid the dynamic ship is as follows.

$$A_{avoid}^i = \left\{ (\alpha, v) \mid v_{obs}^i \cos(\beta - \theta) - v\cos(\alpha - \theta) \geq 0 \right\} \tag{9}$$

where, A_{avoid}^i represents the feasible solution set of USV for avoidance, α represents the optional heading Angle of USV for avoidance, v represents the optional speed of USV for avoidance, v_{obs}^i represents the speed of the USV corresponding to Article i dynamic ship, β represents the Angle between the dynamic vessel speed v_{obs} and the horizontal axis, θ represents the Angle between the linear OP direction and the horizontal axis.

The other collision case can be discussed in two parts. When $v_\theta \neq 0, v_r < 0$ and $r^2 v_\theta^2 \leq R^2 (v_r^2 + v_\theta^2)$, the two boats will collide, and the collision constraint condition is established. The expression of the feasible solution set for USV avoidance is as follows.

$$A_{avoid}^i = \left\{ (\alpha, v) \mid v_{obs}^i \cos(\beta - \theta) - v\cos(\alpha - \theta) \geq 0 \land r^2 v_\theta^2 \leq R^2 (v_r^2 + v_\theta^2) \right\} \tag{10}$$

where, r represents the distance between the center point of the USV and the dynamic ship, R represents the radius of the dynamic ship, v_θ represents the relative speed of the two ships along the straight line OP direction, v_r represents the relative speed of the two ships along the straight line OP direction, and the other expressions have the same meaning as above. When $v_\theta \neq 0$, $v_r < 0$ and $r^2 v_\theta^2 > R^2\left(v_r^2 + v_\theta^2\right)$, the two boats will also collide. At this time, the collision constraint condition, namely the feasible solution set of the USV avoiding the dynamic ship, is expressed as follows.

$$A^i_{\text{crvid}} = \left\{(\alpha, v) \mid v^i_{\text{obs}}\cos(\beta - \theta) - v\cos(\alpha - \theta) < 0 \wedge r^2 v_\theta^2 > R^2\left(v_r^2 + v_\theta^2\right)\right\} \quad (11)$$

The expression of the relative speed v_θ of the two boats in the direction perpendicular to line OP and the relative speed v_r of the two boats in the direction of line OP is updated as follows.

$$\begin{cases} v_r = v^i_{obs}\sin(\beta - \theta) - v_{uvv}\sin(\alpha - \theta) \\ v_\theta = v^i_{obs}\cos(\beta - \theta) - v_{usv}\cos(\alpha - \theta) \end{cases} \quad (12)$$

The meaning of the expression in the formula is the same as above. Combining the above collision analysis results between USV and dynamic ship obstacles, the feasible solution constraint expression of USV's avoidance behavior is obtained as follows.

$$A_{\text{avoid}} = \bigcap_{i=1}^{N} A^i_{\text{cuoid}} \quad (13)$$

where, A_{avoid} is the constraint condition of feasible solution for USV avoidance, and N represents the number of target ships with collision danger.

3 Design of Collision Avoidance Algorithm Based on Behavioral Constraints

3.1 Screening of Dangerous Obstacle Targets

In order to ensure that the USV safely avoid all ships, it is necessary to carry out situation analysis of dynamic targets, screen ships with collision avoidance risks, and avoid obstacles for targets with high risk [12]. The safe collision avoidance range of USV is established, which is divided into three layers of collision avoidance distance: emergency collision avoidance, potential danger and safety encounter.

The USV can obtain the *Arpa* information *DCPA* and *TCPA* of each ship through the perception radar, and the calculation expression is as follows.

$$\begin{cases} DCPA = R_T\sin(\varphi_R - \partial_T - \pi) \\ TCPA = \frac{R_T\cos(\varphi_R - \partial_T - \pi)}{v_R} \end{cases} \quad (14)$$

where, *DCPA* represents the shortest meeting distance between the two vessels, *TCPA* represents the time taken from the current position to the shortest meeting distance, R_T is the distance between the USV and the obstacle, φ_R is the heading Angle of the obstacle relative to the USV, ∂_T is the orientation of the target vessel, and $\overrightarrow{v}_R$ is the movement direction of the target vessel relative to the USV. Whether the target boat needs to avoid obstacles can be determined by the information of *DCPA* and *TCPA*. The judgment process of dangerous dynamic target is shown in Fig. 3.

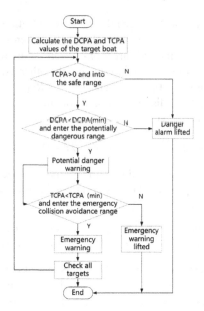

Fig. 3. Dynamic dangerous target determination process.

3.2 Safe Heading and Speed are Solved

On the basis of the screening of dangerous obstacle targets, the target boats with different danger degrees are selected for local collision avoidance, so it is necessary to calculate the safe course and speed value of unmanned vehicle collision avoidance to achieve safe collision avoidance [12, 13]. In the process of determining the safe course and speed, it is necessary to meet the behavioral constraints such as the motion characteristics of the USV and the collision of dynamic obstacles, so as to make the avoidance range of the USV as small as possible. The safe course and speed can be obtained by solving the optimization problem of collision avoidance course and speed.

On the basis of satisfying the motion characteristics and dynamic obstacle collision behavior constraints of the USV, the course and speed of the USV are taken as the optimization objectives, and the local collision avoidance problem is transformed into a single objective optimization problem by the method of weighted average. The course and speed of the collision avoidance are taken as the solutions of a group of collision avoidance behaviors. The optimization objective function of the collision avoidance behavior of USV is expressed as follows.

$$\min J_o(\varphi, v) = \xi_1 \cdot f_h(\varphi) + \xi_2 \cdot g_s(v) \tag{15}$$

where, $J_o(\varphi, v)$ represents the weighted optimization objective function of the safe collision avoidance course and speed of the USV. $f_h(\varphi)$ represents the optimization objective function of the heading Angle, $f_h(\varphi) = |\pi - |\varphi - \varphi_k||/\pi$. φ_k represent the Angle between the position of the unmanned vehicle and the target point in the geodetic coordinate system. $g_s(v)$ is the optimization objective function of speed, $g_s(v) = v/v_{max}$, v_{max} is the maximum speed value of USV. ξ_1 is the weighting coefficient of course objective function; ξ_2 is the weighting coefficient of the speed objective function. In designing the experiment, $\xi_1 = 0.65, \xi_2 = 0.35$.

3.3 The Time of Resuming Navigation is Determined

After solving the safe course and speed of collision avoidance, the USV began to perform the preliminary avoidance of local collision avoidance. When there is no danger of collision, that is, after the target ship avoids the obstacle, the USV needs to resume the original route and continue to sail along the planned path, and then the whole avoidance behavior of the obstacle is completed. There are two ways for the USV to resume sailing, one is to resume the course before collision avoidance, and the other is to resume the original course before collision avoidance [14]. The smaller the deviation from the original course, the more beneficial it is to save fuel consumption [15]. Therefore, it is necessary to determine the resumption time of the USV [16]. In this paper, the resumption of the USV is realized by restoring the route, as shown in Fig. 4.

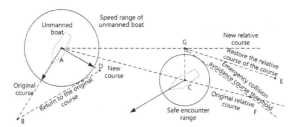

Fig. 4. The reopening process of the USV.

As shown in the figure above, the original route of USV A is in the direction AB. When encountering target boat C, according to the dangerous target judgment process, it can be seen that the boat is in danger of collision, and the USV needs to make a left turn to avoid. According to the new relative course AG as the velocity triangle, the new course of the USV can be obtained as the direction AD. The closest encounter point of the two boats is G-point, and the distance between the two boats is the closest when the USV is at this point, which meets the requirements of safe encounter range, indicating that the collision avoidance route is qualified. When the USV sails to the G-point position, it can resume the original route.

The timing of the resumption of USVs must meet the following conditions:

(1) When $TCPA \leq 0$, it indicates that there is no time for the two boats to meet, then the target boat is a safety boat and there is no risk of collision;

(2) When $TCPA > 0$, and the shortest encounter distance between the two boats is greater than the safe encounter distance, indicating that the encounter distance between the two boats is relatively far, the target boat does not enter the safe distance range of the unmanned boat, then the target boat is a safe boat and there is no collision risk;

According to the figure above, when the USV collision avoidance steering reaches point D, the conditions of the above resumption time have been met, and the resumption operation can be performed. When the USV begins to resume operations at that point, the route between point D and point B needs to be as short as possible. In addition, the opposite course of the two boats after turning needs to be kept outside the safe encounter range to ensure the safety of collision avoidance during the course of the USV.

On the basis of the above optimization objective function and constraint conditions of USV collision avoidance behavior, the safe course and speed are solved by screening the dangerous obstacle target, and the time of re-sailing is determined to complete the entire collision avoidance behavior of USV. In order to ensure the safety of the unmanned boat in the course of avoiding obstacles, the avoidance range is as small as possible and the collision avoidance speed is as slow as possible.

4 Simulation Experiment

In the simulation test of USV collision avoidance, the behavior framework based on moos-ivp is used to build the simulation platform of USV. On the chart, the planned course of USV (blue dashed line), the actual collision avoidance course of USV (green solid line), the model of USV (red hull), the model of obstacles (green circle) and the size of the action range are superimposed. On the basis of the simulation software, the collision avoidance simulation test of USV in multiple obstacle scenes is carried out. Figure 5 shows the simulation trajectory of static obstacles for USV.

In Fig. 5(a) to (d), the initial navigation parameters of the USV are set as follows: the speed is set as $v = 10\,kn$, heading $\varphi_a = 87.92°$, heading $\varphi_b = 83.25°$, heading $\varphi_c = 36.32°$, heading $\varphi_d = 40.59°$. The blue dashed line is the route of the USV, and the green solid line is the actual navigation trajectory of the USV in the collision avoidance process. The obstacles are randomly generated on the route of the USV, and the number is 3–8. The obstacles are red dots, the surrounding red circles represent the size and contour of the obstacles, and the surrounding green circles represent the influence range of the obstacles after the expansion treatment. The static obstacle collision avoidance simulation results using the proposed collision avoidance algorithm are shown in Table 1.

According to Table 1, when there are multiple static obstacles in the environment at a speed of 10kn, the safe meeting distance is 61.181 m to 70.814 m, which can be controlled within the range of 6–9 times of the boat length (boat length 8.075 m). The minimum avoidance range is 13.72°, and the maximum is 19.88°, which can be controlled within 20°. The longest return time is 7.031 s, the shortest is 5.691 s, which can be controlled within 7 s. According to the evaluation function of collision avoidance in Sect. 3.13, the evaluation values of four collision avoidance tests are 0.8316, 0.6996, 0.7411, 0.7048, respectively, which means that the collision avoidance behavior is qualified, has the collision avoidance ability of multiple static obstacles, can achieve small amplitude avoidance of obstacles, and has good collision avoidance safety. Next, the dynamic ship collision avoidance simulation test is carried out, and the simulation trajectory is shown in Fig. 6.

In Fig. 6(a) to (d). The initial navigation parameters of the USV are set as follows: the speed is set as $v_{usv} = 10\,kn$, heading $\varphi_a = 89.52°$, heading $\varphi_b = 42.13°$, heading

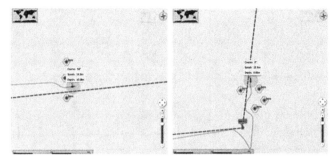

(a) number of obstacles 3 (b) number of obstacles 4

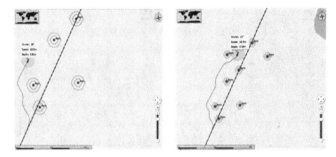

(c) Number of obstacles 5 (d) number of obstacles 8

Fig. 5. Collision avoidance simulation diagram of static obstacle.

Table 1. Simulation results of collision avoidance of static obstacles.

Trial No	Number of obstacles	DCPA(m)	TCPA(s)	Shortest safe meeting distance (m)	Return time of USV (s)	Take shelter Amplitude (°)
1	3	142.598	69.121	64.513	5.691	13.72
2	4	124.138	61.879	61.181	6.125	19.88
3	5	158.291	76.181	70.814	7.031	16.36
4	8	130.981	63.984	67.256	6.993	19.01

$\varphi_c = 8.95°$, heading $\varphi_d = 38.91°$. The blue dashed line is the route of the USV, and the green solid line is the actual navigation trajectory of the USV. The red dotted line is the route of the dynamic vessel.

The initial navigation parameters of the dynamic vessel are set as follows: Encounter scene speed $v_{obs} = 8$kn, heading $\varphi_{obs} = 270°$; Overtaking scene $v_{obs} = 6$kn, heading $\varphi_{obs} = 45°$; Left crossing scene $v_{obs} = 8$kn, heading $\varphi_{obs} = 100°$; Right cross scene $v_{obs} = 8$kn, heading $\varphi_{obs} = 300°$. The simulation results of dynamic vessel collision avoidance are shown in Table 2.

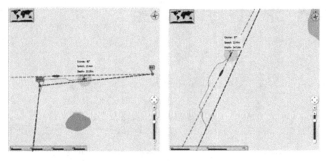

(a) Encounter scenario (b) chase scenario

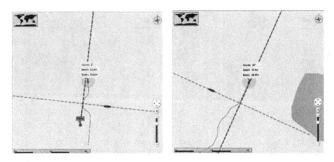

(c) Left cross meet scene (d) right cross meet scene

Fig. 6. Collision avoidance simulation diagram of a dynamic ship.

According to Table 2, under the speed of 10kn, the safe meeting distance for collision avoidance scenarios such as encounter, chase, left cross and right cross is 66.589 m and 72.665 m, which can be controlled within the range of 6–9 times of the boat length (boat length 8.075 m). The minimum avoidance range is 15.32°, the maximum is 20.03°, and the avoidance range can be controlled within 20°. The longest return time is 7.482 s, the shortest is 5.649 s, which can be controlled within 7 s. According to the collision avoidance evaluation function, the evaluation values of four collision avoidance tests are 0.7392, 0.8125, 0.6821, 0.7933, respectively, which means that the collision avoidance behavior is qualified. The algorithm has the ability of collision avoidance in the scene of dynamic ships, and can achieve small amplitude avoidance of dynamic ships, which meets the requirements of motion control in the collision avoidance of USV.

Table 2. Simulation results of collision avoidance for dynamic ships.

Trial No	/	DCPA(m)	TCPA(s)	Shortest safe meeting distance (m)	Return time of USV (s)	Out of the way Amplitude (°)
1	Encounter	121.252	62.589	69.116	6.335	18.26
2	Overtaking	135.282	65.822	70.598	7.482	15.32
3	Left cross	129.829	70.281	72.665	5.649	20.03
4	Right cross	131.281	64.128	66.589	6.929	16.51

5   Conclusion

In order to solve the problem that the existing path planning method is difficult to meet the requirements of its motion control characteristics and the actual track difference is large, this paper designs a local collision avoidance algorithm for USV based on behavior constraints. Through the analysis of the motion characteristics of USV and the calculation of the relevant parameters of collision avoidance, the behavior constraints based on the motion characteristics of USV and dynamic obstacles are established. The optimal collision avoidance behavior of the USV is obtained by solving the optimization problem of the course and speed of collision avoidance, and the value of the course and speed of collision avoidance at the next moment is obtained. An evaluation function is designed to evaluate the collision avoidance index of USV. Through simulation comparison, the proposed method improves the safety of collision avoidance, and can achieve a small range of obstacle avoidance.

References

1. Fu, Z.B., Liu, L. B.: Research on unmanned ship technology and military application. Nat. Defense Sci. Technol. **5**, 42–44+51 (2016). (in Chinese)
2. Yan, R., Pang, S., Sun, H., et al.: Development and missions of unmanned surface vehicle. J. Mar. Sci. Appl. **9**(4), 451–457 (2010)
3. Gu, Y.M.: Overview of technology development of surface unmanned ship fleet. Ship Sci. Technol. **41**(23), 35–38 (2019)
4. Chen, Y.B.: Review of development status and key technologies of unmanned ship. Sci. Technol. Innovation **60**(2), 60–61 (2019). (in Chinese)
5. Wang, X., Song, X., Du, L.: Review and application of unmanned surface vehicle in China. In: Proceedings of the 2019 5th International Conference on Transportation Information and Safety (ICTIS), pp. 1476–1481. IEEE (2019)
6. Wu, B., Wen, Y.Q., Wu, B., et al.: Review and prospect of collision avoidance methods for surface unmanned ships. J. Wuhan Univ. Technol.: Transp. Sci. Eng. Ed. **40**(3), 456–461 (2016). (in Chinese)
7. Zhuang, J.Y., Wan, L., Liao, Y.L., et al.: Research on global path planning of surface unmanned ship based on electronic chart. Comput. Sci. **38**(9), 211–214+219 (2011). (in Chinese)
8. Ou, H., Niu, H., Tsourdos, A., et al.: Development of collision avoidance algorithms for the C-Enduro USV. IFAC Proc. Vol. **47**(3), 12174–12181 (2014)

9. Chang, H.: Global path planning method for USV system based on improved ant colony algorithm. Appl. Mech. Mater. **32**(52), 568–570+785–788 (2014)

10. Laval, B., Bird, J.S., Helland, P.D.: An autonomous underwater vehicle for the study of small lakes. J. Atmos. Oceanic Technol. **17**(1), 69–76 (2000)

11. Song, A., Su, B., Dong, C., et al.: A two-level dynamic obstacle avoidance algorithm for unmanned surface vehicles. Ocean Eng. **170**(15), 351–360 (2018)

12. Wang, X., Yadav, V.: Cooperative UAV formation flying with obstacle/collision avoidance. IEEE Trans. Control Syst. Technol. **15**(4), 672–679 (2007)

13. Moldovan, E., Tatu, S.O., Gaman, T., et al.: A new 94-GHz six-port collision-avoidance radar sensor. IEEE Trans. Microw. Theory Tech. **52**(3), 751–759 (2004)

14. Vandenberg, J., Guy, S.J., Lin, M., et al.: Reciprocal n-body collision avoidance. In: Pradalier, C., Siegwart, R., Hirzinger, G. (eds.) Robotics Research. Springer Tracts in Advanced Robotics, vol 70. Springer, Heidelberg (2011). https://doi.org/10.1007/978-3-642-19457-3_1

15. Meng, R., Su, S.J., Lian, X.F.: Path planning of Mobile Robot based on dynamic fuzzy artificial potential field method. Comput. Eng. Des. **31**(7), 1558–1561 (2010)

16. Wang, H., Wei, Z., Wang, S., et al.: A vision-based obstacle detection system for unmanned surface vehicle. In: Proceedings of the 5th International Conference on Robotics, Automation and Mechatronics (RAM), pp. 364–369. IEEE (2011)

Research on Dynamic Path Planning Algorithm for Unmanned Underwater Vehicles Based on Multi-step Mechanism DDQN

Zheng Wang, Xinyu Qu, Yang Yin$^{(\boxtimes)}$, and Houpu Li

School of Electrical Engineering, Naval University of Engineering, Wuhan 430033, China
602257804@qq.com

Abstract. The dynamic path planning capability of UUV in complex underwater environment is difficult to meet the needs of advanced nature. In this paper, Q(λ) algorithm and DDQN algorithm are combined, the updated iteration strategy of DDQN algorithm is improved, and a MS-DDQN algorithm is designed to obtain more accurate Q value through continuous multi-step interaction and instant reward. According to this algorithm, the input information and reward function of the algorithm are redesigned, and the UUV path planning is divided into two network modules: navigation and obstacle avoidance, which improves the dynamic path planning ability of UUV. The simulation results show that the MS-DDQN algorithm has a high planning success rate and can meet the dynamic operation requirements of UUV.

Keywords: Dynamic path planning · DDQN · Q(λ)

1 Introduction

Path planning is an important part of Unmanned Underwater Vehicles (UUV) task execution, which is mainly divided into static path planning and dynamic path planning [1]. Traditional dynamic path planning algorithms include artificial potential field method [2], A* algorithm [3], fuzzy control algorithm, etc. These algorithms [4] need to analyze all kinds of situations that UUV may encounter in detail. When the environment becomes complex, it will encounter the problem of dimensionality disaster, so it is difficult to apply to engineering practice.

In order to solve the above problems, the ability of deep learning to obtain information and the decision-making ability of reinforcement learning can be combined. Combining the advantages of the two, the deep reinforcement learning method is used to complete the research of UUV dynamic path planning technology. When the commonly used deep reinforcement learning algorithm iterates policy updates, the agent only considers the impact of obtaining the reward value of two adjacent states on the state-action Q value [5], or improves the learning effect by increasing the state weight with higher learning efficiency [6]. These algorithms are all limited by the calculation and update of the current state information. The Q-learning (Q(λ)) algorithm can make the obtained reward affect the state-action Q value of adjacent multi-step states. When applied to the path planning of UUV, it is equivalent to making UUV obtain the ability to perceive the future situation in advance, which will greatly improve the navigation obstacle avoidance ability of UUV. Therefore, the idea of Q(λ) algorithm can be used to improve the Double Deep Q Network (DDQN) algorithm, and the Multi-Step DDQN (MS-DDQN) algorithm for multi-step mechanism is studied. The algorithm has a large number of interactions with the environment, and uses the deep neural network to directly process the high-dimensional original information and approximate the state-action Q-value, so as to avoid the curse of dimensionality in complex working environments and have the ability to predict the future.

2 Dynamic Path Planning Technology for UUV

When executing the dynamic path planning task, UUV mainly detects obstacles and other environmental information through sensors. The UUV path planning technology designed in this paper mainly executes corresponding behavior strategies according to the state space information, so as to update the state space information according to the obtained behavior rewards.

Firstly, the dynamic path planning of UUV is divided into two sub-modules, namely local obstacle avoidance module and global navigation module, which are mainly used to guide UUV to stay away from obstacles and navigate quickly. The structure [7] of the local obstacle avoidance function and the global navigation function are designed respectively. The input state information of the local obstacle avoidance module includes the environmental information detected by the ranging sensor and the relative position information of the UUV. The input state information of the global navigation module is only the relative position information of the UUV, and the control instructions are output after the same information processing.

The instruction selection module determines which module outputs the action command by judging the distance value between the UUV and the nearest obstacle. The threshold value used in this paper is 40. When the distance is less than the threshold value 40, it means that the UUV is very close to the obstacle, and the UUV is executed by the instruction output by the local obstacle avoidance module. Otherwise, it means that the UUV is still a certain distance away from the obstacle, and the global navigation module should be executed to reach the target position at a faster speed [8]. The neural network of the two modules adopts the fully connected structure.

Secondly, in order to improve UUV learning efficiency and learn better strategies, the position relationship between UUV and the target point is added to the state information of the input deep neural network, which mainly includes three information contents: First, the relative coordinate relationship between the current UUV coordinate and the target position coordinate; Second, the Euclidean distance between the current position of UUV and the target position; The third is the Angle between the UUV forward direction vector and the target position direction vector starting from the current UUV coordinate position.

Finally, Rectified Linear Unit-6 (ReLU6) function is selected as the activation function at the front end of the neural network construction, which helps the algorithm to quickly learn the sparse features of data samples. Rectified Linear Unit (ReLU) is used as the activation function at the back end of the network to output the evaluation value of each behavior [9, 10].

3 Algorithm Design

Before designing the algorithm, it is necessary to consider the reward design of the algorithm. Therefore, a kind of continuous combined reward mechanism containing three kinds of reward functions is designed [11, 12], terminal reward and non-terminal reward are redesigned, and another directional reward is designed. The specific reward design is as follows.

3.1 Terminal Rewards

When the UUV reaches the target point, it gets a positive reward, which is expressed as:

$$r_{arr} = 100; \text{ if } d_{r-t} \leq d_{win} \tag{1}$$

where d_{r-t} is the Euclidean distance from UUV to the target point, and d_{win} is the threshold for UUV to reach the target point. When d_{r-t} is less than the set d_{win}, it means that the target point has been reached; otherwise, the target point has not been reached.

When the UUV collids with the obstacle, the negative reward is obtained, which is designed as follows.

$$r_{col} = -100; \text{ if } d_{r-o} \leq d_{col} \tag{2}$$

where d_{r-o} is the Euclidean distance between UUV and the nearest obstacle, d_{col} is the threshold of collision between UUV and obstacle, when d_{r-o} is less than or equal to d_{col}, it means a collision has occurred, otherwise no collision has occurred.

3.2 Non-terminal Rewards

Get a positive reward when the UUV is moving towards the goal point, negative reward otherwise.

$$r_{t\_goal} = c_r \big[d_{r-t}(t) - d_{r-t}(t-1) \big] \tag{3}$$

where $c_r \in (0, 1]$ represents the constant coefficient, which is set to 1 in this paper, and this reward guides the UUV toward the target point.

When the minimum distance between the UUV and the obstacle continues to decrease, the obtained danger reward $r_{dang} \in (0, 1]$ also becomes smaller.

$$r_{dang} = \beta * 2^{d_{min}} \tag{4}$$

where d_{min} is the minimum distance between UUV and the obstacle, β is the coefficient, and the value range of r_{dang} is $(0,1)$.

3.3 Direction Reward

When the absolute value of the Angle between the forward direction of the UUV and the direction of the UUV to the target position is less than 18°, the reward is 1. When the absolute value is greater than 18° and less than 72°, the reward is 0.3.

$$r_{\mathrm{ori}} = \begin{cases} 1, a_{\mathrm{ori}} \leq 18° \\ 0.3, 18° < a_{\mathrm{ori}} \leq 72° \\ 0, \text{ otherwise} \end{cases} \tag{5}$$

where a_{ori} is the absolute value of the Angle between the forward direction vector of the UUV and the direction vector sent by the current coordinate of the UUV to the target position, and the vehicle is guided to the target point faster through this reward.

Take $r_{t\_\mathrm{goal}} + r_{\mathrm{dang}} + r_{\mathrm{ori}}$ as the final reward. This combination of rewards solves the problem of sparse rewards so that UUVs can receive corresponding rewards at each step of the training process. In this way, the UUV can better learn the corresponding strategy.

The MS-DDQN algorithm introduces the idea of Q(λ) [13] into the DDQN algorithm through the method of multi-step goal guidance, and uses the instant reward of continuous multi-step interaction to replace the instant reward of a single moment obtained after executing the behavior of the DDQN algorithm, so as to obtain a more accurate target Q value in training. The reward obtained by the UUV can diffuse the estimated value of the state value of the multi-step interval state backward, which can better guide the UUV to learn quickly and perceive the change of the future state in advance. The output of MS-DDQN target value network is as follows.

$$y_{\mathrm{target}}^{MS-DDQN} = \sum_{i=0}^{i=\lambda-1} \gamma^i r_{i+1} + \gamma^\lambda Q_{target}\left(s_{t+\gamma}, \mathrm{argmax}_a Q(s_{t+\lambda}, a, \theta), \theta'\right) \tag{6}$$

The loss function of MS-DDQN network is as follows.

$$L(\theta) = E\left[y_{\mathrm{target}}^{MS-DDQN} - Q(s, a, \theta)\right)^2\right] \tag{7}$$

The training process of DDQN algorithm is online training. After collecting a sample, it is put into the experience pool, and then sampling from it during training. Using the experience pool can increase the utilization of samples and increase the correlation between samples. MS-DDQN algorithm trains and learns through multi-step target guidance, and the data stored in the experience pool is different from that of the conventional DDQN algorithm. In the conventional DDQN algorithm, the data stored in the experience pool is, and in the MS-DDQN algorithm, it is changed to $\left[s_t, a_t, \sum_{i=0}^{i=\lambda-1} \gamma^i r_{t+i}, s_{t+1}\right]$.

The MS-DDQN algorithm flow is shown in the Fig. 1.

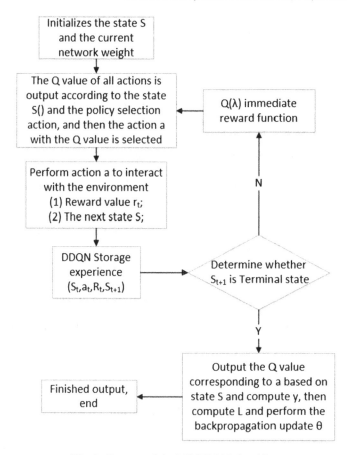

Fig. 1. Process of the MS-DDQN algorithm.

4 Simulation Test

In this section, the obstacle environment, UUV model and dynamic interference are simulated.

4.1 Obstacle Environment

The obstacle environment model refers to the description of obstacles in the environment. The environment model description directly affects the state information of the deep reinforcement learning algorithm input [14].

The geometric method does not need to segment the environment, and the "points", "lines", and "surfaces" of obstacles are used to describe the obstacle information in the environment.

The types of obstacles are divided into ellipse and polygon, and the circular obstacle is also regarded as an ellipse obstacle. The environment model records the vertex coordinates of the upper left corner of the ellipse and the long and short axis of the ellipse.

Triangles, rectangles, and squares are classified as polygonal obstacles, and each vertex coordinate of the polygon is recorded.

4.2 UUV Model

Considering UUV as a particle, in the environment model, the UUV is projected into a circle with a radius of 1 pixel, and multi-beam forward-looking sonar is used as a sensing instrument to detect the environment.

In this paper, multi-beam forward looking sonar is simplified into a plane model. The UUV model is designed to perform only horizontal motion control actions, so a multi-beam forward-looking sonar simulator with a horizontal opening Angle of 180° and a maximum range of 90 is designed and sampled every 5°.

First, the first laser line collected to the left of the UUV forward direction is 0°, and the last laser line collected is 180°. Each time the information detected by laser sonar at time t is measured by (0°, 5°, 10°,..., 180°) are stored in a row vector $s^t = \left[s_1^t, s_2^t, s_3^t, \dots, s_{36}^t \right]$.

If no obstacle is detected, the sonar returns the maximum detection distance; otherwise, the distance to the obstacle is returned. The detected information is normalized by dividing the row vector by the maximum effective detection distance.

The UUV model is set to move forward at a constant speed and can only make discrete turning actions, which are 15° to the left, 30° to the left, straight direction, 15° to the right, and 30° to the right.

4.3 Dynamic Environments

First, assume that the starting position of UUV in the lower left corner of the simulation environment is P_{stant}, the coordinate is $(x_{\text{start}}, y_{\text{start}})$, the motion speed is $v = 0.5$kn, and the direction of the current UUV is $angle$. In the current state, UUV selects a turning action to execute, the turning Angle is $angle_{\text{tran}}$, and the choice range of the turning Angle is $angle_{\text{tran}} \in \left(15^\circ, 30_r^\circ, 0^\circ, 15^\circ, 30_r^\circ \right)$.

Formula (8) represents the Angle of UUV after the execution of the action:

$$angle \leftarrow angle + angle_{\text{tran}} \tag{8}$$

Combining Eq. (8), the UUV coordinates can be obtained as follows.

$$x_{\text{next}} = x_{\text{start}} + \cos(angle) * v; \; y_{\text{next}} = y_{\text{start}} + \sin(angle) * v \tag{9}$$

And then the vector from each edge vector of the obstacle, the beam vector, and the position coordinate of the UUV to each vertex of the obstacle is constructed. The position information between the UUV and the obstacle can be obtained by solving the relative relationship between these vectors.

4.4 Model Training and Simulation Results

In order to verify MS-DDQN algorithm, the above simulation environment model is used as UUV training environment, and the navigation ability of UUV is tested in

Table 1. Setting of hyperparameter

Hyperparameter	Values
Learning rate a	0.002
Discount factor ε	0.9
Experience pool maximum capacity M	20000
Draw the number of training samples N	32
Steps λ	5
Update θ the parameter interval steps C	200

the training environment. Before the training, the relevant hyperparameters in the MS-DDQN algorithm were set accordingly, as shown in Table 1.

To quantitatively evaluate the performance of each algorithm, two metrics are used to evaluate the quality of the navigation model. The first is the success rate, which indicates the proportion of the number of UUV successfully reaching the target position in the total number of training. The other is the average reward value of the training process, which is the quotient of the total reward obtained by the UUV in the training process divided by the number of training rounds. The simulation training curve based on MS-DDQN algorithm is shown in Fig. 3.

It can be concluded from Fig. 2(a) that the success rate curve of MS-DDQN rises rapidly, which indicates that the MS-DDQN algorithm has a high learning efficiency. The reward curve in Fig. 2(b) also supports this view. After 5,000 training sessions, the success rate of MS-DDQN reaching the target position was 84.39%.

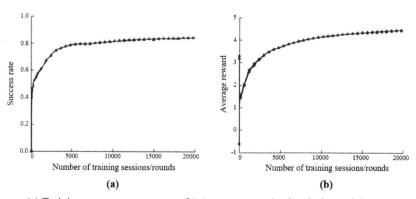

(a) Training process success rate. (b) Average reward value during training.

Fig. 2. Training process success rate and average reward value.

The higher reward value and success rate indicate that the UUV has conducted more training without collision and reaching the target point during the training process, and has stronger obstacle avoidance and navigation functions. In Fig. 2(b), it can be seen that after 10,000 training sessions, the average reward obtained by MS-DDQN is stable at more than 4, and the low reward value means that the UUV has more collisions, which also proves that the UUV based on MS-DDQN has strong navigation ability.

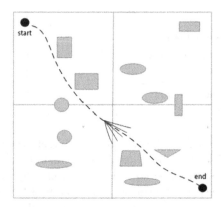

Fig. 3. Dynamic planning trajectory of UUV

The simulation test results can show the planned dynamic path, which is the final path trajectory of MS-DDQN algorithm after training convergence. As shown in Fig. 3, the path trajectory of UUV finally reaches the target position, the planned path is simple and clear, and the whole path has no collision with obstacles, and the algorithm performs well.

5 Conclusion

UUV with autonomous navigation ability can replace humans to complete different maritime operations. In order to improve the performance of UUV path dynamic planning algorithm, a MS-DDQN algorithm is proposed by combining the ideas of DDQN algorithm and Q(λ) algorithm. The improved algorithm changes the input state space, and adds the Euclidean distance between the UUV and the target point and the Angle between the direction of the UUV and the direction of the UUV to the target point as new input information. The improved DDQN algorithm can make the influence of the reward value obtained by the current state of UUV extend backward, sense obstacles in advance and make avoidance actions, so as to improve the accuracy of the algorithm and accelerate its convergence speed.

Through the simulation data, it can be found that MS-DDQN has a higher success rate and reward value, which indicates that the algorithm learns more information in the training process. After testing the trained algorithm model, it can be seen from the test results that the success rate of MS-DDQN algorithm model navigation has reached 100%, and it can be packaged into the UUV system.

References

1. Wu, H., Tian, G.-H., Li, Y., Zhou, F.-Y., Duan, P.: Spatial semantic hybrid map building and application of mobile service robot. Robot. Auton. Syst. **62**(6), 923–941 (2014)
2. Khatib, O.: Real-time obstacle avoidance for manipulators and mobile robots. Auton. Robot Veh. 396–404 (1986)
3. Bao, Y.-L.: A* algorithm for finding the optimal path on vector maps. Comput. Simul. **25**(4), 253–258 (2008)
4. Yang, F., Chen, Y.-Y., Zeng, J.-B., Liu, F., Zhao, X.: Research on obstacle avoidance of UUV formation based on fuzzy control. Ship Sci. Technol. **44**(7), 98–103 (2022)
5. Mnih, V., et al.: Playing atari with deep reinforcement learning. Comput. Sci. (2013)
6. Van Hasselt, H., Guez, A., Silver, D.: Deep reinforcement learning with double q-learning. In: AAAI 2016: Proceedings of the Thirtieth AAAI Conference on Artificial Intelligence, vol. 7, pp. 2094–2100. AAAI, Phoenix, Arizona (2016)
7. Cheng, K., et al.: A heuristic neural network structure relying on fuzzy logic for images scoring. IEEE Trans. Fuzzy Syst. **29**(1), 34–45 (2021)
8. Caldera, S., Rassau, A., Chai, D.: Review of deep learning methods in robotic grasp detection. Multimodal Technol. Interaction **2**(3), 57 (2018)
9. Bai, Y.-Y., Cao, J., Zhang, F.-Y., Zhang, F.-Y., Peng, X.-Y.: Optimization of convolutional neural network structure based on particle swarm optimization. J. Inner Mongolia Univ. (Nat. Sci. Ed.) **50**(1), 88–92 (2019)
10. Jiang, A.-B., Wang, W.-W.: Research on optimization of ReLU activation function. Transducer Microsyst. Technol. **37**(2), 50–53 (2018)
11. Li, Y., Shao, Z., Zhao, Z., Shi, Z., Guan, Y.: Design of reward function in deep reinforcement learning for trajectory planning. Comput. Eng. Appl. **56**(2), 226–232 (2020)
12. Liu, Q., Yan, Q., Hu, D.: A hierarchical reinforcement learning algorithm based on heuristic reward function. In: Xu, H. (ed.) The 2nd IEEE International Conference on Advanced Computer Control (ICACC 2010), vol. 3, pp. 371–376. IEEE, Shenyang (2010)
13. Fu, Q.-M., Liu, Q., Wang, H., Xiao, F., Li, J.: A novel off policy Q(λ) algorithm based on linear function approximation. Chinese J. Comput. **37**(3), 677–686 (2014)
14. Châari, I., et al.: On the adequacy of tabu search for global robot path planning problem in grid environments. Procedia Comput. Sci. **32**, 604–613 (2014)

Dynamic Analysis of Immersion Pump Rotors Considering Fluid Effects

Jinghui Fang[✉], Hailiang Li, and Fujun Wang

ZHEFU Holding Group Co., Ltd., Hangzhou 310013, China
fang_jinghui@zhefu.cn

Abstract. Taking into account the influence of surrounding fluid on the rotor dynamics of immersion pumps, a rotor bearing non rotating component system model is established, and the dynamic characteristic coefficients of the bearings are introduced to establish the system's dynamic equations. Wet mode analysis is conducted, and the shaft response analysis is completed based on this model. Through the above analysis, a reasonable evaluation can be made on the critical speed of the shaft system, system stability, and vibration characteristics. Among them, the rotor bearing system is essentially a nonlinear system, and the radial bearing fluid film force plays a nonlinear stiffness and damping role in the system. However, in practical analysis, according to specific requirements, the model of fluid film force can generally be divided into linear model and nonlinear model based on the magnitude of disturbance around the rotor's static equilibrium position. In response analysis and stability analysis, the liquid film stiffness and damping characteristics are obtained from the relationship between the maximum average and minimum average values of the bearing liquid film force and the speed, respectively. The nonlinear characteristics of the bearing liquid film force are transformed into linear problems for consideration.

Keywords: Pump container · wet mode analysis · nonlinear · linear · response analysis · stability analysis

1 Introduction

1.1 Immersion Pump Applications

Immersion pumps (as shown in Fig. 1) can be widely used in the fields of photo-thermal energy storage, tanker inverting, chemical circuits, nuclear reactors and so on. It has a wide range of application prospects. Especially used in liquid metal nuclear reactor, it effectively solves the problem of high efficiency medium transportation. It is a development direction in the current application field [1].

1.2 The Basic Structure and Characteristics of Immersion Pumps

As shown Fig. 2: immersion pumps have the following characteristics due to the ambient conditions of the arrangement:

L. Pan et al. (Eds.): BIC-TA 2023, CCIS 2062, pp. 418–429, 2024.
https://doi.org/10.1007/978-981-97-2275-4_33

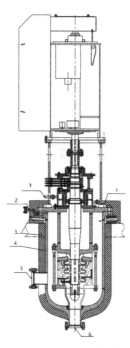

Fig. 1. Schematic arrangement of an immersion pump

(1) Poor bearing capacity of the lower bearing. In order to simplify the structure, the lower radial bearings are used as lubricating medium in immersion pump. Lower radial plain bearing can provide a small bearing capacity, the stability of the system is poor, the rotor bearing system to occur the possibility of system instability than horizontal pumps, pipeline pumps. There is a large difference in the theoretical model and the actual situation. The theoretical model of the radial plain bearing capacity within a certain range, and there is no definitive value [2].

(2) The immersion pump shaft system is very long, and most of the pump shaft is immersed in liquid coolant. When solving for the intrinsic frequency and response, the effect of the surrounding fluid should be taken into account and wet modal analysis should be carried out [3].

(3) Immersion pump spindle and pump motor rotor are mostly connected by flexible coupling, this connection effectively reduces the impact of the motor shaft on the pump shaft, and the effect of the coupling should be considered. This paper calculates the torsional stiffness and lateral stiffness, used to calculate the torsional vibration pattern of the shaft system, as well as the effect of lateral response [4].

(4) Immersion pump acts on the pump support of the container, and the stiffness support of the container and the pump support should be considered.

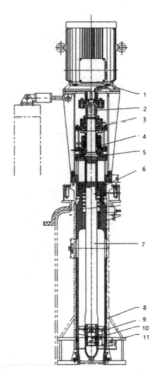

Fig. 2. Immersion Pump Installation Schematic

1.3 Immersion Pump Rotor Dynamics and Vibration Analysis

Self-excited vibration loads in immersion pumps are generated by liquid film forces on the lower radial bearings, which are caused by rotational imbalance of the rotor, hydraulic radial loads, etc.

Dynamic characterization of rotating components of immersion pumps including intrinsic frequency, stability and response analysis for rated operating conditions.

Rotor dynamics analysis is used to ensure that the most unfavorable lateral critical speeds are not generated, nor are unstable sub-synchronous eddy modes.

The damped intrinsic frequency of the spindle should be greater than 120% of the rated speed, or the response amplification factor of the rotating parts should be less than 2.5 [5].

Immersion pumps shall be operated with a vibration peak-to-peak value of less than 0.076 mm, or a vibration limit value mutually agreed upon by A and B [6, 7].

2 Mathematical Modeling

The vibration and stability analysis of immersion pumps should consider the effect of the fluid on the structural system, and the wet modal analysis is used to study the state of the rotor when it is running in the fluid, so it is necessary to consider the fluid-solid coupling effect of the fluid and rotor solid. When the rotor and fluid coupling vibration, taking into account the viscosity of the fluid, the rotor and the fluid will produce friction between the rotor, resulting in energy dissipation, vibration attenuation, in addition to the viscous force due to the fluid and inertial force, the fluid will be attached to the rotor with its rotation, equivalent to the rotor mass increases, the mass distribution changes.

Therefore, in order to take into account, the effect of the fluid on the rotor as it turns, it is necessary to introduce a compressibility fluctuating equation of the fluid:

$$\frac{1}{C^2}\frac{\partial^2 p}{\partial t^2} - \nabla^2 p = 0 \tag{1}$$

where C is the speed of sound in water; p is the fluid dynamic pressure, which be obtained by discretized:

$$Hp + A\dot{p} + E\ddot{p} + \rho B\ddot{r} + q_0 = 0 \tag{2}$$

The equation of motion for a structure in contact with a fluid is:

$$M_s\ddot{r} + C_s\dot{r} + K_s r - B^T p + f_0 = 0 \tag{3}$$

The equations of motion (2) and (3) for the structure-fluid system can be generalized into the following form [8]:

$$\begin{bmatrix} M_s & 0 \\ \rho B & E \end{bmatrix}\begin{Bmatrix} \ddot{r} \\ \ddot{p} \end{Bmatrix} + \begin{bmatrix} C_s & 0 \\ 0 & A \end{bmatrix}\begin{Bmatrix} \dot{r} \\ \dot{p} \end{Bmatrix} + \begin{bmatrix} K_s & -B^T \\ 0 & H \end{bmatrix}\begin{Bmatrix} r \\ P \end{Bmatrix} = \begin{Bmatrix} -f_0 \\ -q_0 \end{Bmatrix} \tag{4}$$

where p, $\dot{p}$, $\ddot{p}$ denote the displacement, velocity and acceleration of the point where the mass of the water body is attached to the fluid-solid coupling surface, M_s, C_s, K_s denote the mass matrix, damping matrix and stiffness matrix of the structure, f_0 is the excitation vector on the structure; q_0 is the excitation vector on the fluid; ρ is the density of the fluid; B, E, A, H are the coefficient matrices associated with the shape function.

This equation can be solved step by step by direct integration methods such as the Newmark β method or Wilson θ Newmark method or Wilson method, etc. For example, the large commercial analysis software ANSYS uses the Newmark β method to solve the differential method of motion.

3   Finite Element Modeling

The above mathematical equations are solved using the Acoustic Modal and Acoustic Harmonic modules of the analysis software ANSYS for immersion pump rotor dynamics and vibration analysis. The pump model and bearing model are introduced below (as shown Fig. 3 and Table 1).

3.1   Pump Model

It is not enough to consider only the stiffness of the rotor and bearing liquid film in the model, but the pool motor pump has a high center of gravity, and the stiffness of the non-rotating parts is weak, so the effects of the non-rotating parts, as well as the supporting parts in the stack, need to be considered. The pool immersion pump is immersed in coolant, and a large part of the pump shaft is immersed in coolant, so the fluid-solid coupling analysis needs to be considered.

The pump shaft and the motor shaft are connected through a flexible coupling, and the radial support stiffness and axial torsional stiffness of the pump shaft and the motor shaft are simulated using a spring unit, and the damping effect of damping is not considered.

The in-stack support stiffness was simulated using multiple single-degree-of-freedom bearing units.

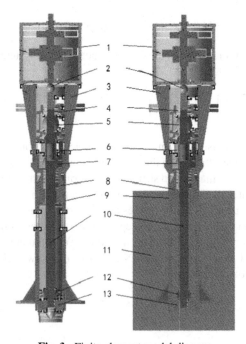

Fig. 3. Finite element model diagram

Table 1. Model simplification and parameter settings for each part of the model.

serial number	framework	model simplification way	Module selection
1	Motor	entity model	Solid187
2	Pump shaft	entity model	Solid187
3	Impellers	entity model	Solid187
4	Pump body	entity model	Solid187
5	Pump cover	entity model	Solid187
6	Shielding layer filler	quality point	Mass21
7	Motor bracket	entity model	Solid187
8	Shaft seal	quality point	Mass21
9	Bearing support	entity model	Solid187
10	Upper bearing	spring	Combin-214
11	Lower bearing	spring	Combin-214
12	Coupling	Springs, mass points	Mass21 Combin-14
13	Pump Support	entity model	Solid187
14	In-stack support stiffness	spring	Combin-14
15	Fluids	fluids	Fluid221

3.2 Bearing Model

Due to the nonlinear nature of bearing performance with bearing position and speed at the bearing-rotor system, the bearing performance is evaluated in this paper for two different bearing load cases to assess the rotor vibration response and the stability of the rotor bearing system (as shown Fig. 4).

Mass imbalance forces and hydraulic radial loads, etc. on the bearing to produce bearing loads, the phase angle of the various loads is different in the bearing superimposed on the total load generated is different.

If the bearing stiffness is small relative to the shaft stiff-ness, then the deformation of the bearing is large relative to the deformation of the shaft, so the damping produced by the bearing is used to reduce the vibration of the shaft, and when the bearing stiffness is better than the shaft stiffness, then the deformation of the shaft is larger than the deformation of the bearing, in this case, even if the bearing produces a large damping, the damping force produced by the bearing will be small because of the small displacement at the bearing. So, the bearing stiffness is much larger than the bending stiffness of the shaft, and there will be a large amplification factor for the vibration of the shaft near the critical speed [10].

Therefore, the average value of the predictable maxi-mum radial bearing load is used to calculate the bearing stiffness and damping coefficients for the vibration response analysis. Similarly, the average value of the predictable minimum radial bearing load is used to calculate the bearing stiffness and damping for considering the stability of

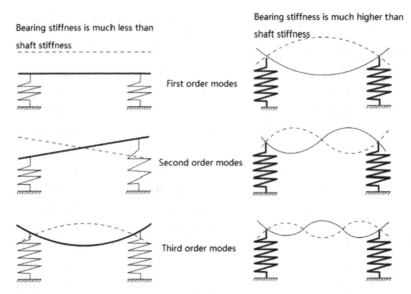

Fig. 4. Bearing stiffness relative to shaft stiffness to obtain modal vibration patterns.

the rotor-bearing system, because light load bearing is the most demanding condition for stability. Note that the stiffness and damping of the bearing are speed-dependent coefficients (as shown Table 2 and Table 3).

Table 2. Bearing liquid film stiffness and damping at maximum bearing loads.

Working condition	Lower radial plain bearings	
	Stiffness N/m	Damping N.s/m
Low speed 300 r/min	$K_{xx} = 4.31E6$	$C_{xx} = 1.91E5$
	$K_{yy} = 6.43E6$	$C_{yy} = 5.19E5$
	$K_{xy} = -1.84E6$	$C_{xy} = 1.42E5$
	$K_{yx} = 8.10E6$	$C_{yx} = 1.42E5$
Half speed 495 r/min	$K_{xx} = 1.14E7$	$C_{xx} = 2.29E5$
	$K_{yy} = 2.51E7$	$C_{yy} = 8.96E5$
	$K_{xy} = -1.76E6$	$C_{xy} = 2.26E5$
	$K_{yx} = 2.30E7$	$C_{yx} = 2.26E5$
Full speed 990 r/min	$K_{xx} = 4.31E7$	$C_{xx} = 3.16e5$
	$K_{yy} = 1.62E8$	$C_{yy} = 2.08e6$
	$K_{xy} = 5.58E6$	$C_{xy} = 4.34e5$
	$K_{yx} = 1.05E8$	$C_{yx} = 4.34e5$

Table 3. Bearing liquid film stiffness and damping at minimum bearing loads.

Working condition	Lower radial plain bearings	
	Stiffness N/m	Damping N.s/m
Low speed 300 r/min	Kxx = 8.62E+05	Cxx = 3.18E+04
	Kyy = 1.29E+06	Cyy = 8.65E+04
	Kxy = −3.68E+05	Cxy = 2.37E+04
	Kyx = 1.62E+06	Cyx = 2.37E+04
Half speed 495 r/min	Kxx = 2.28E+06	Cxx = 3.82E+04
	Kyy = 5.02E+06	Cyy = 1.49E+05
	Kxy = −3.52E+05	Cxy = 3.77E+04
	Kyx = 4.60E+06	Cyx = 3.77E+04
Full speed 990 r/min	Kxx = 8.62E+06	Cxx = 5.27E+04
	Kyy = 3.24E+07	Cyy = 3.47E+05
	Kxy = 1.12E+06	Cxy = 7.23E+04
	Kyx = 2.10E+07	Cyx = 7.23E+04

4 Evaluation Methods (Models, Criteria) for Axial Stability

The liquid film of the bearing generates radial and tangential forces on the rotor. These loads are complex and have a nonlinear relationship with the operating position of the rotor, but for small perturbations, the rotor displacement and velocity can be considered linear at this point.

Bearing cross-stiffness coefficient plays a role in promoting vortex, when this role is greater than the stabilizing effect of damping, then the vortex energy is increasing, so that the center of the hinder journal motion becomes a dispersed motion, ultimately leading to the instability of the system.

Vortex energy:

$$E_{DS} = F_Y V_Y = -K_{YX} X \dot{Y} \tag{5}$$

Dissipation of energy:

$$E_S = F_Y V_Y = -\left(C_{YY} \dot{Y}\right)\dot{Y} = -C_{YY}\dot{Y}^2 \tag{6}$$

The rotor bearing system is stabilized when $E_S > E_{DS}$, the rotor bearing system is stabilized, and when $E_S < E_{DS}$, the rotor bearing system is unstable (as shown Fig. 5).

In the immersion pump finite element analysis model, the stability criterion uses "four stiffness, four damping coefficients" under the minimum average load of the bearing, this coefficient can be obtained through the method of linearization of the perturbation, and then set up the bearing-rotor system model, with the dynamic characteristic coefficients obtained in the previous, the establishment of the system's dynamics equations; and at the same time, use the Routh-Hurwitz stability discrimination method to determine the stability of the system. Routh-Hurwitz stability discrimination method is used to determine the stability of the system (as shown Table 4).

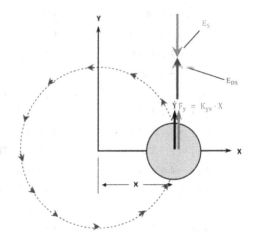

Fig. 5. Schematic diagram of bearing stability

Complex modal analysis is performed and if the real part of the eigenvalue is less than 0, the modal stability of this order is obtained, conversely, the system is unstable.

Table 4. Stability analysis results of the immersion pump.

Working condition	Modal	Damped frequency/Hz	Stability/Hz
Low speed 300 r/min	1	4.70	−2.9481e−6
	2	4.73	−1.3865e−5
	3	10.03	−1.0754e−4
	4	10.07	−16.514
	5	20.19	−0.59533
	6	20.88	−9.0922e−2
Half speed 495 r/min	1	4.69	−2.9481e−6
	2	4.72	−1.3865e−5
	3	10.02	−1.0754e−4
	4	10.08	−16.514
	5	22.26	−0.59533

(continued)

Table 4. (*continued*)

Working condition	Modal	Damped frequency/Hz	Stability/Hz
	6	22.78	−9.0922e−2
Full speed 990 r/min	1	4.70	−2.3863e−6
	2	4.73	−2.9481e−6
	3	10.12	−1.3865e−5
	4	10.15	−1.0754e−4
	5	24.13	−0.59533
	6	24.58	−9.0922e−2

Table 5. Modal analysis results at different speeds.

Number of revolutions per minute	Spindle intrinsic frequency/Hz
low speed 300 r/min	20.185
	20.876
half speed 495 r/min	22.262
	22.784
full speed 990 r/min	24.13
	24.584
maximum speed 1188 r/min	24.525
	25.011

5   Study and Analysis of the Influencing Elements of the Critical Speed of the Axle System

Immersion pumps for wet critical speed analysis, taking into account the operating conditions of the pumping liquid in the rotor in the operating clearance of the additional support role and damping effect, and take into account the permissible bearing stiffness and damping effect of these factors, the rotor critical speed calculated.

If the standard only makes requirements for dry critical speed, it is obvious that it cannot meet the requirements. Pool-type fast reactor in the wet modal calculation at the same time, using the standard requirements for the dry critical speed, for damped wet critical speed evaluation, the calculation is closer to the actual [5] (as shown Table 5).

Based on the calculations a Campbell diagram is drawn as shown below (Fig. 6):

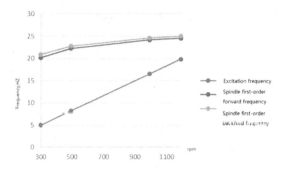

Fig. 6. Campbell diagram.

Based on the calculations, it is clear that the pump does not experience critical speeds in the interval from low speed to 120% of the rated speed.

6 Study and Analysis of Various Influencing Elements of the Harmonic Response Vibration of the Shaft System

The response analysis of the unbalanced force includes the response curve, the position of the critical speed as well as the amplification factor and the isolation region, and the rotor deflection curve at the position of the critical speed [7, 10] (as shown Fig. 7).

Further research on the unbalanced response of the rotor system should focus on two aspects: one is to establish a mechanical model that conforms as much as possible to the actual rotor structure and operation state and calculate the response characteristics of such large and complex rotor systems, and the other is to study the response characteristics caused by various types of nonlinear excitations in more depth. For the former, the use of mature commercial analysis software such as ANSYS can achieve good results.

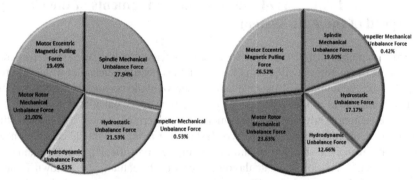

(a) At the upper bearing section rotor.   (b) At the rotor in the hydrostatic bearing section.

Fig. 7. Weight share of each unbalance force on the amplitude response at 990 r/min operating condition.

7 Conclusion

When the influence of the transport medium is accounted for, only dry modal analysis is performed, which does not satisfy the requirement of calculation accuracy, especially for heavy metal fluids of liquid metal reactor coolant.

ANSYS analysis software calculates the Campbell diagram, only the gyroscopic effect caused by the positive and negative in motion to analyze and consider, without considering the impact of the bearing stiffness, need to be manually animated Campbell, to get a more realistic critical speed.

Evaluation of stability on the logarithmic decay rate requirements, the requirements of different specifications are not the same, specific problems can be analyzed specifically, and further experimental verification.

References

1. Yu, J., Jia, B.: Reactor Thermal Hydraulic-s. Tsinghua University Press (2011)
2. Sun, L.: Stability study of vertical rotor bearing system. Harbin Institute of Technology (2010)
3. Zou, Y., Zhao, D.: Computation of eigenvalues of acoustic-structural coupled vibration for underwater structures. J. Ship Mech. **8**(2), 109–120 (2004)
4. Chen, W., Song, Q., Hong, J.: Influence of coupling stiffness on centrifuge speed stability. Equipment Environ. Eng. **12**(5) (2015)
5. API: Centrifugal Pumps for Petroleum, Petrochemical and Natural Gas Industries, American Petroleum Institute, Washington, DC, API Standard No. 610 (2004)
6. API: Axial and Centrifugal Compressors and Expander-compressors for Petroleum, Chemical, and Gas Industry Service, Downstream Segment, 7th ed., American Petroleum Institute, Washington, DC, API Standard No. 617 (2002)
7. API: Sealless Centrifugal Pumps for Petroleum, Petrochemical, and Gas Industry Process Service, American Petroleum Institute, Washington, DC, API Standard No. 685 (2011)
8. Zhang, A., Dai, S.: Fluid-Structure Interaction Dynamics. National Defense Industry Press (2011)
9. Timothy, W.D., Amir, A.Y., Paul, A.: The effect of tilting pad journal bearing dynamic models on the linear stability analysis of an 8-stege compressor. J. Eng. Gas Turbines Power **134**(5) (2012)
10. API: Rotordynamic Tutorial: Critical Speeds, Unbalance Response, Stability, Train Torsionals, and Rotor Balancing, 2nd edn. Balancing. American Petroleum Institute, Washington, DC. API Recommended Practice 684 (2003)

Research on USV Heading Control Algorithm Based on Model Predictive Control

Fei Long[1], Zeyu Zhang[1], Qizhen Ge[1], Yeliang Xia[1], and Guanglong Zeng[2(✉)]

[1] Wuhan Institute of Marine Electric Propulsion, Wuhan 430064, China
[2] Wuhan University of Technology, Wuhan 430070, China
z183362517@163.com

Abstract. To solve the course control problem of unmanned surface ships, a course controller based on prediction function is designed, considering the inherent steering constraints of unmanned surface ships and the disturbance of stroke, wave and current during navigation. Based on prediction function, the controller adopts the strategy of prediction before control to overcome the constraints and interference in course control. In the online optimal control calculation, the potential influence of steering constraints and external interference on course control is considered, and the timely compensation of navigation interference is realized. Finally, the course controller based on prediction function is verified by simulation experiment and compared with the traditional PID controller. Through observation and analysis of the test results, it is proved that the new controller is effective in solving the heading control problem.

Keywords: Model prediction · USV · Heading control · Environmental disturbances

1 Introduction

Unmanned surface vessels (USVs) hold significant military and civilian value due to their characteristics of low risk, high maneuverability, deploy-ability, and environmental adaptability [1–3]. Currently, USVs have been widely employed in civilian applications, encompassing tasks such as offshore meteorological monitoring, real-time alerts, shallow-water mapping, environmental monitoring, and port surveys. Despite some successful applications, their usage scenarios remain relatively limited. Many potential scenarios urgently need further exploration, such as the water pollution issues faced by our country. In the context of water pollution control and regulation, USVs demonstrate enormous application potential. By continuously sampling and detecting the water quality of rivers, oceans, and other water bodies, USVs can monitor surrounding pollution sources around the clock, providing reliable data support for water quality governance. Additionally, traditional methods of channel mapping and measurements face challenges when involving specialized personnel and navigating through shallow or narrow waterways. The introduction of USVs can expand the measurement range, adapt to different water depths, reduce the involvement of specialized personnel, thereby improving mapping efficiency. USVs can also be applied to high-intensity, low-safety tasks such as channel maintenance, underwater topography mapping, and water-based search and rescue, bringing significant benefits to the socio-economic landscape. These applications not only contribute to increased operational efficiency but also play a crucial role in responding to emergency situations [4, 5].

Unmanned boat heading control, as a prominent research focus within unmanned boat motion control, has witnessed a proliferation of control technologies. Concerning the algorithm for unmanned boat heading control, the traditional PID control algorithm proposed by Minorsky [6] has been widely utilized due to its excellent robustness. However, because the traditional PID heading control algorithm is unable to resist the disturbances encountered by unmanned boats in water, such as wind and waves, experts and scholars worldwide have put forth various methods to enhance and modify the traditional PID heading control algorithm. These modifications include improvements in the parameter tuning aspect of traditional PID control, integration of advanced algorithmic technologies such as neural networks, genetic algorithms, and generalized predictive methods. In many practical applications of unmanned boat heading control, the feasibility of these enhanced algorithms has been validated.

H Jing proposed a ship PID control algorithm and an auxiliary indicator TTAR algorithm to address the issue of overlapping desired heading angles for unmanned boats [7], achieving decentralized control of the unmanned boat's trajectory. Experimental data indicates that the control range of the unmanned boat increased by 27% and 19.5%, demonstrating practical advantages. To address the heading control speed and robustness issues in the navigation environment of small unmanned surface vessels (USVs) in inland water, Yan D designed a propulsion cabin-based USV heading control system using a bipolar fuzzy controller [8]. This system can meet the requirements of large-angle turning control and small-angle heading maintenance. Simulation results show that, compared to traditional PID control, this control system is more suitable for inland navigation conditions.

USV inherently face steering constraints, and during their navigation, they encounter disturbances such as wind and waves. To address these issues, this paper, leveraging the unmanned boat motion model, employs a Model Predictive Control (MPC) algorithm to design a heading controller. This approach aims to achieve effective control over the heading of USV.

2 USV Control Mathematical Model Construction

2.1 Coordinate System Establishment

The motion of the USV is studied using two Cartesian coordinate systems: the fixed coordinate system and the body-fixed coordinate system, as illustrated in Fig. 1.

In the analysis of the arbitrary motion of Unmanned Surface Vehicles (USV) in three-dimensional space from the perspective of rigid body mechanics, it can be considered as a superposition of two components: the first part involves translational motion with respect to the reference point 'o' along the three coordinate axes denoted as u, v, and w. Here, u is the surge velocity, v is the sway velocity, and w is the heave velocity. The second part involves rotational motion about the reference point 'o' around the three coordinate axes denoted as p, q, and r. Here, p is the roll angular velocity, q is the pitch

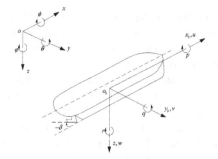

Fig. 1. Fixed coordinate system and body-fixed coordinate system.

angular velocity, and r is the yaw angular velocity. These kinematic quantities of the two components constitute the six degrees of freedom mathematical description of the motion of the USV, and their positive directions are determined by the right-hand screw rule. The motion of the USV in the fixed coordinate system can be described using its position and attitude. When performing force (moment) analysis on the USV in the body-fixed coordinate system, the forces along the three axes are represented by X, Y, and Z, while the moments about the three axes are represented by K, M, and N.

According to the above definition, the arbitrary motion of the six degrees of freedom in the USV can be represented by the following three vectors in space:

$$\eta = \left[x_0, y_0, z_0, \phi, \theta, \psi\right]^T \tag{1}$$

$$v - [u, v, w, p, q, r]^T \tag{2}$$

$$\tau = [X, Y, Z, K, M, N]^T \tag{3}$$

where, η represents the spatial position and attitude angles of the USV in the fixed coordinate system, v represents the linear velocity and angular velocity of the USV in the body coordinate system, and τ represents the various forces and torques acting on the USV in the body coordinate system.

The six degrees of freedom motion state of the USV can be described either in the body coordinate system using the velocity vector $[u, v, w]^T$ and angular velocity vector $[p, q, r]^T$, or in the fixed coordinate system using the derivatives of the position vector $[\dot{x}_0, \dot{y}_0, \dot{z}_0]^T$ and the derivatives of the Euler attitude angle vector $[\dot{\phi}, \dot{\theta}, \dot{\psi}]^T$. The choice between the two representations can be made based on specific requirements. The transformation between these vectors in the two coordinate systems can be achieved through the following conversion:

$$\begin{bmatrix} \dot{x}_0 \\ \dot{y}_0 \\ \dot{z}_0 \end{bmatrix} = T_V \begin{bmatrix} u \\ v \\ w \end{bmatrix}, \begin{bmatrix} \dot{\phi} \\ \dot{\theta} \\ \dot{\psi} \end{bmatrix} = T_\Omega \begin{bmatrix} p \\ q \\ r \end{bmatrix} \tag{4}$$

where, T_V and T_Ω are transformation matrices, represented using the Euler attitude angles of the USV as:

$$T_V = \begin{bmatrix} \cos\theta\cos\psi & \sin\phi\sin\theta\cos\psi - \cos\phi\sin\psi & \cos\phi\sin\theta\cos\psi + \sin\phi\sin\psi \\ \cos\theta\sin\psi & \sin\phi\sin\theta\cos\psi + \cos\phi\sin\psi & \cos\phi\sin\theta\cos\psi - \sin\phi\sin\psi \\ -\sin\theta & \sin\phi\cos\phi & \cos\phi\cos\theta \end{bmatrix} \tag{5}$$

$$T_\Omega = \begin{bmatrix} 1 & \sin\phi\tan\theta & \cos\phi\tan\theta \\ 0 & \cos\phi & -\sin\phi \\ 0 & \sin\phi\sec\theta & \cos\phi\sec\theta \end{bmatrix}, T_\Omega = \begin{bmatrix} 1 & \sin\phi\tan\theta & \cos\phi\tan\theta \\ 0 & \cos\phi & -\sin\phi \\ 0 & \sin\phi\sec\theta & \cos\phi\sec\theta \end{bmatrix} \tag{6}$$

2.2 USV Six Degrees of Freedom Space Motion Equation

Assuming the mass of the USV is m, the momentum is H, and the angular momentum relative to the center of mass G is L, according to the principles of theoretical mechanics, the absolute rate of change of rigid body momentum and angular momentum (i.e. in the fixed coordinate system) is equal to the sum of the applied total external force and the total external torque at that instant [9], as follows:

$$\frac{dH}{dt} = F_E, \frac{dL}{dt} = T_E \tag{7}$$

By taking the derivatives of the momentum and angular momentum expressions mentioned above, we obtain the following:

$$\begin{aligned} \frac{dH}{dt} &= m\frac{\delta U}{\delta t} + m\frac{\delta \Omega}{\delta t} \times R_G + \Omega \times (mU + m\Omega \times R_G) \\ \frac{dT}{dt} &= J\frac{\delta \Omega}{\delta t} + \Omega \times J\Omega + R_G \times m\frac{\delta U}{\delta t} + R_G \times (\Omega \times mU) \end{aligned} \tag{8}$$

where, U is the linear speed of the ship's motion, Ω is the angular speed of the ship's rotation, J is the moment of inertia matrix, and R_G is the position coordinate vector of the ship's center G in the body coordinate system. Expanding in the body coordinate system, we obtain:

$$\begin{aligned} m\left[\dot{u} - vr + wq - x_G\left(q^2 + r^2\right) + y_G(pq - \dot{r}) + z_G(pr + \dot{q})\right] &= X \\ m\left[\dot{v} - wp + ur - y_G\left(p^2 + r^2\right) + z_G(qr - \dot{p}) + x_G(pq + \dot{r})\right] &= Y \\ m\left[\dot{w} - uq + vp - z_G\left(p^2 + q^2\right) + x_G(qr - \dot{q}) + y_G(qr + \dot{p})\right] &= Z \end{aligned} \tag{9}$$

$$\begin{aligned} J_{xx}\dot{p} + J_{xy}\dot{q} + J_{xz}\dot{r} + \left(J_{xz}p + J_{yz}q + J_{zz}r\right)q - \left(J_{xy}p + J_{yy}q + J_{yz}r\right)r + \\ m\left[y_G(\dot{w} + vp - uq) + z_G(-\dot{v} - ur + wp)\right] = K \\ J_{yx}\dot{p} + J_{yy}\dot{q} + J_{yz}\dot{r} + \left(J_{xx}p + J_{yx}q + J_{xz}r\right)r - \left(J_{xz}p + J_{yz}q + J_{zz}r\right)p + \\ m\left[z_G(\dot{u} + wp - ur) + x_G(-\dot{w} - vp + uq)\right] = M \\ J_{zx}\dot{p} + J_{zy}\dot{q} + J_{zz}\dot{r} + \left(J_{xy}p + J_{yy}q + J_{xy}r\right)p - \left(J_{xx}p + J_{xy}q + J_{zy}r\right)q + \\ m\left[x_G(\dot{v} + ur - wp) + y_G(-\dot{u} - wq + vr)\right] = N \end{aligned} \tag{10}$$

The above two equations represent the translational and rotational equation sets of the general equations of motion for ship spatial movement. In this work, the origin *o* of the body coordinate system of the USV is set at the center G of the hull. It is also assumed that the shape and mass of the USV are symmetric, resulting in:

$$x_G = y_G = z_G = 0$$
$$J_{xy} = J_{yx} = J_{yz} = J_{zy} = 0 \tag{11}$$

Then the USV six-degree-of-freedom space motion equations can be obtained:

$$m(\dot{u} - vr + wq) = X$$
$$m(\dot{v} - wp + ur) = Y$$
$$m(\dot{w} - uq + vp) = Z$$
$$(J_{xx} + J_{xz})\dot{p} + (J_{zx}p + J_{zz}r)q - J_{yy}qr = K$$
$$J_{yy}\dot{q} + (J_{xx}p + j_{zx}r)r - (J_{zx}p + J_{zz}r)p = M$$
$$(J_{zz} + J_{zx})\dot{r} - (J_{xx}p + J_{zx}r)q + J_{yy}pq = N \tag{12}$$

where, m is the mass of USV; J_x, J_{yy} and J_{zz} are the moments of inertia of USV with respect to o_x, o_y and o_z axes in the servo system, respectively. J_{xz} and J_{zx} are the moments of inertia of USV with respect to the o_{xz} plane. When the USV device condition is constant, these physical quantities are constant, forming a mass inertia matrix. In order to more clearly reflect the physical meaning of the equations of motion and facilitate the establishment of mathematical models, the above equation is converted into the form of the following matrix equation:

$$M^{-1}\begin{bmatrix} X - m(wq - vr) \\ Y - m(ur - wp) \\ Z - m(vp - uq) \\ K - (J_{zz} - J_{yy})qr - J_{zx}pq \\ M - (J_{xx} - J_{zz})pr - J_{zx}(r^2 - p^2) \\ N - (J_{yy} - J_{xx})pq + J_{zx}qr \end{bmatrix} = \begin{bmatrix} \dot{u} \\ \dot{v} \\ \dot{w} \\ \dot{p} \\ \dot{q} \\ \dot{r} \end{bmatrix} \tag{13}$$

where, the mass inertia matrix M is:

$$M = \begin{bmatrix} m & 0 & 0 & 0 & 0 & 0 \\ 0 & m & 0 & 0 & 0 & 0 \\ 0 & 0 & m & 0 & 0 & 0 \\ 0 & 0 & 0 & J_{xx} & 0 & J_{xz} \\ 0 & 0 & 0 & 0 & J_{yy} & 0 \\ 0 & 0 & 0 & J_{zx} & 0 & J_{zz} \end{bmatrix} \tag{14}$$

where, $[\dot{u}, \dot{v}, \dot{w}, \dot{p}, \dot{q}, \dot{r}]^T$ is the acceleration vector of USV; $[X, Y, Z, K, M, N]^T$ is the component vector of the external torque subjected to the USV, including the hydrodynamic force, propulsive force, and environmental force.

The above formula is the basis of USV 6-DOF motion studied in this project, and also the theoretical framework of USV motion mathematical model.

3 Path Following Control of Unmanned Surface Vehicles Based on Model Predictive Control

3.1 Classic Model Predictive Control Algorithm

The basic principle of MPC is illustrated in Fig. 2. At time step k, based on the current measured output y(h) and the reference trajectory y_d, the optimal predictive control input sequence $\hat{y}(k + i)$ within the prediction horizon P is solved. This sequence minimizes a cost function defined by the predictive output and the reference trajectory. The first control input from the optimal predictive control input sequence is applied as the system control input at time step k. After applying this input to the system, the computation process at time step k is repeated at time step k + 1 to obtain the optimal system input at time step k + 1. This rolling optimization strategy ensures that each step's input is based on the currently calculated optimal value. The control performance of MPC is influenced by modeling accuracy, prediction horizon, and the efficiency of solving the optimal objective function. During the control process, it is necessary to choose appropriate simplified models, control parameters, and objective functions based on specific requirements to meet the needs of different control systems.

The predictive control method is based on solving the optimal control problem at arbitrary sampling intervals. The algorithm can be represented by the following key principles:

(1) Utilize a recognizable model to predict the future outputs of the system.
(2) Optimize an objective function, typically a function of prediction error and control benefits, to calculate control signals.
(3) Apply the signals obtained from the optimization calculation to implement control.
(4) Repeat the above three steps in a loop.

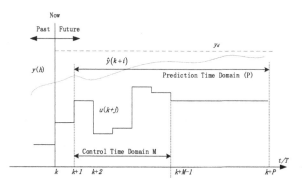

Fig. 2. MPC Basic Principles Diagram.

The main idea of Model Predictive Control, along with its basic structure, can be represented by Fig. 3. The cost function in predictive control is typically a function of prediction errors. In most cases, variations in control signals are also included in the cost function. By minimizing the cost function, the optimization aims to improve both prediction error and control signal variations. The cost function can be single-step or multi-step, meaning that the prediction errors and control signal variations may be considered in the cost function at a specific future time or aggregated over several future time steps.

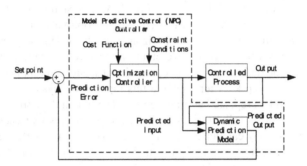

Fig. 3. Model Predictive Controller.

3.2   Heading Controller Based on Predictive Function Control

The overall block diagram of the Heading Control System for the surface unmanned boat based on Predictive Function Control (PFC) is shown in Fig. 4. In this system, the input signal for the heading control system is the target heading angle Wd of the unmanned boat, and the output signal is the actual heading angle W of the unmanned boat. The input signal for the heading controller based on Predictive Function Control is the target heading angle Wd and the actual heading angle W of the unmanned boat, and the output signal is the control input for the boat's rudder angle A.

The control process of the Predictive Function Controller in the heading control system for the surface unmanned boat involves the following steps:

(1) The unmanned boat prediction model calculates the predicted output of the unmanned boat's heading angle within the prediction time domain based on the current rudder angle control input.
(2) The predicted output values within the prediction time domain are corrected based on the deviation between the current actual heading angle output of the unmanned boat and the predicted output values.
(3) The optimization controller calculates the optimal control sequence for the boat's rudder angle control within the control time domain based on the optimization performance indicator function and the current target heading of the unmanned boat. The optimal control sequence for the boat's rudder angle control is then used to determine the rudder angle control signal applied to the unmanned boat at the current moment.
(4) The rolling optimization strategy continues the above calculation process (as shown in Fig. 4), repeating steps (1)–(3).

According to the above control process, the design of this heading controller mainly involves determining the prediction model of the unmanned boat, defining the optimization performance indicators, formulating the feedback correction strategy, and selecting parameters for the prediction time domain and control time domain. The following will provide a detailed explanation of the design process for the heading controller of the unmanned boat based on Predictive Function Control from these aspects.

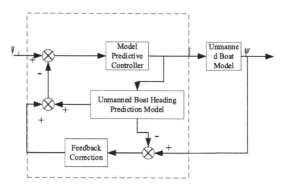

Fig. 4. Diagram of Unmanned Boat Heading Control System Based on Predictive Function Control (PFC).

1) Prediction Model

The prediction model for the unmanned boat heading control system refers to a model that accurately describes the relationship between the inputs and outputs of the heading control system. In the study of unmanned boat heading control, the Nomoto equation, which can describe the maneuvering characteristics of unmanned boats, is commonly used as the prediction model. The Nomoto equation is represented as follows:

$$T\ddot{\psi} + \dot{\psi} = K\delta \tag{15}$$

where, K represents the turning index of the unmanned boat, T represents the tracking index of the unmanned boat. δ is the rudder angle of the unmanned boat, and ψ is the heading angle of the unmanned boat.

Taking the sampling time as T_s, discretizing Eq. 15 yields the difference equation representation of the unmanned boat prediction model:

$$\psi(k) = e^{-\frac{T_s}{T}}\psi(k-1) + K\left(1 - e^{-\frac{T_s}{T}}\right)\delta(k-1) \tag{16}$$

In Eq. 16, the yawing index K of the unmanned boat and the following index T of the unmanned boat can be obtained through zigzag maneuvering experiments conducted on existing experimental boats with a 10°/10° pattern. The basic parameters of the experimental boat in this paper are shown in Table 1.

An experimental zigzag (10°/10°) test was conducted on a calm lake surface. Utilizing the calculation method mentioned in reference [30], the unmanned boat's experimental data was fitted, resulting in the values of the parameters K and T for the experimental

Table 1. The primary parameters of the unmanned boat

Major parameters	Length	Vertical Distance	Draught	Molded Depth	Displacement	Maximum speed	Cruising Speed
Numerical value	8.075	8	0.6	1.15	3.2	12	6

boat as follows: $K = 0.49$, $T = 1.94$. When the unmanned boat is navigating on the sea, it is susceptible to disturbances such as wind and waves. These disturbances, denoted as F, are equivalently added as rudder angle disturbances to the prediction model. Thus, Eq. 15 is modified as follows:

$$T\ddot{\psi} + \dot{\psi} = K(\delta + f_a) \tag{17}$$

2) Feedback Correction

The feedback correction component in the Predictive Function Control (PFC) unmanned boat heading controller primarily involves correcting the predicted heading output values within the prediction time domain given by the unmanned boat prediction model. This correction aims to make the predicted values closer to the actual output values. At time k, the error between the predicted output heading angle $\psi_m(k)$ of the unmanned boat prediction model and the actual system output heading angle $\psi(k)$ is used to correct the predicted output heading angle $\hat{\psi}(k+i)$ within the future prediction time domain. The error expression is given by:

$$e(k + i) = \psi(k) - \psi_m(k) \tag{18}$$

The expression for the corrected predicted output heading angle $\hat{\psi}(k+i)$ of the unmanned boat prediction model within the future prediction time domain is then given by:

$$\hat{\psi}(k + i) = \psi_m(k + i) + e(k + i) \tag{19}$$

3) Roll Optimization

The heading controller based on Predictive Function Control utilizes rolling optimization to determine the optimal rudder angle input quantity at each moment. The criterion for obtaining this rudder angle control quantity is the controller's tracking effectiveness for the preset heading. Additionally, the selection of the rudder angle control quantity, while considering the maximization of heading tracking, takes into account the changes in the rudder angle consistent with the actual steering of the unmanned boat.

Therefore, a standard function J is constructed, as shown in Eq. 20. It calculates the weighted average of the square of heading tracking error within the prediction time domain and the square of the increment in rudder angle control quantity within the control time domain. The minimization of the standard function J is pursued to obtain the optimal rudder angle input control quantity for the unmanned boat. Adjusting the weighting factors in the standard function reflects the importance of heading error and rudder angle increment in solving the optimal rudder angle control quantity within the control time domain.

Typically, while satisfying the constraints on the unmanned boat's rudder angle, more emphasis is placed on heading control effectiveness. Therefore, in Eq. 20, the weighting factor a_i for heading error should be set larger than the weighting factor b_i for rudder angle control quantity error.

$$J = \frac{1}{2}\left\{\sum_{i=1}^{H_p} a_i[\psi(k+i) - \psi_r(k+i)]^2 + \sum_{i=1}^{H_c} b_i[\delta(k+i-1) - \delta(k+i-2)]^2\right\}$$
(20)

where, ψ is the actual output heading angle of the heading controller, ψ_r is the reference heading angle of the unmanned boat, H_p is the prediction time domain of Predictive Function Control, H_c is the control time domain of Predictive Function Control, satisfying $H_c \leq H_p$, a_i is the weighting factors for heading error, b_i is the weighting factor for rudder angle control increment.

Furthermore, based on the control characteristics of the unmanned boat itself, the model predictive heading controller must take into account rudder angle constraints during the process of outputting the unmanned boat's rudder angle control quantity. The rudder angle constraint used in this paper for the unmanned boat model is:

$$-35° \leq \delta \leq 35°$$
(21)

The problem of obtaining the optimal rudder angle control quantity at time k in rolling optimization can be transformed into a local optimal control problem, as represented by Eq. 21. The first control value of the optimal control sequence obtained at time k is then applied as the optimal rudder angle control quantity at time k in the unmanned boat heading control system, achieving the heading control of the unmanned boat.

$$\min J = \frac{1}{2}\left\{\sum_{i=1}^{H_p} a_i[\psi(k+i) - \psi_r(k+i)]^2 + \sum_{i=1}^{H_c} b_i[\delta(k+i-1) - \delta(k+i-2)]^2\right\}$$
$$\text{s.t. } \delta_{\min} \leq \delta \leq \delta_{\max}$$
(22)

4 Simulation Experiment and Analysis

The system simulation of the designed course control system is carried out in matlab simulation environment to verify the feasibility of the controller. The system simulation mainly involves two aspects: course keeping and course tracking. The algorithm designed in this paper is compared with the PID control algorithm, and the specific parameters are set as follows:

(1) The initial parameters of the unmanned ship are set as follows: speed $v = 1$ m/s, course Angle $\psi = 0°$.
(2) PID heading controller parameter Settings: $K_p = 0.5, K_i = 0.00005, K_d = 0.9$.

(3) MPC heading controller parameter setting: forecast time domain $H_p = 10$; Control time domain $H_c = 2$. In the objective function J, course error weighting factor $a_i = 1, i = 1, 2, \cdots H_p$, and rudder Angle control increment weighting factor $b_i = 0.1, i = 1, 2, \cdots H_c$.

4 1 Course Keeping Effect

Under the ideal condition without interference, the course keeping effect of two unmanned ship course controllers is simulated. Given the constant input of the target heading Angle $\psi_d = 10°$, the obtained heading control effect is shown in Fig. 5, and the change of rudder Angle control quantity is shown in Fig. 6.

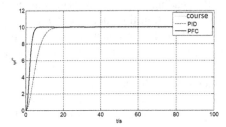
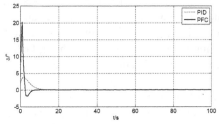

Fig. 5. Course keeping rendering (no interference).

Fig. 6. Course maintains rudder Angle control (no interference).

As can be seen from the simulation results of Fig. 5 and Fig. 6, both the MPC-based heading controller and the PID-based heading controller can reach the given target heading Angle without overshoot under ideal conditions without interference. The response speed of the MPC-based heading controller is faster than that of the PID-based heading controller, but its rudder Angle control changes greatly.

In the course control simulation with external interference, the disturbance term is replaced by a white noise multiplied by the second-order wave function. The mean value of the white noise signal in the interference signal is 2, and the power spectral density is 0.5. The second-order wave function expression is shown as follows, and the parameters are as follows: gain constant $K_\omega = 0.42$, dominant wave frequency $\omega_0 = 0.606$, damping coefficient $\zeta = 0.3$.

$$h(s) = \frac{K_\omega s}{s^2 + 2\zeta\omega_0 s + \omega_0^2} \tag{23}$$

The interference signal in the simulation experiment is shown in Fig. 7. In the case of external interference, the control effect of the two heading controllers is shown in Fig. 8, and the corresponding change of rudder Angle control quantity is shown in Fig. 9.

According to the simulation results shown in Fig. 8 and Fig. 9, it can be seen that when external wind and wave interference exists in the course control system of unmanned ship, the control effect of predictive function control is better than that of classical PID control. As can be seen from the course tracking curve of Fig. 8, the course control system based on predictive function control can track the target course faster, and the course

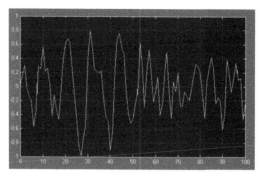

Fig. 7. Wind and waves interfere with the signal.

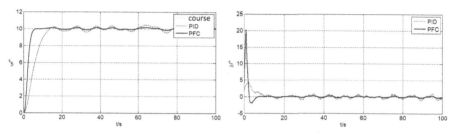

Fig. 8. Curse keeping rendering (exist interference).

Fig. 9. Course maintains rudder Angle control (exist interference).

jitter is smaller when tracking the target course. As can be seen from the change curve of rudder Angle control in Fig. 9, the change of output rudder Angle control quantity of course controller based on prediction function control is smoother.

4.2 Heading Tracking Performance

In an ideal scenario without disturbances, simulate the heading tracking performance of the two heading control systems. The heading tracking simulation experiment sets the target heading angle for the unmanned boat as a sine signal with an amplitude of and a frequency of 0.02 rad/s. The heading tracking performance obtained from the two controllers in the simulation experiment is illustrated in Fig. 10, and the rudder angle control variation curves are shown in Fig. 11.

To further analyze the control performance of the PID-based heading controller and the MPC-based heading controller, Table 2 provides the average heading tracking error for both controllers. From Fig. 12 and 13, as well as Table 2, it can be observed that the MPC-based heading controller achieves stable heading tracking performance with higher tracking speed and accuracy compared to the traditional PID heading controller.

In the presence of external disturbances, simulate the heading tracking performance of the two heading controllers with the target heading angle input for the unmanned boat as a sine signal with an amplitude of and a frequency of 0.02 rad/s. The disturbance signal continues to be the wind and wave disturbance as described by Eq. 23, and the

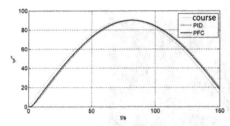

Fig. 10. Heading Tracking Performance (No Disturbances).

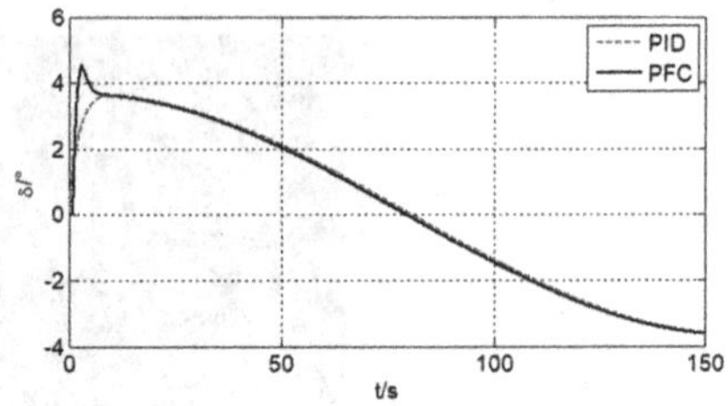

Fig. 11. Heading Tracking Rudder Angle Control Input (No Disturbances).

Table 2. Heading Tracking Error

Controller	Average Tracking Error
PID	4.9180°
MPC	3.4923°

disturbance signal is illustrated in Fig. 7. The obtained heading tracking performance is depicted in Fig. 12, and the rudder angle control variation is shown in Fig. 13.

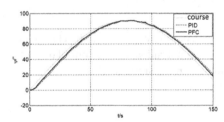
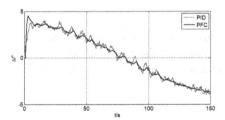

Fig. 12. Heading Tracking Performance (Exist Disturbances).

Fig. 13. Heading Tracking Rudder Angle Control Input (Exist Disturbances).

To further analyze the control performance of the PID-based heading controller and the MPC-based heading controller, Table 3 provides the heading tracking error for both controllers in the presence of disturbances. From Fig. 12 and 13, as well as Table 3, it can be observed that the MPC-based heading controller achieves stable heading tracking performance even in the presence of disturbances, with higher tracking speed and accuracy compared to the traditional PID heading controller.

Table 3. Heading Tracking Error (with disturbances)

Controller	Average Tracking Error
PID	4.9384°
MPC	3.5001°

5 Conclusion

Building upon the principles of Model Predictive Control (MPC) and aligning with the requirements of unmanned boat heading control, this paper introduces a waterborne unmanned boat heading controller based on Model Predictive Control. Through comparative simulation experiments, the proposed heading controller is evaluated for heading maintenance and tracking performance against a waterborne unmanned boat heading controller based on the classical PID algorithm. The results validate that the designed heading controller in this paper exhibits superior heading maintenance and tracking effectiveness.

References

1. Ma, T.Y., Yang, S.L., Wang, T.T., Xin, L., Chen, Y.: An outline of current status and development of the multiple USV cooperation system. Ship Sci. Technol. **36**(6), 7–13 (2014)
2. Liu, T., Pang, B., Zhang, L., Yang, W., Sun, X.: Sea surface object detection algorithm based on YOLO v4 fused with reverse depthwise separable convolution (RDSC) for USV. J. Mar. Sci. Eng. **9**(7), 753 (2021)
3. Jin, J., Liu, D., Wang, D., Ma, Y.: A practical trajectory tracking scheme for a twin-propeller twin-hull unmanned surface vehicle. J. Mar. Sci. Eng. **9**(10), 1070 (2021)
4. Qin, J., Guo, C.: Path following control of unmanned surface vessel with unknown ocean currents disturbances. In: Deng, Zhidong (ed.) Proceedings of 2021 Chinese Intelligent Automation Conference. LNEE, vol. 801, pp. 56–64. Springer, Singapore (2022). https://doi.org/10.1007/978-981-16-6372-7_7
5. Liu, X., Guo, T., Song, D., Li, J., Wang, X.: Modeling and predictive control of an automatic launch and recycling system for USV based on NAR neural network. Mar. Technol. Soc. J. **56**(2), 73–87 (2022)
6. Sørensen, A.J., Fossen, T.I., Strand, J.P.: Design of Controllers for Positioning of Marine Vessels. In: The Ocean Engineering Handbook (2000)
7. Hui, J.: Research on route dispersion control algorithm of multi-unmanned boat. Ship Sci. Technol. (2018)
8. Yan, D., Xiao, C., Wen, Y.: Pod propulsion small surface USV heading control research. In: International Ocean and Polar Engineering Conference (2016)
9. Huang, S., Liu, W., Luo, W., Wang, K.: Numerical simulation of the motion of a large scale unmanned surface vessel in high sea state waves. J. Mar. Sci. Eng. **9**(9), 982 (2021)
10. Lawrynczuk, M., Nebeluk, R.: Beyond the quadratic norm: computationally efficient constrained nonlinear MPC using a custom cost function. ISA Trans. **134**, 336–356 (2023)
11. Oliveira-Silva, E., de Prada, C., Montes, D., Navia, D.: Economic MPC with modifier adaptation using transient measurements. Comput. Chem. Eng. **173**, 108205 (2023)
12. Zhong, Z., de Rio-Chanona, E.A., Petsagkourakis, P.: Tube-based distributionally robust model predictive control for nonlinear process systems via linearization. Comput. Chem. Eng. **170**, 108112 (2023)

Author Index

Printed in the United States
by Baker & Taylor Publisher Services